Robust Intelligence and Trust in Autonomous Systems

Ranjeev Mittu • Donald Sofge • Alan Wagner
W.F. Lawless
Editors

Robust Intelligence and Trust in Autonomous Systems

Editors
Ranjeev Mittu
Naval Research Laboratory
Washington, DC, USA

Donald Sofge
Naval Research Laboratory
Washington, DC, USA

Alan Wagner
Georgia Tech Research Institute
Atlanta, GA, USA

W.F. Lawless
Paine College
Augusta, GA, USA

ISBN 978-1-4899-7666-6 ISBN 978-1-4899-7668-0 (eBook)
DOI 10.1007/978-1-4899-7668-0

Library of Congress Control Number: 2015956336

Springer New York Heidelberg Dordrecht London
© Springer Science+Business Media (outside the USA) 2016

Chapters 1, 2, 10, and 12 were created within the capacity of an US governmental employment. US copyright protection does not apply.

This work is subject to copyright. All rights are reserved by the Publisher, whether the whole or part of the material is concerned, specifically the rights of translation, reprinting, reuse of illustrations, recitation, broadcasting, reproduction on microfilms or in any other physical way, and transmission or information storage and retrieval, electronic adaptation, computer software, or by similar or dissimilar methodology now known or hereafter developed.
The use of general descriptive names, registered names, trademarks, service marks, etc. in this publication does not imply, even in the absence of a specific statement, that such names are exempt from the relevant protective laws and regulations and therefore free for general use.
The publisher, the authors and the editors are safe to assume that the advice and information in this book are believed to be true and accurate at the date of publication. Neither the publisher nor the authors or the editors give a warranty, express or implied, with respect to the material contained herein or for any errors or omissions that may have been made.

Printed on acid-free paper

Springer Science+Business Media LLC New York is part of Springer Science+Business Media (www.springer.com)

Preface

This book is based on the Association for the Advancement of Artificial Intelligence (AAAI) Symposium on "The Intersection of Robust Intelligence (RI) and Trust in Autonomous Systems"; the symposium was held at Stanford March 24–26, 2014. The title of this book reflects the theme of the symposium. Our goal for this book is to further address the current state of the art in autonomy at the intersection of RI and trust and to more fully examine the existing research gaps that must be closed to enable the effective integration of autonomous and human systems. This research is particularly necessary for the next generation of systems, which must scale to teams of autonomous platforms to better support their human operators and decision makers.

The book explores the intersection of RI and trust across multiple contexts and among arbitrary combinations of humans, machines, and robots. To help readers better understand the relationships between artificial intelligence (AI) and RI in a way that promotes trust among autonomous systems and human users, this edited volume presents a selection of the underlying theories, computational models, experimental methods, and possible field applications. While other books deal with these topics individually, this book is unique in that it unifies the fields of RI and trust and frames them in the broader context of effective integration for human-autonomous systems.

The volume begins by describing the current state of the art for research in RI and trust presented at Stanford University in the Spring of 2014 (copies of the technical articles are available from AAAI at http://www.aaai.org/Library/Symposia/Spring/ss14-04.php; a link to the presentation materials and photographs of participants is at https://sites.google.com/site/aaairobustintelligence/).

After the introduction, chapter contributors elaborate on key research topics at the heart of effective human-systems integration. These include machine learning, Big Data, workload management, human-computer interfaces, team integration and performance, advanced analytics, behavior modeling, training, and test and evaluation, the latter known as V&V (i.e., verification and validation).

The contributions to this volume are written by world-class leaders from across the field of autonomous systems research, ranging from industry to academia and to government. Given the diversity of the research in this book, we strove to thoroughly examine the challenges and trends of systems that exhibit RI; the fundamental implications of RI in developing trusted relationships among humans, machines, and robots with present and future autonomous systems; and the effective human systems integration that must result for trust to be sustained.

A brief summary is presented below of the AAAI Symposium in the Spring of 2014.

AAAI-2014 Spring Symposium Organizers

Jennifer Burke, Boeing: jennifer.l.burke2@boeing.com
Alan Wagner, Georgia Tech Research Institute: Alan.Wagner@gtri.gatech.edu
Donald Sofge, Naval Research Laboratory: don.sofge@nrl.navy.mil
William F. Lawless, Paine College: wlawless@paine.edu

AAAI-2014 Spring Symposium: Keynote Speakers

- Suzanne Barber, barber@mail.utexas.edu, AT&T Foundation Endowed Professor in Engineering, Department of Electrical and Computer Engineering, Cockrell School of Engineering, U Texas
- Julie L. Marble, julie.marble@navy.mil, Program Officer: Hybrid human computer systems at Office of Naval Research, Washington, DC
- Ranjeev Mittu, ranjeev.mittu@nrl.navy.mil, Branch Head, Information Management & Decision Architectures Branch, Information Technology Division, US Naval Research Laboratory, Washington, DC
- Hadas Kress-Gazit, hadaskg@cornell.edu, Cornell University; High-Level Verifiable Robotics
- Satyandra K. Gupta, skgupta@umd.edu, Director, Maryland Robotics Center, University of Maryland
- Dave Ferguson, daveferguson@google.com, Google's Self-Driving Car project, San Francisco
- Mo Jamshidi, mo.jamshidi@usta.edu, University of Texas at San Antonio, Lutcher Brown Endowed Chair and Professor, Computer and Electrical Engineering
- Dirk Helbing, dirk.helbing@gess.ethz.ch, http://www.futurict.eu; ETH Zurich

Symposium Program Committee

- Julie L. Marble, julie.Marble@jhuapl.edu, cybersecurity, Johns Hopkins Advanced Physics Lab, MD
- Ranjeev Mittu, ranjeev.mittu@nrl.navy.mil, Branch Head, Information Management & Decision Architectures Branch, Information Technology Division, U.S. Naval Research Laboratory, Washington, DC
- David Atkinson, datkinson@ihmc.us, Senior Research Scientist, Institute of Human-Machine Cognition (IHMC)
- Jeffrey Bradshaw, jbradshaw@ihmc.us; Senior Research Scientist, Institute of Human-Machine Cognition (IHMC)
- Lashon B. Booker, booker@mitre.org, The MITRE Corporation
- Paul Hyden, paul.hyden@nrl.navy.mil, Naval Research Laboratory
- Holly Yanco, holly@cs.uml.edu, University of Massachusetts Lowell
- Fei Gao, feigao@MIT.EDU.MIT
- Robert Hoffman, rhoffman@ihmc.us, Senior Research Scientist, Institute of Human-Machine Cognition (IHMC)
- Florian Jentsch, florian.Jentsch@ucf.edu, Department of Psychology and Institute for Simulation & Training, *Director*, Team Performance Laboratory, University of Central Florida
- Howell, Chuck, howell@mitre.org, Chief Engineer, Intelligence Portfolio, National Security Center, The MITRE Corporation
- Paul Robinette, probinette3@gatech.edu, Graduate Research Assistant, Georgia Institute of Technology
- Munjal Desai, munjaldesai@google.com
- Geert-Jan Kruijff, gj@dfki.de, Senior Researcher/Project Leader, Language Technology Lab, DFKI GmbH, Saarbruecken, Germany

This AAAI symposium sought to address these topics and questions:

- How can robust intelligence be instantiated?
- What is RI for an individual agent? A team? Firm? System?
- What is a robust team?
- What is the association between RI and autonomy?
- What metrics exist for robust intelligence, trust, or autonomy between individuals or groups, and how well do these translate to interactions between humans and autonomous machines?
- What are the connotations of "trust" in various settings and contexts?
- How do concepts of trust between humans collaborating on a task differ from human-human, human-machine, machine-human, and machine-machine trust relationships?
- What metrics for trust currently exist for evaluating machines (possibly including such factors as reliability, repeatability, intent, and susceptibility to catastrophic failure), and how may these metrics be used to moderate behavior in collaborative teams including both humans and autonomous machines?

- How do trust relationships affect the social dynamics of human teams, and are these effects quantifiable?
- What validation procedures could be used to engender trust between a human and an autonomous machine?
- What algorithms or techniques are available to allow machines to develop trust in a human operator or another autonomous machine?
- How valid are the present conceptual models of human networks? Mathematical models? Computational models?
- How valid are the present conceptual models of autonomy in networks? Mathematical models? Computational models?

Papers at the symposium specified the relevance of their topic to AI or proposed a method involving AI to help address their particular issue. Potential topics included (but were not limited to) the following:

Robust Intelligence (RI) topics:

- Computational, mathematical, conceptual models of robust intelligence
- Metrics of robust intelligence
- Is a model of thermodynamics possible for RI (i.e., using physical thermodynamic principles, can intelligent behavior be addressed in reaction to thermodynamic pressure from the environment?)?

Trust topics:

- Computational, mathematical, conceptual models of trust in autonomous systems
- Human requirements for trust and trust in machines
- Machine requirements for trust and trust in humans
- Methods for engendering and measuring trust among humans and machines
- Metrics for deception among humans and machines
- Other computational and heuristic models of trust relationships, and related behaviors, in teams of humans and machines

Autonomy topics:

- Models of individual, group, and firm autonomous system behaviors
- Mathematical models of multitasking in a team (e.g., entropy levels overall and by individual agents, energy levels overall and by individual agents)

Network topics:

- Constructing, measuring, and assessing networks (e.g., the density of chat networks among human operators controlling multi-unmanned aerial vehicles)
- For networks, specify whether the application is for humans, machines, robots, or a combination, e.g., the density of inter-robot communications

After the symposium was completed, the book and the symposium took on separate lives. The following individuals were responsible for the proposal submitted to Springer after the symposium, for the divergence between the topics of the two, and for editing the book that has resulted.

Washington, DC, USA Ranjeev Mittu
Washington, DC, USA Donald Sofge
Atlanta, GA, USA Alan Wagner
Augusta, GA, USA W.F. Lawless

Contents

Chapter 1
Introduction

RanjeevMittu, Donald Sofge, AlanWagner, and W. F. Lawless

1.1 The Intersection of Robust Intelligence (RI) and Trust in Autonomous Systems

The Intersection of Robust Intelligence (RI) and Trust in Autonomous Systems addresses the current state-of-the-art in autonomy at the intersection of Robust Intelligence (RI) and trust, and the research gaps that must be overcome to enable the effective integration of autonomous and human systems. This is particularly true for the next generation of systems, which must scale to teams of autonomous platforms to better support their human operators and decision makers. This edited volume explores the intersection of RI and trust across multiple contexts among autonomous hybrid systems (where hybrids are arbitrary combinations of humans, machines and robots). To better understand the relationships between Artificial Intelligence (AI) and RI in a way that promotes trust between autonomous systems and human users, this edited volume explores the underlying theory, mathematics, computational models, and field applications.

To better understand and manage RI with AI in a manner that promotes trust in autonomous agents and teams, our interest is in the further development of theory, network models, mathematics, computational models, associations, and field

R. Mittu • D. Sofge
Naval Research Laboratory, 4555 Overlook Ave SW, Washington, DC 20375, USA
e-mail: ranjeev.mittu@nrl.navy.mil; donald.sofge@nrl.navy.mil

A. Wagner
Georgia Tech Research Institute, 250 14th Street NW, Atlanta, GA 30318, USA
e-mail: Alan.Wagner@gtri.gatech.edu

W.F. Lawless (✉)
Paine College, 1235 15th Street, Augusta, GA 30901, USA
e-mail: WLawless@paine.edu

© Springer Science+Business Media (outside the USA) 2016

R. Mittu et al. (eds.), *Robust Intelligence and Trust in Autonomous Systems*,
DOI 10.1007/978-1-4899-7668-0_1

applications at the intersection of RI and trust. We are interested not only in effectiveness with a team's multitasking or in constructing RI networks and models, but in the efficiency and trust engendered among interacting participants.

Part of our symposium in 2014 sought a better understanding of the intersection of RI and trust for humans interacting with other humans and human groups (e.g., teams, firms, systems; also, the networks among these social objects). Our goal is to use this information with AI to not only model RI and trust, but also to predict outcomes from interactions between autonomous hybrid groups (e.g., hybrid teams in multitasking operations).

Systems that learn, adapt, and apply their experience to the problems faced in an environment may be better suited to respond to new and unexpected challenges. One could argue that such systems are "robust" to the prospect of a dynamic and occasionally unpredictable world. We expect the systems that exhibit this type of robustness to afford to those who interact with the system a greater degree of trust. For instance, an autonomous vehicle which, in addition to driving to different locations by itself, can also warn a passenger of locations where it should not drive, might likely be viewed as more robust than a similar system without such a warning capability. But would it be viewed as more trustworthy? This workshop endeavored to examine such questions that lay at the intersection of robust intelligence and trust. Problems such as these are particularly difficult because they imply situational variations that may be hard to define.

The focus of our workshop centered on how robust intelligence impacts trust in the system and how trust in the system makes it more or less robust. We explored approaches to RI and trust that included, among others, intelligent networks, intelligent agents, and multitasking by hybrid groups (i.e., arbitrary combinations of humans, machines and robots).

1.2 Background of the 2014 Symposium

Robust intelligence (RI) has not been easy to define. We proposed an approach to RI with artificial intelligence (AI) that may include, among other approaches, the science of intelligent networks, the generation of trust among intelligent agents, and multitasking among hybrid groups (humans, machines and robots). RI is the goal of several government projects to explore the intelligence as seen at the level of humans, including those directed by NSF (2013); the US Army (Army 2014) and the USAF (Gluck 2013). DARPA (2014) has a program on physical intelligence that is attempting to produce the first example of "“intelligent” behavior under thermodynamic pressure from their environment." Carnegie Mellon University (CMU 2014) has a program to build a robot that can execute "complex tasks in dangerous ... environments." IEEE (2014) has the journal *Intelligent Systems* to address various topics on intelligence in automation including trust; social computing; health; and, among others, coalitions that make the "effective use of limited resources to achieve complex and multiple objectives." From another perspective, IBM has built a program that beat the reigning world champion at chess in 1997; another program that won at the game of Jeopardy in 2011 (du Sautoy 2014); and an intelligent operations center for the management of cities, transportation, and water (IBM 2014). Multiple other ways may exist to define or approach RI, and to measure it.

In an attempt to advance AI with a better understanding and management of RI, our interest is in the theory, network models, mathematics, computational models, associations, and field applications of RI. This means that we are interested in not only effectiveness with multitasking or in constructing RI networks and models, but in the efficiency and trust engendered among the participants during interactions.

Part of the goal in this symposium was to find a better understanding of RI and the autonomy it produces with humans interacting with other humans and human groups (e.g., teams, firms, systems; also, networks among these social objects). Our ultimate goal is to use this information with AI to not only model RI and autonomy, but also in the predictions of the outcomes from interactions between hybrid groups that interdependently generate networks and trust.

For multitasking with human teams and firms, interdependence is an important element in their RI: e.g., the Army is attempting to develop a robot that can produce "a set of intelligence-based capabilities sufficient to enable the teaming of autonomous systems with Soldiers" (Army 2014); and ONR is studying robust teamwork (ONR 2013). But a team's interdependence also introduces uncertainty, fundamentally impacting measurement (Lawless et al. 2013).

Unlike conventional computational models where agents act independently of neighbors, where, for example, a predator mathematically consumes its prey or not as a function of a random interaction process, interdependence means that agents dynamically respond to the bi-directional signals of actual or potential presence of other agents (e.g., in states poised to fight or flight), a significant increase over conventional modeling complexity; as an example of interdependence in Yellowstone's National Park (Hannibal 2012):

> aspen and other native vegetation, once decimated by overgrazing, are now growing up along the banks ... [in part] because elk and other browsing animals behave differently when wolves are around. Instead of eating down to the soil, they take a bite or two, look up to check for threats, and keep moving. [This means that the] greenery can grow tall enough to reproduce.

That the problem of interdependence remains unsolved, mathematically and conceptually, precludes hybrid teams based on artificial intelligence from processing information like human teams operating under interdependent challenges and perceived threats.

At this AAAI Symposium, we explored the various aspects and meanings of robust intelligence, networks and trust between humans, machines and robots in different contexts, and the social dynamics of networks and trust in teams or organizations composed of autonomous machines and robots working together with humans. We sought to identify and/or develop methods for structuring networks and engendering trust between agents, to consider the static and dynamic aspects of behavior and relationships, and to propose metrics for measuring outcomes of interactions.

1.3 Contributed Chapters

Chapter 2 is titled "*Towards modeling the behavior of autonomous systems and humans for trusted operations.*" Its authors are Gavin Taylor, Ranjeev Mittu, Ciara Sibley and Joseph Coyne. The first author is with the U.S. Naval Academy; and authors Mittu, Sibley and Coyne are with the Naval Research Laboratory. In this chapter, the authors have studied the promise offered to the Department of Defense by autonomous robot and machine systems to improve its mission successes and to protect its valuable human users; but this promise has been countered by the increased complexity and workloads that have been placed on human supervisors by these systems. In this new era, as autonomy increases, the trust humans place in these systems becomes an important factor. Trust may depend on knowing whether anomalies exist in these systems; whether the anomalies that do exist can be managed; and whether these anomalies further affect the limitations of the human supervisors (acknowledged but not studied in this chapter). Using a mathematical manifold that captures a platform's trajectories to represent the tasks to be performed by an unassisted and unmanned autonomous system, the authors propose an example that exploits the errors generated for alarms and system analyses. The authors point out the existing research questions (e.g., user interaction patterns) and challenges that must also be addressed, including the best way for users to interact with autonomy; the optimized formal models of human decision-making; the modeling of active decision contexts; and the adaptation of concept drift techniques from the machine learning community.

Chapter 3 is titled "*Learning trustworthy behaviors using an inverse trust metric*"; its authors are Michael W. Floyd and Michael Drinkwater with Knexus Research Corporation, Springfield, Virginia; and David W. Aha, with the Navy Center for Applied Research in Artificial Intelligence, Naval Research Laboratory, Washington, DC. The authors present an algorithm by which a robot measures its own trustworthiness—an inverse trust metric—and uses this information to adapt its behavior, ideally becoming more trustworthy. They use case-based reasoning to gauge whether or not some previously used behavior or set of behaviors is likely to be trustworthy in the current environment. They present simulation experiments demonstrating the use of this "inverse trust metric" in a patrol scenario. The authors begin their chapter by assuming that those robots that can be trusted to perform important tasks may become helpful to human teams, but only if the humans can trust the robot as a member of a team to perform its assigned tasks as expected. But how to determine trust on the fly is a difficult problem. Instead of asking users how much they trust an autonomous agent, the authors use "inverse trust". They estimate the "inverse trust" for their concept as judged by the robot when determining its own performance while working for multiple human operators, or in front of multiple human operators. Using simulation, the authors demonstrate the superiority of their case-based reasoning approach to random learning for robot team members assisting a human team. The authors conclude that more robotic uncertainty in performance impedes trust.

Chapter 4 is titled "*'Trust V'—Building and measuring trust in autonomous systems*"; its authors are Gari Palmer, Anne Selwyn and Dan Zwillinger with the Raytheon Corporation. The authors develop a framework based on the system V framework to codify how trust is built into a new system and how a system should respond in order to maintain trust. Their framework is a life-cycle model which adds trust components. Testing and evaluation ensure that the trust components are functional. During operational use these trust components allow the user to query the system to better understand (and trust) its operation. This framework is becoming more important as autonomous systems become more prominent; e.g., the Department of Defense has made autonomy one of its research priorities. Autonomous systems, those in use today and anticipated in the future, will need both system trust (i.e., when their specifications have been met) and operational trust (when the user's expectations have been met). Automated systems are more easily trusted than autonomous systems. But trusting complex automated systems requires rigorous Test & Evaluation (T&E) and Verification & Validation (V&V) processes. While similar processes are likely to be used to establish trust for autonomous systems, new methods set within these processes must address the unique attributes of autonomy, like adaptation to situations, or self-organization within situations. Using their framework, the authors identify specific methods for engendering trust in automated and autonomous systems, where systems range from automated to autonomous systems as endpoints. The authors give the example of a prototyped method that has been shown to enable trust. This framework supports the insertion of new methods to generate and measure operational trust in existing and future autonomous systems.

Chapter 5 is titled "*Big Data analytic paradigms—From principle component analysis to deep learning*"; its authors are Mo Jamshidi, Barney Tannahill and Arezou Moussavi with the Autonomous Control Engineering (ACE) Laboratory at The University of Texas, San Antonio (Tannahil is also from the Southwest Research Institute, or SwRI). This chapter presents an overview of Artificial Neural Networks (ANNs) ranging from multi-layer networks to recent advances related to deep architectures including auto-encoders and restricted Boltzmann machines (RBMs). Large sets of data (numerical, textural and image) have been accumulating at a rapid pace from multiple sources in all aspects of society. Advances in sensor technology, the Internet, social networks, wireless communication, and inexpensive memory have all contributed to the explosion of "Big Data" as this phenomenon has come to be known. Big Data is produced in many ways in today's interdependent global economy. Social networks, system of systems (SoS), and wireless systems are only some of the contributors to Big Data. Instead of a hindrance, many researchers have come to consider Big Data as a rich resource for future innovations in science, engineering, business and other potential applications. But the flood of data has to be managed and controlled before useful information can be extracted. For the extraction of information to be useful, recent efforts have developed a promising approach known as "Data Analytics". This approach uses statistical and computational intelligence tools like principal component analysis (PCA), clustering, fuzzy logic, neuro-computing, evolutionary computation, Bayesian networks and other tools to reduce the size of Big Data. One of these tools, Deep Learning, is described by the development and use of neural networks in the machine learning community that has allowed for the extraordinary results recently obtained for digital speech, imagery, and natural language processing tasks. The authors present an example of Neural Networks using the data collected from a wind farm to demonstrate Data Analytics.

Chapter 6 is titled "*Artificial brain systems based on neural network discrete chaotic dynamics. Toward the development of conscious and creative robots*"; its author, Vladimir Gontar, prepared his chapter at the Biocircuits Institute, University of California in San Diego, while on a sabbatical; he has since returned to his affiliation at the Ben-Gurion University of the Negev in Israel. He is working on new theory and mathematical models of the human brain based on first principles for neural networks that model biochemical reactions to simulate consciousness. Consciousness is a hard problem. From Marcus and his colleagues (2014), although "no consensus" exists, current research tends to address how "systems might bridge from neuronal networks to symbolic cognition". Gontar's approach is similar. In contrast to regular information processes approximated linearly as a function of the energy available, he models information exchanges between neurons and neural networks based on the infinitesimally small energies needed to change chaotic systems. He compares the example of a mandala drawn by an artist matched step-by-step with one drawn by his chaos equations, concluding that this is how consciousness may be addressed computationally.

Chapter 7, on the "*Modeling and control of trust in human-robot collaborative manufacturing*", is authored by Behzad Sadrfaridpour, Hamed Saeidi, Jenny Burke, Kapil Madathil and Yue Wang with Clemson University and the Boeing Company. The authors explore trust in the context of Human-Robot Collaboration (HRC) on the factory floor. To measure and gauge the improvement in a system on the factory floor, they use a time-series model of trust, a model of a robot's performance to tie its speed to flexibility, and a model of a person that includes fatigue. They present a series of experiments which investigate how the robot and the human adapt to each other's changing performance and how these changes impact trust. HRC already exists on factory floors today, opening a new realm of manufacturing with robots in real-world settings. There, humans and robots work together by collaborating as coworkers. HRC plays a critical role in safety, productivity, and flexibility. Human-to-robot trust determines the human's acceptance and allocation of autonomy to a robot that in turn decides the efficiency of the task performed and the human's workload. Using Likert scales and time-series models of performance to measure trust, the authors studied trust in a robot in the laboratory subjectively and objectively under three control modes of the robot; viz., the robot placed under manual, autonomous and collaborative control conditions. Human operator control was used in the manual condition; a neural network was used for intelligent control in the autonomous condition; and a mixed control was used in the collaborative condition. For this study, the authors did not find strong support for the autonomous mode. They also showed that under the collaborative mode, human-to-robot trust will be improved since the human has more control over the robot speed while the robot is adapting to the human speed as well.

Chapter 8 is titled "*Investigating human-robot trust in emergency scenarios: Methodological lessons learned*"; its authors Paul Robinette, Alan Wagner and Ayanna Howard are with the Georgia Institute of Technology; they conclude that trust has an elusive, subjective meaning depending on the context and the culture of the perceiver and the bias introduced by a questioner, especially in emergency scenarios. Being that few research protocols exist to study human-robot trust (HRT), the authors devised their own protocol to include risk on the part of both the human and the robot. Overall, they conclude that studies of HRT are inherently problematic, even though HRT has been studied as computational cognition; neurological change; and, among other studies, in the probability distributions of an agent's actions. They like Lee and See's claim that trust is an attitude associated with the goals sought under uncertainty and vulnerability. The authors performed experiments using crowdsourcing techniques. They found that the word phrasing of a narrative significantly affected decisions; that anchoring biases also had significant effects; and that unsuccessful robot leaders did not always dissuade their human followers. The latter finding presents a significant challenge to researchers to design robots in a way so that the robots communicate clearly with humans, so that humans do not overly-trust robots when they should not, and so that crowdsourcing for testing hypotheses provide generality and empirical evaluations if coupled with complementary methods (viz., narratives and simulated scenarios).

In Chap. 9, titled "*Designing for robust and effective teamwork in human-agent teams*", the authors Fei Gao, M.L. Cummings and Erin Solovey are with the Massachusetts Institute of Technology. The authors examine the impact of team structure, task uncertainty, and information-sharing tools on team coordination and performance. They present several information sharing tools which allow users to update others with regard to their status thus reducing work duplication and infrequent communication. The authors investigated the impact on human-agent teams of team structure, task uncertainty, and information-sharing, including coordination and performance. From their perspective, in the future, search and rescue, command and control, and air traffic control operators will be working in teams with robot teammates. But teams involve tasks that individual humans cannot do at all or are inefficient at doing. The authors contrasted organizational structures based on divisional teams, where self-contained redundancy governs under high uncertainty to make them more robust; and functional teams, where uncertainty is low and predictability is high. They discussed team situational awareness, where each member's contributions to and impacts on team tasks must be predictable and appreciated. The authors also discussed the costs of coordination and communication; and that these costs and duplication could be reduced with information-sharing tools, while increasing robustness for divisional teams. The authors found that information sharing tools allowed users to communicate more effectively.

The author of Chap. 10, Kristin Schaefer, is with the U.S. Army Research Laboratory. In her article on "*Measuring Trust in Human Robot Interactions: Development of the 'Trust Perception Scale-HRI'* ", she studied the importance of trust in human-robot interaction and teaming as robotic technologies continue to improve their functional capability, robust intelligence, and autonomy. The author explores the development of a unifying survey scale to measure a human user's trust in a robotic system and in Human Robot Interaction (HRI) settings. She presents a series of related experiments leading to the creation of a 40 item survey which she argues measures trust across multiple conditions and robot domains. In this chapter, the author has summarized her PhD research to produce a reliable and valid subjective measure of the trust humans have of robots, the *Trust Perception Scale-HRI*. She performed an extensive literature review of trust in the interpersonal, automation and robot domains to determine if specific attributes accounted for human-robot trust. Schaefer developed an initial pool of items, tested it with human subjects, analyzed the results with a mental model of a robot, and reduced the number of items based on statistical and Subject Matter Expert (SME) content-validation procedures. This resulted in her 42 item scale plus a 14 item shorter scale derived from the feedback by her SMEs. She then used computer simulated human-robot interaction experimentation for a two-part task-based validation process to determine if the scale could measure a change in survey scores and measure the construct of trust. She first demonstrated that the scale measured a change pre-post interaction and across two reliability conditions (100% reliable feedback versus 25% reliable feedback) during a supervisory human-robot target detection task. This was followed by a second validation experiment using a Same-Trait approach during a team Soldier-robot navigation task. Her finalized 40-item scale performed well in both cases, and provided support for additional benefits when used in the HRI domain, above and beyond results achieved when using a previously developed automation-specific trust scale.

Chapter 11 is titled "*Methods for developing trust models for intelligent systems*"; its authors are Holly A. Yanco, Munjal Desai, Jill L. Drury and Aaron Steinfeld with the University of Massachusetts Lowell (Yanco and formerly Desai), The MITRE Corporation (Drury and Yanco), and Carnegie Mellon University (Steinfeld). The number of robots in use across the width of society, including in industry, with the military, and on the highways, is increasing rapidly, along with an expansion of their abilities to operate autonomously. Benefits from autonomy are increasing rapidly along with concerns about how well these systems can be, should be, and are being trusted. Human automation interaction (HAI) research is crucial to the further expansion of intelligence, but also its disuses and abuses. The research by the authors is designed to understand and model the factors that affect intelligent systems. The chapter begins with a review of prior research in the development of trust models, including surveys and experiments. Then the authors discuss two methods for investigating trust and creating trust models: surveys and robot studies.

They also produce 14 guidelines as well as an overall model of trust and the factors that increase and decrease trust. Finally, the authors review their conclusions and discuss the path forward.

In Chap. 12, titled, "*The intersection of robust intelligence and trust: Hybrid teams, firms & systems*", the authors are W.F. Lawless with Paine College and Donald Sofge with the Naval Research Laboratory; they are developing the physics of interdependent relations among social agents to reflect uncertainty arising from these relationships but also the power of social groups to solve difficult problems. Interdependence depends on the existence of alternative (bistable) interpretations of social reality. Interdependence makes social situations non-linear and non-intuitive, making interdependence a difficult problem to address. But if this problem can be solved, unlike today when robots work as individual agents, it will allow humans, machines and robots to work together in teams by multitasking to solve problems that only human teams can now solve. On the other hand, as interdependence increases across a group, its chances increase that it can make a mistake. Traditional models of interdependence consist primarily of traditional game theory. But game theory's solution of this problem relies heavily on increasing cooperation, thereby increasing static interdependence, further increasing the likelihood of a mistake. To avoid mistakes, the authors argue for a competitive situation similar to a Nash equilibrium, where the two sides engage in a nonlinear competition for neutrals (independent agents) to determine the winning argument at one point in time; mathematically, the result is a limit cycle as one side wins, but then that side falls behind in the next argument when the limits to its "solution" become apparent. The result is a method that increases social welfare. The authors describe how this may work in human-machine-robot environments.

References

Army (2014) Army robotics researchers look for into the future, D. Mcdally, RDECOM public affair, from http://www.army.mil/molule/article/?p=137837

CMU (2014) http://www.darpa.mil/Our_Work/TTO/Programs/DARPA_Robotics_Challenge/Track_A_Participants.aspx

Darpa (2014) http://www.darpa.mil/Our_Work/DSO/Programs/Physical_Intelligence.aspx

du Sautoy M (2014) The guardian (2012, 3/31), "AI robot: how machine intelligence is evolving". http://www-03.ibm.com/software/products/us/en/intelligent-operations-center/

Gluck K (2013) Robust decision making in integrated human-machine systems, U.S. Air Force BAA: 13.15.12.B0909

Hannibal ME (2012) Why the beaver should thank the wolf, New York Times. http://www.nytimes.com/2012/09/29/opinion/the-world-needs-wolves.html

IBM (2014) Smarter Cities. http://www.ibm.com/smarterplanet/us/en/smarter_cities/overview/

IEEE (2014) http://www.computer.org/csdl/mags/ex/2013/01/index.html

Lawless WF, Llinas J, Mittu R, Sofge DA, Sibley C, Coyne J, Russell S (2013) Robust Intelligence (RI) under uncertainty: mathematical and conceptual foundations of autonomous hybrid (human-machine-robot) teams, organizations and systems. Struct Dyn 6(2)
Marcus G, Marblestone A, Dean T (2014) The atoms of neural computation. Science 346:551–552
NSF (2013) Robust Intelligence (RI); National Science Foundation. http://www.nsf.gov/funding/pgm_summ.jsp?pims_id=503305&org=IIS
ONR (2013) Command Decision Making (CDM) & Hybrid Human Computer Systems (HHCS) Annual Program Review, Naval Research Lab, Washington, DC, 4–7 June 2013

Chapter 2
Towards Modeling the Behavior of Autonomous Systems and Humans for Trusted Operations

Gavin Taylor, Ranjeev Mittu, Ciara Sibley, and Joseph Coyne

2.1 Introduction

Unmanned systems will perform an increasing number of missions in the future, reducing the risk to humans, while increasing their capabilities. The direction for these systems is clear, as a number of Department of Defense roadmaps call for increasing levels of autonomy to invert the current ratio of multiple operators to a single system (Winnefeld and Kendall 2011). This shift will require a substantial increase in unmanned system autonomy and will transform the operator's role from actively controlling elements of a single platform to supervising multiple complex autonomous systems. This future vision will also require the autonomous system to monitor the human operator's performance and intentions under different tasking and operational contexts, in order to understand how she is influencing the overall mission performance.

Successful collaboration with autonomy will necessitate that humans properly calibrate their trust and reliance on systems. Correctly determining reliability of a system will be critical in this future vision since automation bias, or overreliance on a system, can lead to complacency which in turn can cause errors of omission and commission (Cummings 2004). On the other hand, miscalibrated alert thresholds and criterion response settings can cause frequent alerts and interruptions (high false alarm rates), which can cause humans to lose trust and underutilize a system (i.e., ignore system alerts) (Parasuraman and Riley 1997). Hence, it is imperative that not only does the human have a model of normal system behavior in different

G. Taylor (✉)
United States Naval Academy, Annapolis, MD, USA
e-mail: taylor@usna.edu

R. Mittu • C. Sibley • J. Coyne
Naval Research Laboratory, Washington, DC, USA
e-mail: ranjeev.mittu@nrl.navy.mil; ciara.sibley@nrl.navy.mil; joseph.coyne@nrl.navy.mil

© Springer Science+Business Media (outside the USA) 2016
R. Mittu et al. (eds.), *Robust Intelligence and Trust in Autonomous Systems*,
DOI 10.1007/978-1-4899-7668-0_2

contexts, but that the system has a model of the capabilities and limitations of the human. The autonomy should not only fail transparently so that the human knows when to assist, but autonomy should also predict when the human is likely to fail and be able to provide assistance. The addition of more unmanned assets with multi-mission capabilities will increase operator demands and may challenge the operator's workload just to maintain situation awareness. Autonomy that monitors system (including human) behavior and alerts users to anomalies, however, should decrease the task load on the human and support them in the role of supervisor.

Noninvasive techniques to monitor a supervisor's state and workload (Fong et al. 2011; Sibley et al. 2011) would provide the autonomous systems with information about the user's capabilities and limitations in a given context, which could provide better prescriptions for how to interact with the user. However, many approaches to workload issues have been based on engineering new forms of autonomy assuming that the role of the human will be minimized. For the foreseeable future, however, the human will have at least a supervisory role within the system; rather than minimizing the actions of the human and automating those actions the human can already do well, it would be more efficient to develop a supervisory control paradigm that embraces the human as an agent within the system and leverages on her capabilities and minimizes the impact of her limitations.

In order to best develop techniques for identifying anomalous behaviors associated with the complex human-autonomous system, models of normal behaviors must be developed. For the purpose of this paper, an anomaly is not just a statistical outlier, but rather a deviation that prevents mission goals from being met, dependent on the context. Such system models may be based on, for example, mission outcome measures such as objective measures of successful mission outcomes with the corresponding behaviors of the system. Normalcy models can be used to detect whether events or state variables are anomalous, i.e., probability of a mission outcome measure that does not meet a key performance parameter or other metric.

The anomalous behavior of complex autonomous systems may be composed of internal states and relationships that are defined by platform kinematics, health and status, cyber phenomena and the effects caused by human interaction and control. Once the occurrence and relationships between abnormal behaviors in a given context can be established and predicted, our hypothesis is that the operational bounds of the system can be better understood. This enhanced understanding will provide transparency about the system performance to the user to enable trust to be properly calibrated with the system, making the prescriptions for human interaction that follow to become more relevant and effective during emergency procedures.

A key aspect of using normalcy models for detecting abnormal behaviors is the notion of context; and behaviors should be understood in the context in which they occur. In order to limit the false alarms, effectively integrating context is a critical first step. Normalcy models must be developed for each context of a mission, and used to identify potential deviations to determine whether such deviations are anomalous (i.e., impact mission success). Proper trust calibration would be assisted through the development of technology that provides the user with transparency about system behavior. This technology will provide the user with information about how the system is likely to behave in different contexts and how the user should best respond.

We present an approach for modeling anomalies in complex system behavior; we do not address modeling human limitations and capabilities in this paper, but recognize that this is equally important in the development of trust in collaborative human-automation systems.

2.2 Understanding the Value of Context

The role of context is not only important when dealing with the behavior of autonomous systems, but also quite important in other areas of command and control. Today's warfighters operate in a highly dynamic world with a high degree of uncertainty, compounded by competing demands. Timely and effective decision making in this environment is challenging. The phrase "too much data—not enough information" is a common complaint in most Naval operational domains. Finding and integrating decision-relevant information (vice simply data) is difficult. Mission and task context is often absent (at least in computable and accessible forms), or sparsely/poorly represented in most information systems. This limitation requires decision makers to mentally reconstruct or infer contextually relevant information through laborious and error-prone internal processes as they attempt to comprehend and act on data. Furthermore, decision makers may need to multi-task among competing and often conflicting mission objectives, further complicating the management of information and decision making.

Clearly, there is a need for advanced mechanisms for the timely extraction and presentation of data that has value and relevance to decisions for a given context. To put the issue of context in perspective, consider that nearly all national defense missions involve Decision Support Systems (DSS)—systems that aim to decrease the cycle time from the gathering of data to operational decisions. However, the proliferation of sensors and large data sets are overwhelming DSSs, as they lack the tools to efficiently process, store, analyze, and retrieve vast amounts of data. Additionally, these systems are relatively immature in helping users recognize and understand important contextual data or cues.

2.3 Context and the Complexity of Anomaly Detection

Understanding anomalous behaviors within the complex human-autonomous system requires an understanding of the context in which the behavior is occurring. Ultimately, when considering complex, autonomous systems comprised of multiple entities, the question is not what is wrong with a single element, but whether that anomaly affects performance of the team and whether it is possible to achieve the mission goals in spite of that problem. For example, platform instability during high winds may be normal, whereas the same degree of instability during calm winds may be abnormal. Furthermore, what may appear as an explainable deviation may actually be a critical problem if that event causes the system to enter future states that prevent the satisfaction of a given objective function. The key distinction is that in certain settings, it may be appropriate to consider anomalies as those situations that effect outcomes, rather than just statistical outliers. In terms of the team, the question becomes which element should have to address the problem (the human or the autonomy).

The ability to identify and monitor anomalies in the complex human-autonomous system is a challenge, particularly as increasing levels of autonomy increase system complexity and, fundamentally, human interactions inject significant complexity via unpredictability into the overall system. Furthermore, anomaly detection within complex autonomous systems cannot ignore the dependencies between communication networks, kinematic behavior, and platform health and status.

Threats from adversaries, the environment, and even benign intent will need to be detected within the communications infrastructure, in order to understand its impact to the broader platform kinematics, health and status. Possible future scenarios might include cyber threats that take control of a platform in order to conduct malicious activity, which may cause unusual behavior in the other dimensions and corresponding states. The dependency on cyber networks means that a network provides unique and complete insight into mission operations. The existence of passive, active, and adversarial activities creates an ecosystem where "normal" or "abnormal" is dynamic, flexible, and evolving. The intrinsic nature of these activities results in challenges to anomaly detection methods that apply signatures or rules that have a high number of false positives. Furthermore, anomaly detection is difficult in large, multi-dimensional datasets and is affected by the "curse of dimensionality." Compounding this problem is the fact that human operators have limited time to deal with complex (cause and effect) and/or subtle ("slow and low") anomalies, while monitoring the information from sensors, and concurrently conducting mission planning tasks. The reality is that in future military environments, fewer operators due to reduced manning may make matters worse, particularly if the system is reliant on the human to resolve all anomalies!

Below we describe research efforts underway in the area of anomaly detection via manifolds and reinforcement learning.

2.3.1 Manifolds for Anomaly Detection

A fundamental challenge in anomaly detection is the need for appropriate metrics to distinguish between normal and abnormal behaviors. This is especially true when one deals with nonlinear dynamic systems where the data generated contains highly nonlinear relationships for which Euclidean metrics aren't appropriate.

One approach is to employ a nonlinear "space" called a manifold to capture the data, and then use the natural nonlinear metric on the manifold, in particular the Riemannian metric, to define distances among different behaviors.

We view the path of an unmanned system as a continuous trajectory on the manifold and recognize any deviations due to human inputs, environmental impacts, etc. Mathematically, we transform the different data types into a common manifold-valued data so that comparisons can be made with regard to behaviors.

For example, a manifold for an unmanned system could be 12 dimensional, composed of position, pitch, roll, yaw, velocities of the position coordinates, and angular velocities of the pitch, roll and yaw. This 12-dimensional model captures any platform (in fact any moving rigid object's) trajectories under all possible environment conditions or behaviors. This manifold is the tangent bundle, TM of $SO(3) \times \Re^3$. Here $SO(3)$ denotes the set of all possible rotations of the unmanned system which is a Lie group, and $\Re^3$ the set of all translations of the platform. Since rotations and translations do not commute, this is not a direct product of $SO(3)$ with $\Re^3$. The product between $SO(3)$ and $\Re^3$ is a "Semi-Product" $\ltimes$. Non-linear key geometric, dynamical and kinematic characteristics are represented using TM. This manifold model is able to encapsulate the unique structure of the environment, effects of human behaviors, etc. through continuous parameterizations and coherent relationships.

Once we have this manifold model and its Riemannian metric, it is possible to define concepts of geodesic neighborhood and other appropriate measurements and map those to mission cost. Such a mapping is done by designing a weighted cost function with dynamical neighborhoods around a trajectory of the platform. For example, if the weather is good in the morning, the neighborhood is smaller than it would be with bad weather. This innovative manifold method could be used to dynamically identify normal or abnormal behaviors occurring during a mission, taking into consideration whether a mission could be successfully achieved under a given cost constraint. We also have the freedom to adjust normal neighborhoods if a mission suddenly changes while en-route. Our model is robust and captures complicated dynamics of unmanned systems and is able to encapsulate very high dimensional data using only a 12 dimensional configuration space.

The algorithms use continuous parameterizations and coherent relationships and are scalable. Our manifold-based methods provide new techniques to combine qualitative (platform mechanics) and quantitative (measured data) methods and are able to handle large, nonlinear dynamic data sets.

2.4 Reinforcement Learning for Anomaly Detection

The military and commercial communities increasingly rely on autonomous systems to augment their capabilities. For example, power plants feature automatic monitoring and safety features, and the military increasingly employs unmanned systems in denied or politically sensitive theaters. However, it is rare for these systems to be truly autonomous; generally, fine-grained decisions are made by computer (such as an autopilot), while coarse-grained decisions (such as mission tasking or flight plans) are made by human operator. Thus, autonomy is not a binary on-off switch, but instead lies on a continuum. As technology progresses, it is expected that daily operations will shift along this continuum such that more and more decisions will be made by autonomous machine. This point in the continuum is referred to as supervised autonomy, in which human supervisors are tasked with maintaining awareness and identifying and responding to negative surprises, potentially for multiple systems simultaneously. In this scenario, it is important that accurate measures of mission progress be communicated quickly and succinctly, so that supervisors can direct their attention properly, particularly when overtasked.

In this section we make several points. First, we believe the field of Reinforcement Learning can provide the tools necessary to communicate this type of information. Second, we believe automated feature selection for Reinforcement Learning provides an intuitive way to identify features of interest in autonomous systems. Finally, we believe Reinforcement Learning may provide several additional benefits for negative anomaly detection and human decision-making analysis.

2.4.1 Reinforcement Learning

The purpose of this section is to describe the field of Reinforcement Learning, so that the new applications can be better understood. We begin by describing the problem intuitively, before introducing the mathematical background. Imagine, for some new unknown system, a human is presented with telemetry data entirely describing every aspect of the system (position, orientation, velocity, sensor readings, control surface position, etc.) for every second of a large number of missions. Given this data, it is possible to imagine the human learning what the system is capable of doing (for example, when learning about an unmanned aircraft, the human may learn that altitude gain can be achieved by setting particular control surfaces to certain settings, but there is a maximum climb rate which cannot be exceeded). Now imagine an expert assigns a "reward" to each observation; a state in which the mission is completed would receive a high reward, while a state in which the system is damaged would receive a negative reward. A logical question, then, is given some new state, what rewards can we expect to receive in the future? Are there certain actions we can take to maximize this expected future rewards? These questions are very important in a supervised context; if we expect to receive many high rewards, no intervention is required. However, if we are in a state with low expected rewards, retasking may be necessary. Reinforcement Learning provides algorithms for performing this predictive analysis.

Mathematically, the problem is described as a Markov Decision Process (MDP). An MDP is a tuple $M = (\mathcal{S}, \mathcal{A}, \mathcal{P}, R, \gamma)$, where $\mathcal{S}$ is the set of possible states the agent could inhabit, $\mathcal{A}$ is the set of actions the agent can take, $\mathcal{P}$ is a transition kernel describing the probability of transitioning between two states given the performance of an action (mathematically, $p(s'|s,a)$, where $s, s' \in \mathcal{S}$ and $a \in \mathcal{A}$), R is the reward function, and $\gamma \in (0, 1)$ is the discount factor, which is used to describe how much we prefer rewards in the short term to rewards in the long term. The expected future rewards from a given state is defined as the *optimal value* of that state, where the optimal value function is defined by the Bellman equation

$$V^*(s) = R(s) + \gamma \max_a \int_{\mathcal{S}} p(s'|s,a)V^*(s')ds'. \tag{2.1}$$

This function defines the optimal value of a state as the expected, discounted sum of rewards from this state forward. In large or complex domains, it is impossible to calculate V^* exactly, forcing instead the construction of an approximation. A huge number of approaches to perform this approximation exist, each with their own benefits and drawbacks.

Classically, an accurate value function is useful for creating autonomy; if choosing between two states, the decision-making agent should choose the one with higher value. This turns long-term planning into a series of greedy, short-term decisions. Because value function approximation techniques are usually general, and can often be applied to any MDP, advancements in Reinforcement Learning lead to better autonomy in any domain.

This same generality is a drawback in practical application, however. While it is theoretically pleasing to begin with no assumptions or prior knowledge about the agent being observed, this is not usually the case. For example, in the case of a new UAV, the known dynamics of the UAV can usually be used to generate a better autopilot than would be generated by Reinforcement Learning, which ignores this tremendously useful prior information.

Therefore, the remainder of this section focuses not on the traditional applications of a value function for autonomy generation, but instead on new applications of this function in a supervised autonomy context.

2.4.2 Supervised Autonomy

In a supervised autonomy context, we assume the agent under most circumstances is capable of performing its task unassisted; the problem then becomes one of helping a possibly-overtasked human supervisor identify cases which do require human intervention. For example, consider the case of a human tasking multiple UAVs at a time to accomplish a variety of missions. Perhaps one of the systems being watched is making slower-than-expected progress, and so another system will have to be tasked to complete its mission before the opportunity expires. Or, more dramatically, perhaps a malfunction occurs and human intervention is required to land the UAV away from the runway with as little damage as possible, in a location easy for recovery. In these cases, the identification of states from which we expect to receive low rewards is an appropriate way to identify moments which are unlikely to result in mission success.

Intuitively, the value of a state can be thought of as a one-dimensional measure of predicted success. In scenarios where supervisors must keep situational awareness on several independent agents, the distillation of this information into something quickly digestible and understandable is essential. Because the value function is built based on what is expected to occur, given past behavior and an optimal autopilot, a small or negative value function may be an extremely useful indicator of possible mission failure.

We note that the level of accuracy necessary for providing this feedback to a supervisor or performing this analysis is far less than the level of accuracy an agent requires to actually choose actions based entirely on the value function.

2.4.3 Feature Identification and Selection

This viewpoint also may provide ancillary benefits. Mathematically, many approximation schemes take the form of a linear approximation, where $V(s) \approx \Phi(s)w$, for all $s \in \mathscr{S}$. In this approximation, $\Phi(s)$ is a set of feature values for state s, and w is a vector which linearly weights these features. The quality of the approximation depends in large part on the usefulness of these features. Therefore, techniques which can select the few features from a large dictionary which result in the highest quality approximation can provide a great benefit, and are an active area of study (Kolter and Ng 2009; Johns et al. 2010; Mahadevan and Liu 2012; Liu et al. 2012). Interestingly for this domain, the identification of features which are highly correlated with future success and failure may provide an intuitive means of understanding the domain and in the construction of early-warning alarms.

One such technique for performing automated feature selection while calculating an approximate value function is L_1-Regularized Approximate Linear Programming (RALP) (Petrik et al. 2010). RALP is unique among other feature-selection approaches in that it has tightly bounded value function approximation error, even when using off-policy samples, and when the domain features a great deal of noise (Taylor and Parr 2012).

RALP works by solving the following convex optimization problem:

$$\begin{aligned} &\min_w \rho^T \Phi w \\ &\text{s.t. } \sigma^r + \gamma \Phi(\sigma^{s'}) w \leq \Phi(\sigma^s) w \;\; \forall \sigma \in \Sigma \\ &\|w\|_{-1} \leq \psi. \end{aligned} \tag{2.2}$$

Φ is the feature matrix, Σ is the set of all samples, where σ is one such sample, in which the agent started at state σ^s, received reward σ^r, and transitioned to state $\sigma^{s'}$. ρ is a distribution, which we call the state-relevance weights, in keeping with the (unregularized) Approximate Linear Programming (ALP) terminology of de Farias and Van Roy (2003). $\|w\|_{-1}$ is the L_1 norm of the vector consisting of all weights excepting the one corresponding to the constant feature.

This final constraint, which contributes L_1 regularization, provides several benefits. First, regularization in general ensures the linear program is bounded, and produces a smoother value function. Second, L_1 regularization in particular produces a sparse solution, producing automated feature selection from an over-complete feature set. Finally, the sparsity results in few of the constraints being active, speeding the search for a solution by a linear program solver, particularly if constraint generation is used.

Most relevantly to this problem, the selected features are those most useful in predicting the future receipt of rewards. Practically, these can be interpreted as the few features, from a potentially large candidate set, which carry the strongest signal predicting success or failure; while this definitely improves prediction accuracy, it may also be useful in related tasks in which an intuitive prediction of future autonomous performance will be useful, such as in scenarios of transfer learning.

2.4.4 Approximation Error for Alarming and Analysis

We also propose an exciting direction of research in using the error in approximation as a useful signal in flight analysis and alarming. Consider again the Bellman equation of Eq. (2.1), and a single observation, consisting of a state, a received reward, and the next state (an observation can be denoted (s, r, s'), where $s, s' \in \mathcal{S}$, and $r = R(s)$). Also assume the existence of some approximation $\hat{V}$, such that $\hat{V}(s) \approx V^*(s)$ for all $s \in \mathcal{S}$. If the approximation is exactly correct, and everything goes as expected, then $\hat{V}(s) = r + \gamma\hat{V}(s')$ for all such observations. If this equality does not hold, then we have non-zero *empirical Bellman error*

$$BE(s) = r + \gamma\hat{V}(s') - \hat{V}(s). \tag{2.3}$$

When the Bellman error is non-zero, there are several possible explanations.

The first explanation is that the approximation to the value function was incorrect. This explanation has classically received the most attention, as researchers tried to tune their predictions to be as accurate as possible. However, even if the value function $\hat{V} = V^*$ were calculated exactly, in a non-deterministic environment there would still be non-zero Bellman error.

A second explanation is that the prediction would have been accurate, but the controller (autopilot or human) chose an action which was not optimal. We can expect this to happen consistently to some degree. This is because optimality is defined as choosing the action that maximizes the sum of rewards; however, existing autopilot and human decision making is performed using an approach independent of Reinforcement Learning, and ignorant of the designed reward function. Realistically, all approaches are targeting some approximation of optimal performance, and it is unreasonable to expect them to precisely agree on optimal decision making. However, we argue that decision making resulting from these differing approaches should not differ too much, and that unexpected large drops in the value function would be interesting in performing objective performance assessment, such as in human training and flight analysis.

A third explanation is that the approximation was accurate, but something unexpected happened during the state transition. A positive example would be if a tailwind pushed the aircraft closer to a target location than would normally be expected; in this case, the value of the next state would be larger than expected. More interestingly for supervised autonomy, however, would be in identifying unexpected drops in value, or negative anomalies. It is perhaps obvious this would be useful in identifying sudden, dramatic drops, as might result from an event such as mechanical failure. More promising, however, is the potential for identifying prolonged periods of underperformance which might indicate a "low-and-slow" anomaly which might otherwise be difficult to detect. These types of anomalies occur when a small error impacts an output over a long period of time. These errors are difficult to detect as they typically do not cross any predefined error boundaries.

To our knowledge, the Bellman error has never been used to perform this type of analysis.

2.4.5 *Illustration*

To illustrate the insights of Sects. 2.4.2 and 2.4.4, we use data from two different domains. First, we use a simulated domain often used for benchmarking value function approximation algorithms. This domain provides an easily-predicted environment to demonstrate the application of reinforcement learning in an easily-visualized way. Second, we use real world data collected by the in-flight computers of landing TigerShark UAVs to demonstrate the technique's efficacy even in less-predictable, real-world scenarios.

2.4.5.1 Synthetic Domain

We use a common Reinforcement Learning benchmark domain, in which an autonomous agent learns to ride a bicycle (Randløv and Alstrøm 1998; Lagoudakis and Parr 2003). In this domain, the bicycle begins facing north, and must reach a goal state 1 km to the east. To be precise, a state in the space consists of the bicycle's angle from perfectly upright, the angular velocity and angular acceleration with respect to the bicycle and the ground, the handlebar angle and angular velocity of the handlebar, and the angle of the goal state from the current direction of travel. Actions include leaning left or right, turning the handlebars left or right, or doing nothing at all. The reward function is a shaping function based on progress made towards the goal.

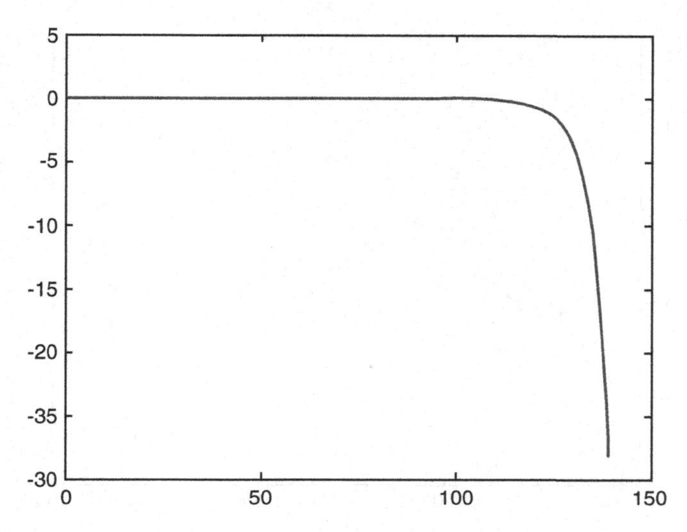

Fig. 2.1 Value as a function of time

Samples to learn from are drawn by beginning in an upright position, and taking random actions until failure. Given a large enough set of such samples, the agent can learn a value function, which it then uses to balance and successfully navigate through the space on a high percentage of trials. To do this, the approximate value function was constructed using RALP, using features composed of a large number of monomials defined on dimensions of the state space.

As our illustration, we consider one run, in which the bicyclist fails to remain upright despite access to a good previously-calculated value function. As the agent moves through the state space, the value function of its current state changes as a function of time. Figure 2.1 graphs this value until failure. Notice that as the bicycle begins to oscillate and eventually collapse, the expectation begins to drop, as it becomes less and less likely the agent will be able to recover. It is easy to imagine an alarm system set off in the beginning stages of this process, alerting a supervisor of the coming failure.

Given the same run, we also plot the absolute value of the empirical Bellman error of Eq. (2.3) as a function of time in Fig. 2.2. Notice the predictions were extremely accurate, and Bellman error very low, until shortly before the bicycle began to fall. In analyzing the performance of the agent, this information provides insight into when the initial mistakes were made which would, several time steps in the future, result in a failed mission; in the analysis of the agent's performance, this is crucial information.

If, however, the interest is not in analyzing past performance, but is in identifying anomalies which may require human intervention in real time, the Bellman error illustrated in Fig. 2.2 would again be extremely useful. It is plainly visually obvious when unexpected behavior began, and it is equally clear it correlates with the

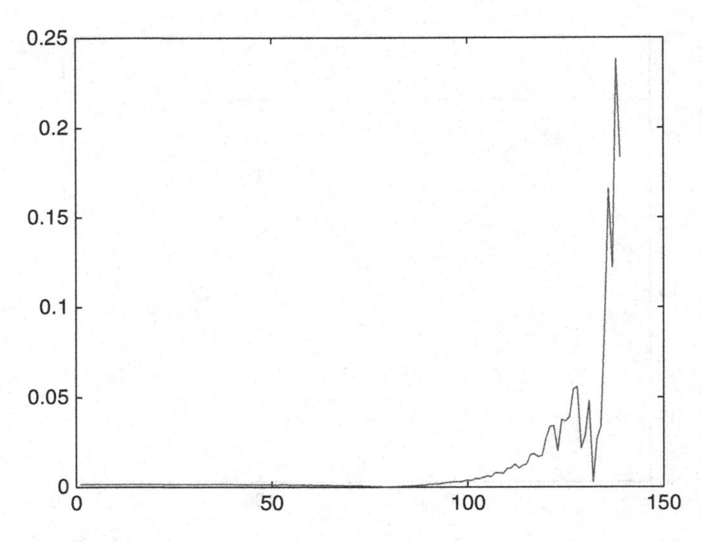

Fig. 2.2 Empirical Bellman error as a function of time

beginnings of the task failure. In particular, we would like to emphasize the anomalies begin to appear well before the bicycle collapses, allowing it to serve as an early warning system.

2.4.5.2 Real-World Domain

The TigerShark Unmanned Aerial Vehicle is a remotely-operated surveillance aircraft with a 22-foot wingspan and weighing 260 pounds, currently in use in theater operations. It can perform autonomous waypoint navigation, or be controlled manually. For this domain, we focus on autonomous alarming for TigerShark landings.

Telemetry data from 1267 landing attempts were collected, for a total of approximately 6.2 million individual samples, which were collected every 1000 ms. Landings were performed both by human pilots and the autopilot. The samples themselves are made up of 192 different sensor readings from the UAV control and sensor surfaces, as well as sensor data taken on the ground. Aside from flight status information like altitude, latitude, longitude, ground speed, etc., the data includes information about the controls coming from the pilot or autopilot, and the connection between the ground station and the UAV.

During landing, pilots have instructions to not allow their roll to extend past a certain angle, to follow a prescribed glide slope, and to keep their aircraft as close to the extended center line of the runway as possible. Therefore, in addition to the sensor readings, features also included the glide slope, the distance from the desired glide slope squared, the distance from the center line squared, the roll squared, and the reward awarded to the state.

The reward function was a shaping function based around the error from the desired glide slope, the distance from the center line of the runway, and the absolute value of the roll. In addition, the reward function contained additional penalties if the aircraft left the extended center third of the runway, or performed too large of a roll. By rule, either of the latter two cases should result in a decision to abort the landing.

This reward function differs from the one in the bicycle domain in a very significant way. In the synthetic domain, the actions taken by the autonomous agent were taken with the purpose of maximizing the rewards received. In the real-world domain, the actions taken are simply those of a pilot attempting to land an aircraft. As such, we can expect a much larger Bellman error in most states, as the value function is making a prediction on the erroneous assumption the pilot is aware of, and is trying to maximize, the reward function. Nevertheless, we will see the Bellman error is nonetheless helpful.

From the set of all landing samples, 2000 were randomly selected, and a value function computed using RALP. We then applied this value function to two separate flights. In the first flight, though the pilot eventually succeeded in landing the plane, the path was erratic and the landing should have been aborted due to an inability to stay within the allowed center third of the runway, and inability to maintain the roll within allowed parameters.

In Fig. 2.3 we see the value function of this landing. The shaded area is an area of special interest, as it is quite low, and produces the lowest value of the landing. We zoom on this area for both the value function (Fig. 2.4) and the Bellman error (Fig. 2.5). First, we note that both the value function and the Bellman error are

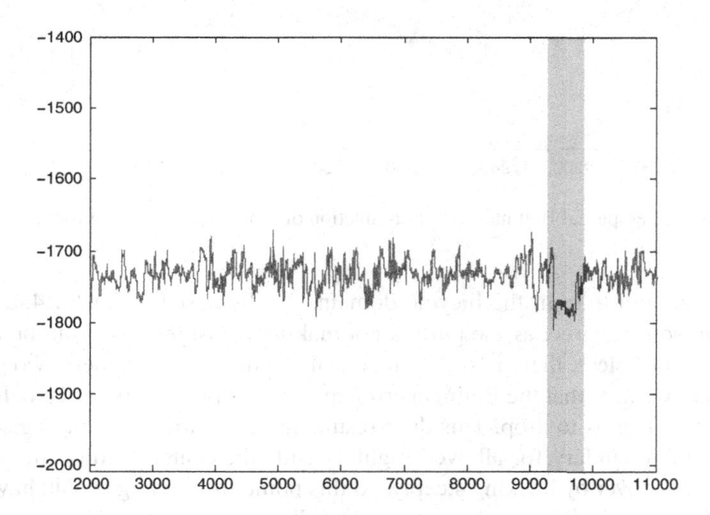

Fig. 2.3 Value as a function of time on unsuccessful flight

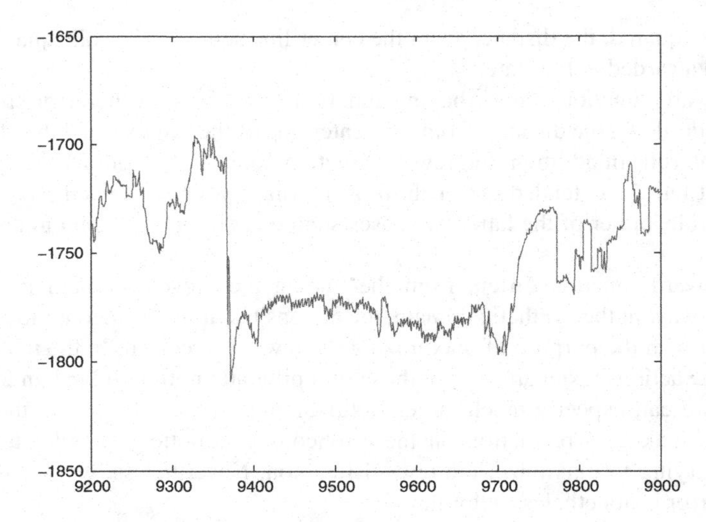

Fig. 2.4 Zoomed value as a function of time on unsuccessful flight

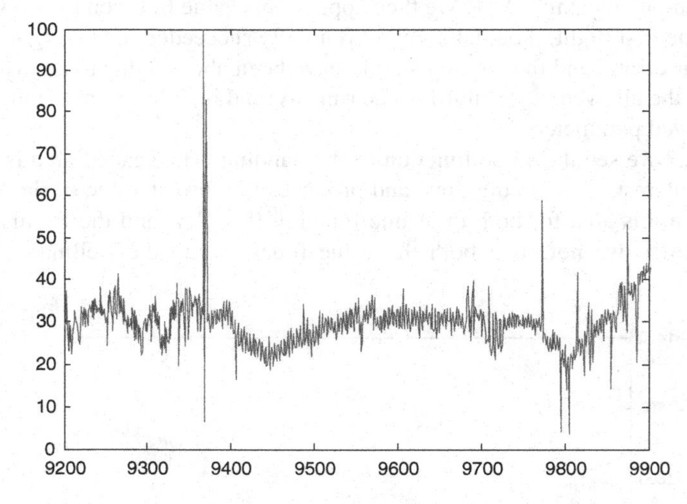

Fig. 2.5 Zoomed empirical Bellman error as a function of time on unsuccessful flight

much less smooth than in the bicycle domain. As discussed in Sect. 2.4.4, this is expected to some degree, as the pilot is not making decisions based on the reward function. Nevertheless, there is still a great deal of information in these two graphs. In particular, we note that the Bellman error spikes greatly in the timestep before the value function begins to drop. This drop results from the aircraft swinging so wide as to cross the boundary for allowed flight, outside the center third of the runway, and trying to recover by banking steeply. At this point, the landing should have been aborted. The spike in Bellman error preceding this drop illustrates the potential for

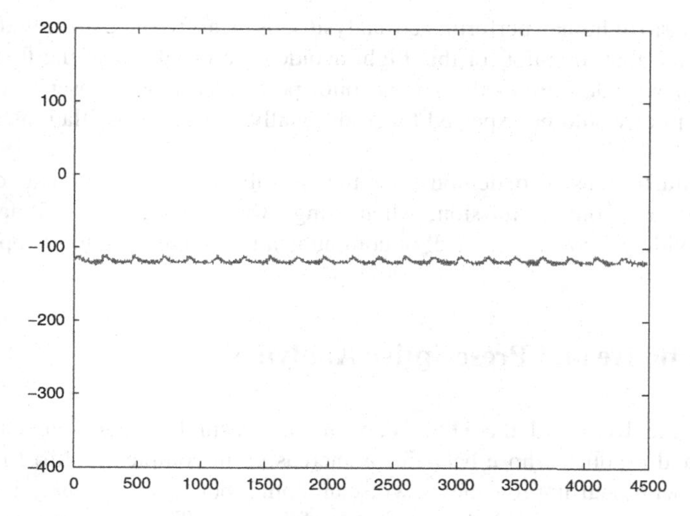

Fig. 2.6 Value as a function of time on successful flight

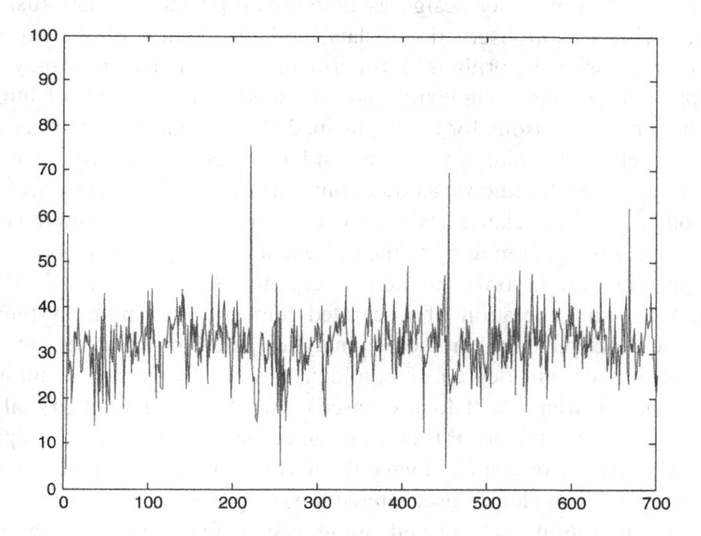

Fig. 2.7 Empirical Bellman error as a function of time on successful flight

a Bellman error-based alarm. In contrast, Figs. 2.6 and 2.7 are the result of a flight without any such negative occurrences (Fig. 2.7 is zoomed to provide the same scale as Fig. 2.5). Note that the two value functions (Figs. 2.3 and 2.6) are scaled identically on the y axis, though the bad value function is much, much lower, as you would expect from a flight struggling to stay on course. Second, we note the value function for the good flight is much less erratic. The Bellman error, meanwhile, remains generally low, with no significant spikes to set off an alarm.

In a scenario of human performance analysis, it is clear from the magnitude of the value function that the pilot for this flight avoided the problems of the first flight; the small magnitude implies the second pilot performed much closer to the way an optimal pilot would be expected to. Additionally, the small oscillations imply a steadier flight.

These illustrations also demonstrate the usefulness of the Bellman error in alarming an autonomous mission; when things went wrong, the Bellman error spiked, providing a concise channel of communication of the error to the operator.

2.5 Predictive and Prescriptive Analytics

It is clear the DoD and the U.S. Navy are increasingly reliant on autonomy, machines and robotics whose behavior is increasing in complexity. Most research indicates operational improvements with autonomy, but autonomy may introduce errors (Manzey et al. 2012) that impact performance. These errors result from many factors, including faulty design assumptions especially in data fusion aids, stochasticity with sensor/observational data, and the quality of the information sources fed into fusion algorithms. Furthermore, additional factors may include greater sophistication and complexity and the subsequent inability of humans to fully comprehend the reasons for decisions made by the automated system (i.e., a lack of transparency). In spite of these known faults, some users rely on autonomy more than is appropriate, known as autonomy "misuse" (Parasuraman and Riley 1997). Another bias associated with human-automation collaboration is disuse, where users underutilize autonomy to the detriment of task performance.

We conjecture that in order to help overcome issues associated with misuse/disuse, the next generation of integrated human-autonomous systems must build upon the descriptive and predictive analytics paradigm of understanding and predicting, with a certain degree of confidence, what the complex autonomous system has done and what it will do next based on what is considered normal for that system in a given context. Once this is achievable, it will enable the development of models that proactively recommend what the user should do in response in order to achieve a prescriptive model for user interactions (Fig. 2.8).

By properly presenting current and future system functioning to the user, and capturing user interactions in response to such states, we believe more effective human-automation collaboration and trust calibration can be established. The key question is "how are the best user interactions captured?"

2.6 Capturing User Interactions and Inference

Transparency in how a system behaves should enable the user to calibrate their level of trust in the system. However, there are still significant challenges that remain

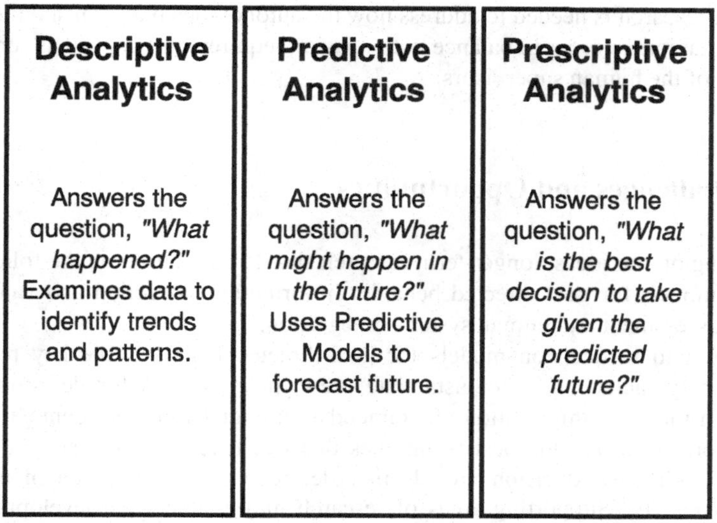

Descriptive Analytics

Answers the question, *"What happened?"* Examines data to identify trends and patterns.

Predictive Analytics

Answers the question, *"What might happen in the future?"* Uses Predictive Models to forecast future.

Prescriptive Analytics

Answers the question, *"What is the best decision to take given the predicted future?"*

Fig. 2.8 Different forms of analytics

with regard to capturing and understanding the human dimensions of supervisory control in order to provide prescriptions for interaction. We envision several longer term challenges related to the notion of prescriptive analytics, specifically how best to understand and model the information interaction behaviors of the user. These information seeking behaviors may be in reference to the potential anomalies in the system, in relation to what is provided by the on-board sensors, etc. and may require the development of the following capabilities:

- Adequately capturing users' information interaction patterns (and subsequently user information biases)
- Reasoning about information interaction patterns in order to infer decision making context; for example, the work being done by researchers within the Contextualized Attention Metadata community and the Universal Interaction Context Ontology (Rath et al. 2009) might serve as a foundation
- Instantiating formal models of decision making based on information interaction behaviors (potentially using cognitive architectures)
- Leveraging research from the AI community in plan recognition to infer which decision context (model) is active, and which decision model should be active
- Recognizing decision shift based on work that has been done in the Machine Learning community with "concept drift," and assessing how well this approach adapts to noisy data and learns over time
- Incorporating uncertainty and confidence metrics when fusing information and estimating information value in relation to decision utility
- Using models of cognition and decision making (and task performance) to drive behavior development and interface development

Lastly, research is needed to address how the autonomous platform should adapt to user behaviors in order to balance both mission requirements as well as servicing the needs of the human supervisors.

2.7 Challenges and Opportunities

Elaborating on our ideas, longer-term research should be focused on the following: decision models for goal-directed behavior, information extraction and valuation, decision assessment and human systems integration.

With regard to decision models for goal-directed behavior, the key research question may include how to instantiate prescriptive models for decision making, which integrate information recommendation engines that are context-aware. Furthermore, what are the best techniques that can broker across, generalize, or aggregate individual decision models in order to enable application in broader mission contexts? Supporting areas of research may include the development of similarity metrics that enable the selection of the appropriate decision model for a given situation, and intuitive decision model visualizations.

The notion of information extraction and valuation would involve locating, assessing, and enabling, through utility-based exploitation, the integration of high-value information within decision models, particularly in the big data realm. This is a particular research challenge due to heterogeneous data environments when dealing with unmanned systems. In addition, techniques that can effectively stage relevant information along the decision trajectory (while representing, reducing and/or conveying information uncertainty) would enable a wealth of organic data to be maximally harvested.

In reference to decision assessment, research needs to address what are the most effective techniques for modeling decision "normalcy," in order to identify decision trajectories that might be considered outliers and detrimental to achieving successful outcomes in a given mission context. Furthermore, techniques that proactively induce the correct decision trajectory to achieve mission success are also necessary. Metrics for quantifying decision normalcy in a given context can be used to propose alternate sequences of decisions or induce the exact sequence of decisions. This would require pre-staging the appropriate information needed to support the evaluation of decisions, potentially improving the speed and accuracy of decision making.

Lastly, with regard to human systems integration, the key challenges are in understanding, modeling and integrating the human state (workload, fatigue, experience) as well as the human decision making component as an integral part of the aforementioned areas. Specific topics include: representing human decision-making behavior computationally; accounting for individual differences in ability and preferences; assessing human state and performance in real-time (during a mission) in order to facilitate adaptive automation; mathematically capturing the human assessment of information value, risk, uncertainty, prioritization, projection and insight; and computationally representing human foresight and intent.

2.8 Summary

The development of robust, resilient, and intelligent systems requires the calibration of trust by humans when working with autonomous platforms. We contend that this can be enabled through a capability which allows system operators to understand anomalous states within the system of systems, which may lead to failures and hence impact system reliability. Likewise, the autonomy should understand the decision making capabilities and other limitations of the humans in order to proactively provide the most relevant information given the user's task or mission context.

This position paper has discussed the need for anomaly detection in complex systems in order to promote a human supervisor's understanding of system reliability. This is a challenging problem due to the increasing sophistication and growing number of sensor feeds in such systems which creates challenges for conducting big data analytics. Technical approaches that enable dimensionality reduction and feature selection should improve anomaly detection capabilities. Furthermore, building models that account for the context of each situation should improve the understanding of what is considered an anomaly. Additionally, we argue that anomalies are more than just statistical outliers, but should also be based upon whether they hinder the ability of the system to achieve some target end state. Understanding anomalies, we believe, should inform, and make more effective, the user's interaction with system. The interaction may include learning more about the anomaly through some form of query, command and control of the situation, entering into some emergency control procedure, etc.

Numerous research questions remain about the most effective interactions between human and autonomy. We believe the following research areas require further exploration in order to build more robust and intelligent systems. First, researchers should seek to capture users' interaction patterns (and subsequently user information biases) and reasoning about interaction patterns in order to infer decision making context. The work being done by researchers within the Contextualized Attention Metadata community and the Universal Interaction Context Ontology (Rath et al. 2009) might serve as a foundation for this approach. Second, instantiating formal models of human decision making based on interaction behaviors would lead to autonomous recognition of human capabilities and habits. Third, leveraging research from the AI community in plan recognition would allow for the inference of active decision contexts (model), and decision model selection. Fourth, adapting work that has been done in the Machine Learning community with concept drift, to recognize decision shifts and assessing how well this approach adapts to noisy data. Finally, it is necessary to incorporate uncertainty and confidence metrics when fusing information and estimating information value in relation to decision utility.

In order to build trusted systems which include a human component performing supervisory control functions, it is vital to understand the behaviors of the autonomy as well as the human (and his/her interaction with the autonomy). This should provide a holistic approach to building effective collaborative human-automation systems, which can operate with some level of expectation and predictability.

Acknowledgements The collaboration between the Naval Research Lab and the US Naval Academy was supported by the Office of Naval Research, grant number N001613WX20992.

References

Cummings M (2004) Automation bias in intelligent time critical decision support systems. In: AIAA 3rd intelligent systems conference

de Farias DP, Van Roy B (2003) The linear programming approach to approximate dynamic programming. Oper Res 51(6):850–865

Fong A, Sibley C, Coyne J, Baldwin C (2011) Method for characterizing and identifying task evoked pupillary responses during varying workload levels. In: Proceedings of the annual meeting of the human factors and ergonomics society

Johns J, Painter-Wakefield C, Parr R (2010) Linear complementarity for regularized policy evaluation and improvement. In: Lafferty J, Williams CKI, Shawe-Taylor J, Zemel R, Culotta A (eds) Advances in neural information processing systems, vol 23, pp 1009–1017

Kolter JZ, Ng A (2009) Regularization and feature selection in least-squares temporal difference learning. In: Bottou L, Littman M (eds) Proceedings of the 26th international conference on machine learning, Omnipress, Montreal, Canada, pp 521–528

Lagoudakis MG, Parr R (2003) Reinforcement learning as classification: leveraging modern classifiers. In: Proceedings of the twentieth international conference on machine learning, vol 20, pp 424–431

Liu B, Mahadevan S, Liu J (2012) Regularized off-policy TD-learning. In: Proceedings of the conference on neural information processing systems (NIPS)

Mahadevan S, Liu B (2012) Sparse Q-learning with mirror descent. In: Conference on uncertainty in artificial intelligence

Manzey D, Reichenbach J, Onnasch L (2012) Human performance consequences of automated decision aids. J Cogn Eng Decis Mak 6:57–87

Parasuraman R, Riley V (1997) Humans and automation: use, misuse, disuse, abuse. Hum Factors 39:230–253

Petrik M, Taylor G, Parr R, Zilberstein S (2010) Feature selection using regularization in approximate linear programs for Markov decision processes. In: Proceedings of the 27th international conference on machine learning

Randløv J, Alstrøm P (1998) Learning to drive a bicycle using reinforcement learning and shaping. In: Proceedings of the 15th international conference on machine learning, pp 463–471

Rath A, Devaurs D, Lindstaedt S (2009) UICO: an ontology-based user interaction context model for automatic task detection on the computer desktop. In: CIAO '09: proceedings of the 1st workshop on context, information and ontologies

Sibley C, Coyne J, Baldwin C (2011) Pupil dilation as an index of learning. In: Proceedings of the annual meeting of the human factors and ergonomics society

Taylor G, Parr R (2012) Value function approximation in noisy environments using locally smoothed regularized approximate linear programs. In: de Freitas N, Murphy K (eds) Conference on uncertainty in artificial intelligence. Catalina island, California, pp 835–842

Winnefeld JA, Kendall F (2011) Unmanned systems integrated roadmap FY2011-2036. US Department of Defense. http://www.acq.osd.mil/sts/docs/Unmanned%20Systems%20Integrated%20Roadmap%20FY2011-2036.pdf

Chapter 3
Learning Trustworthy Behaviors Using an Inverse Trust Metric

Michael W. Floyd, Michael Drinkwater, and David W. Aha

3.1 Introduction

The addition of a robot to a human team can be beneficial if the robot improves the team's sensory capabilities, performs new tasks, or allows for operation in harsh environments (e.g., rough terrain or dangerous situations). This may allow the team to better achieve their goals, improve team productivity, or reduce the risk to humans. In some situations, it may not be possible to achieve team goals or guarantee human safety without the robot. However, in order to adequately use the robot the human teammates will need to trust it.

The need to trust a robot teammate is especially true when the robot operates in an autonomous or semi-autonomous manner. In these situations, a human operator would issue a command or delegate a task to the robot and the robot would act on its own to complete its assignment. A lack of trust in the robot could result in the operator underutilizing the robot (i.e., not assigning it tasks it is capable of completing), excessively monitoring the robot's actions, or not using the robot at all (Oleson et al. 2011). Any of these issues could result in an increased workload for human teammates or the possibility of the team being unable to achieve their goal.

M.W. Floyd (✉) • M. Drinkwater
Knexus Research Corporation, Springfield, VA, USA
e-mail: michael.floyd@knexusresearch.com; michael.drinkwater@knexusresearch.com

D.W. Aha
Naval Research Laboratory (Code 5514), Navy Center for Applied Research in AI, Washington, DC, USA
e-mail: david.aha@nrl.navy.mil

© Springer Science+Business Media (outside the USA) 2016

R. Mittu et al. (eds.), *Robust Intelligence and Trust in Autonomous Systems*,
DOI 10.1007/978-1-4899-7668-0_3

One possibility would be to design a robot that is guaranteed to operate in a trustworthy manner. However, this may be impractical if the robot is expected to handle changes in operators, environments, or mission contexts. These changes would make it impractical to elicit a complete set of rules for trustworthy behavior. Additionally, the way in which an operator measures its trust in the robot may be user-dependent, task-dependent, or time-varying (Desai et al. 2013). For example, if the robot received a command to navigate between two locations in an urban environment, one operator might prefer the task be performed as quickly as possible whereas another might prefer the task be performed as safely as possible (e.g., not driving down a road with heavy automobile traffic or potholes). Each of these operators has distinct preferences for how the robot should perform the task, which may conflict, and these preferences will influence how trustworthy they find the robot's behavior. Even if these preferences were known in advance, a change in context could influence the operator's preferences and what is considered trustworthy behavior. The operator who preferred the task be performed quickly would likely change their preference if the robot was transporting hazardous material, whereas the operator who preferred safety would likely change their preference in an emergency situation.

For a robot to behave in a trustworthy manner regardless of the operator, environment, or context, it must be able to evaluate its trustworthiness and adapt its behavior accordingly. The workload of the human teammates or time-critical nature of the team's mission may make in difficult to get explicit feedback from the operator about the robot's trustworthiness. Instead, the robot will use information from the standard interactions it has with the operator (i.e., being assigned tasks and performing those tasks). Such an estimate, which we refer to as an *inverse trust estimate*, differs from traditional computation trust metrics in that it measures how much trust another agent has in the robot rather than how much trust the robot has in another agent. Additionally, the inverse trust metric does not directly measure trust, since the information necessary to compute such a metric is internal to the operator, but instead estimates trust based on observable factors that are known to influence trust. In this chapter we examine how a robot can estimate the trust an operator has in it, adapt its behavior to become more trustworthy, and learn from previous adaptations so it can find trustworthy behaviors more quickly in the future.

In the remainder of this chapter we describe our inverse trust estimate and how the robot uses it to adapt its behavior. In Sect. 3.2 we examine related work on human-robot trust and adapting to user preferences. We define the robot's behavior and the aspects that it can modify in Sect. 3.3. Section 3.4 presents the inverse trust metric and Sect. 3.5 describes how that metric is used by the robot to guide behavior adaptation. An evaluation of trust-guided behavior adaption, in a simulated robotics domain, is provided in Sect. 3.6 and reports evidence that it can efficiently adapt the robot's behavior to align with the operator's preferences. Concluding remarks and potential areas for future work are presented in Sect. 3.7.

3.2 Related Work

Traditional computational trust metrics are used to measure the trust an agent should have in other agents (Sabater and Sierra 2005). The agent determines another agent's trustworthiness based on prior interactions or using feedback from peers (Esfandiari and Chandrasekharan 2001). However, these metrics are not applicable when attempting to determine how trustworthy an agent is in the eyes of another agent. The primary reason for this is because the agent will not have all of the other agent's internal reasoning information available to it (e.g., outcomes of past interactions, peer feedback, past experiences, internal model of trust). Instead, the agent will need to acquire a subset of this information and use that to infer trust. In the remainder of this section we will examine factors influencing trust in human-robot interaction and how agents can adapt their behavior to humans.

3.2.1 Human-Robot Trust

Factors that influence human-robot trust can be grouped into three main categories (Oleson et al. 2011): robot-related factors (e.g., performance, physical attributes), human-related factors (e.g., engagement, workload, self-confidence), and environmental factors (e.g., group composition, culture, task type). While numerous factors have been found to influence human-robot trust (e.g., Li et al. 2010; Kiesler et al. 2008; Biros et al. 2004), a meta-analysis of numerous studies found the strongest indicator of trust is the robot's performance (Hancock et al. 2011). Similarly, user's identified performance as being among the most important factors they considered in relation to automated cars and medical diagnosis (Carlson et al. 2014).

Kaniarasu et al. (2012) have examined the topic of inverse trust and use an online, performance-based measure to identify decreases in trust. Their measurement is based on the number of times a human takes control of the robot or warns the robot it is behaving poorly. They have extended this work to also identify increases in trust, but it requires direct feedback from the operator at regular intervals (Kaniarasu et al. 2013). Saleh et al. (2012) have also proposed a measure of inverse trust using a set of expert-authored rules. However, if the robot does not have access to direct feedback or predefined rules, these metrics would not be appropriate to use.

3.2.2 Behavior Adaptation

Shapiro and Shachter (2002) discuss why an agent with a reward function that is similar to the utility of the user is desirable; it ensures the agent acts in the interests of the user. The agent may need to perform behavior that appears to be sub-optimal if it better aligns with the preferences of the user. Their work involves identifying the underlying influences of the user's utility and modifying the agent's reward function accordingly. This is similar to our own work in that the agent is willing to behave sub-optimally in order to align with the user's preferences, but our robot is not given an explicit model of the user's reasoning process.

Conversational recommender systems (McGinty and Smyth 2003) use interactions with a user to tailor recommendations to the user's preferences. These systems make initial recommendations and then iteratively improve the recommendations through a dialog with the user. As the system learns the user's preferences through feedback, a model of the user is continuously refined. In addition to learning a user preference model, conversational recommender systems can also learn preferences for how the dialog and interactions should occur (Mahmood and Ricci 2009). Similarly, search systems have been developed that update their behavior based on a user's preferred search results (Chen et al. 2008). These systems use information from the links a user clicks to infer their preferences and update search rankings accordingly.

Learning interface agents assist users when performing a task (e.g., e-mail sorting (Maes and Kozierok 1993), schedule management (Horvitz 1999), note taking (Schlimmer and Hermens 1993)). These systems observe how users perform certain tasks and learn their preferences. Since these agents are meant to be assistive, rather than autonomous, they are able to interact with the user to get additional information or verify if an action should be performed. Similar to conversational recommender systems, learning interface agents are designed to be assistive with one specific task. In contrast, our robot does not know in advance the specific task it will be performing so it cannot bias itself toward learning preferences for that task.

Preference-based planning (Baier and McIlraith 2008) involves incorporating user preferences into automated planning tasks. These preferences are usually defined a priori but there has also been work to learn the planning preferences (Li et al. 2009). This approach learns the probability of a user performing actions based on the previous plans that user has generated. In our work, that would be equivalent to the operator controlling the robot and providing numerous demonstrations of the task. Such an approach would not be practical if there were time constraints or the operator did not have a fully constructed plan for how to perform the task. For example, the operator might have general preferences for how the robot should move between two locations without knowing the exact route it should take.

Our work also has similarities to other areas of learning during human-robot interaction. When a robot learns from a human, it is often beneficial for the robot to understand the environment from the perspective of the human (Berlin et al. 2006). Breazeal et al. (2009) have examined how a robot can learn from a cooperative human teacher by mapping its sensory inputs to how it estimates the human is viewing the environment. This allows the robot to learn from the viewpoint of the teacher and possibly discover information it would not have noticed from its own viewpoint. This is similar to our own work since it involves inferring information about the reasoning of a human. However, like preference-based planning, it involves observing a teacher demonstrate a specific task.

3.3 Agent Behavior

We assume that the robot, in addition to being autonomous or semi-autonomous, has the ability to control and modify some aspects of its behavior. This could include selecting among comparable algorithms (e.g., switching the path planning algorithm it uses), modifying parameter values, or changing the data it uses (e.g., using an environment map from an alternate source). We call these the *modifiable components* of the robot's behavior.

We define each modifiable component i to have a set of selectable values $\mathscr{C}_i$. If the robot has m modifiable components, its current behavior B is a tuple containing the currently selected value c_i for each modifiable component ($c_i \in \mathscr{C}_i$):

$$B = \langle c_1, c_2, \ldots, c_m \rangle$$

By changing one or more of these components, the robot can immediately influence its behavior by switching from its current behavior B to a new behavior B_{new}. These changes can occur multiple times over the course of operation, resulting in a sequence of behaviors $\langle B_1, B_2, \ldots, B_n \rangle$ that have been used. Since the robot is motivated to perform trustworthy behavior, behavior changes will occur because the current behavior B was found to be untrustworthy (or it anticipates the behavior will be untrustworthy given a change in the team's goals or mission context), at which time we want the robot to perform what it believes to be a more trustworthy behavior.

3.4 Inverse Trust Estimate

Traditional trust metrics, which measure how much trust an agent has in other agents, use information related to previous interactions with those agents or feedback from others to compute trustworthiness (Sabater and Sierra 2005). This information is likely internal to the agent and would not be accessible to other agents (both the agent whose trustworthiness is being measured and external observers). In a robotics context, the robot would not have access to the information the operator uses to measure the robot's trustworthiness. If the robot wanted to estimate its own trustworthiness, it would need a method to access the operator's beliefs.

One option would be to elicit explicit feedback from the operator about the trustworthiness of the robot, either at run-time (Kaniarasu et al. 2013) or after the task has been completed (Jian et al. 2000; Muir 1987). However, this might not be practical if the operator does not have time to provide feedback (e.g., a time-critical mission with a heavy operator workload) or there would be a significant delay in providing feedback (e.g., at the end of a multi-day search and rescue mission). Similarly, the operator might not provide accurate information when self-reporting on the robot's trustworthiness, either intentionally (e.g., not wanting to be overly critical of the robot) or unintentionally (e.g., denial). In these situations, even though the operator cannot explicitly tell the robot how trustworthy it is, it would still be beneficial for the robot to *infer* its trustworthiness.

Without full knowledge about how the operator measures trust or the necessary internal information to actually compute the trust value, the robot needs to rely on observable *evidence* of trust. As we discussed previously, there are numerous factors that have been found to influence a human's trust in a robot (Oleson et al. 2011). If the robot can directly observe some of these factors, it can attempt to estimate its own trustworthiness. However, some factors may not be easily observable or have clear models of how they influence trust (e.g., the physical appearance of the robot).

One factor that is observable to the robot and has been found to be the strongest indicator of human-robot trust is the robot's performance (Hancock et al. 2011; Carlson et al. 2014). The inverse trust estimate we present is based on the robot's performance and uses the number of times the robot completes an assigned task, fails to complete a task, or is interrupted while performing a task. The robot assumes that the operator will be satisfied with any completed tasks (i.e., the robot is performing well) and unsatisfied when tasks are failed or must be interrupted (i.e., the robot is performing poorly[1]). Task completion and interruptions have been found to align with changes in operator trust (based on user feedback (Kaniarasu et al. 2013) and post-run surveys (Kaniarasu et al. 2012)), so it serves as a viable option for the robot to estimate its own trustworthiness.

Our inverse trust estimate monitors whether the user's trust in the robot is increasing, decreasing, or remaining constant while the current behavior B' is being used. We estimate this value as follows:

$$Trust_{B'} = \sum_{i=1}^{n} w_i \times cmd_i,$$

where there were n commands issued to the robot while it was using the behavior B'. If the ith command ($1 \leq i \leq n$) was interrupted or failed it will decrease the trust estimate and if it was completed successfully it will increase the trust estimate ($cmd_i \in \{-1, 1\}$). The ith command also receives a weight w_i which denotes the relative importance of that command (e.g., a command that resulted in poor performance would likely be given less weight than a command that resulted in the robot injuring a human). While our inverse trust estimate uses a simple step function to represent the current estimate of trust, a more complex (or cognitively plausible) function could be used that more closely aligns with the operator's actual

[1]An interruption could also be a result of the operator identifying a more important task for the robot to perform or failures could be the result of unachievable tasks. The robot works under the assumption that those situations occur rarely and most failures/interruptions are a result of poor performance.

trust. However, the additional computational complexity of such a function might not provide additional benefits if, like with our robot, we seek general trends in trustworthiness rather than an exact trust value.

3.5 Trust-Guided Behavior Adaptation

The robot uses the inverse trust estimate to infer if its current behavior is trustworthy, untrustworthy, or it does not yet know. Since the trust estimate is being updated over time (after each success, failure, or interruption) the robot continuously monitors the estimate and compares it to two threshold values: the trustworthy threshold (τ_T) and the untrustworthy threshold (τ_{UT}).

If the trust estimate is between the two thresholds ($\tau_{UT} < Trust_{B'} < \tau_T$), the robot will not make any conclusions and will continue to monitor its trustworthiness. However, if the trust estimate reaches the trustworthy threshold ($Trust_{B'} \geq \tau_T$), the robot will conclude it has found a sufficiently trustworthy behavior. The robot will continue to use its current behavior, since it is believed to be trustworthy, but may continue to measure trustworthiness in case any changes occur (e.g., a new operator or new mission goals). Finally, if the trust estimate falls to or below the untrustworthy threshold ($Trust_{B'} \leq \tau_{UT}$), the robot will conclude that its current behavior is untrustworthy and should be changed. In this situation, the robot will perform behavior adaptation to switch to a new behavior.

Figure 3.1 shows an example of the robot's estimate of the trustworthiness of its current behavior. The robot initially starts from a baseline value, since it does not know if the behavior is trustworthy or untrustworthy, and updates the estimate as new evidence becomes available. In this example, the robot was issued five tasks to perform. The robot completed the first two tasks successfully (as indicated by the increases in the trust estimate), failed or was interrupted during the third task

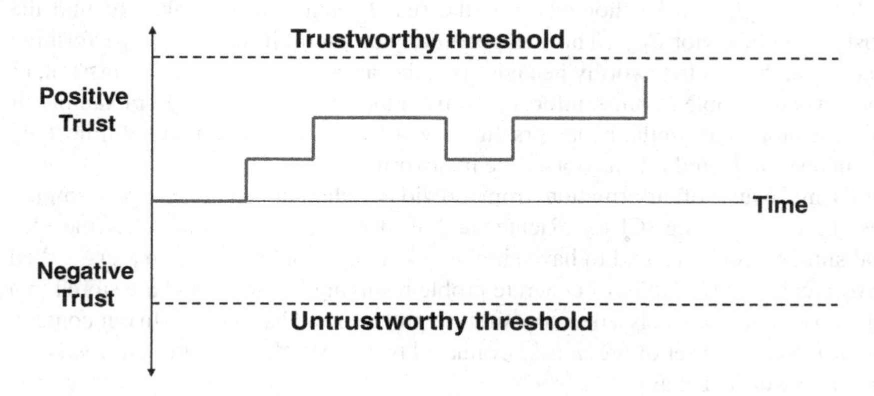

Fig. 3.1 An example of the robot's trust estimate after being issued five tasks

(as indicated by the decrease in the trust estimate), and then successfully completed the fourth and fifth tasks. The robot's trust estimate is trending upwards, but since it is still between the two thresholds it cannot conclude if its behavior is trustworthy or untrustworthy.

3.5.1 Evaluated Behaviors

When a behavior B is found to be untrustworthy (i.e., the trust estimate reached the untrustworthy threshold), it is stored as an *evaluated pair* E that also contains the time t it took the behavior to be labeled as untrustworthy:

$$E = \langle B,\ t \rangle$$

The motivation for storing the time it took to label a behavior as untrustworthy, instead of only storing the behavior itself, is that it allows for a comparison between untrustworthy behaviors. This permits a relative level of untrustworthiness so that we can say one behavior is closer to being trustworthy than another. A behavior B'' that reaches the untrustworthy threshold more quickly than another behavior B'' ($t' < t''$) is defined to be less trustworthy than the other. This is based on the assumption that if a behavior took longer to reach the untrustworthy threshold then it was either performing some trustworthy actions, was not failing as quickly, or was appearing to behave trustworthy for longer periods of time.

The robot maintains a set $\mathcal{E}_{past}$ of previously evaluated behaviors. This set, which is initially empty, is extended as the robot evaluates more behaviors. If the robot has found n behaviors to be untrustworthy then $\mathcal{E}_{past}$ will contain n evaluated behaviors ($\mathcal{E}_{past} = \{E_1, E_2, \ldots,\ E_n\}$). However, if the robot determines that a behavior B_{final} is trustworthy (i.e., the trustworthy threshold was reached), that behavior will not be added to the set and the robot will not change its behavior.

The set $\mathcal{E}_{past}$ can be thought of as the search path that was taken to find the trustworthy behavior B_{final}. This can potentially be useful if the robot is performing a new search for a trustworthy behavior (i.e., because of a new operator, mission, or context) and is able to reuse information from the previous search. For example, if two operators find similar behaviors untrustworthy in a similar amount of time, they might also find similar behaviors to be trustworthy.

To make use of information from previous behavior adaptation, we employ case-based reasoning (CBR) (Richter and Weber 2013). CBR embodies the idea that similar problems tend to have similar solutions. Problem-solution pairs, called *cases*, represent examples of concrete problem solving instances and are stored in a *case base*. Each case C is a pair containing a problem and its solution. In our context, the *problem* is the set of previously evaluated behaviors $\mathcal{E}_{past}$ and the *solution* is the final trustworthy behavior B_{final}:

$$C = \left\langle \mathcal{E}_{past},\ B_{final} \right\rangle$$

The case base, which is initially empty, grows each time a new case is created. Since each case represents a single problem-solving episode (i.e., finding a trustworthy behavior for an operator in a given context), the case base represents all of the problem solving experience that the robot has collected.

3.5.2 Behavior Adaptation

Behavior adaptation, which we have only described abstractly to this point, is performed when the currently evaluated behavior reaches the untrustworthy threshold and the robot needs to select a new behavior to perform. The new behavior B_{new} is selected as a function of the set of previously evaluated behaviors $\mathcal{E}_{past}$ and the robot's case base CB:

$$B_{new} = selectBehavior\left(\mathcal{E}_{past}, CB\right)$$

The *selectBehavior* function (Algorithm 3.1) searches for a case C_i in CB with a set of evaluated behaviors that is most similar to $\mathcal{E}_{past}$. The motivation for this is that if they have similar problems then they might have similar solutions, so the robot can adapt its behavior by switching to the final behavior stored in C_i.

Algorithm 3.1: Selecting a new behavior
Function: $selectBehavior\left(\mathcal{E}_{past}, CB\right)$ **returns** B_{new}

```
1  bestSim ← 0; B_best ← ∅;
2  foreach C_i ∈ CB do
3      if C_i.B_final ∉ E_past then
4          sim_i ← sim(E_past, C_i.E_past);
5          if sim_i > bestSim then
6              bestSim ← sim_i;
7              B_best ← C_i.B_final;
8  if B_best = ∅ then
9      B_best ← modifyBehavior(E_past);
10 return B_best;
```

The algorithm iterates through each case in the case base (line 2) and checks to see if the case's final behavior has already been evaluated (line 3). This check is done to ensure that behaviors that have already been found to be untrustworthy are not evaluated again. The sets of evaluated behaviors of the remaining cases are compared to the robot's current set of evaluated behaviors using a similarity metric (line 4). The most similar case's final behavior is stored (lines 5–7) and returned to the robot (line 10). This behavior is immediately used by the robot and the robot begins measuring the trustworthiness of that behavior. If no similar cases were found (i.e., the case base is empty or the final behaviors of all cases have already been evaluated), the *modifyBehavior* function is used to select the next behavior to perform (line 9).

The *modifyBehavior* function selects an evaluated behavior E_{max} that took the longest to reach the untrustworthy threshold ($\forall E_i \in \mathcal{E}_{past}, E_{max}.t \geq E_i.t$). A random walk (without repetition) is performed to find a behavior B_{new} that requires the minimum number of changes to $E_{max}.B$ and has not already been evaluated ($\forall E_i \in \mathcal{E}_{past}, B_{new} \neq E_i.B$). This is based on the assumption that E_{max} is the least untrustworthy of the evaluated behaviors and a slight change might lead to a more trustworthy behavior. If all possible behaviors have been evaluated and found to be untrustworthy, the robot will stop adapting its behavior and use $E_{max}.B$.

Algorithm 3.1 relies on calculating the similarity between two sets of evaluated behaviors (line 4). This similarity (Algorithm 3.2) is complicated by the fact that the sets may vary in size. This occurs because the number of evaluated behaviors in each case is dependent on how long the search took in that instance. Similarly, there is no guarantee that the same behaviors were evaluated in each set. To account for this, the similarity function looks at the overlap between the two sets and ignores behaviors that have only been evaluated in one set. The algorithm goes through each evaluated behavior in the first set (line 2) and finds the most similar evaluated behavior E_{max} in the second set (line 3). The similarity between two behaviors is a function of the similarity of each behavior component:

$$sim\left(B_1, B_2\right) = \frac{1}{m}\sum_{i=1}^{m} sim\left(B_1.c_i, B_2.c_i\right),$$

where the similarity function for each behavior component will depend on its specific type. For example, a behavior component that represents a binary parameter value would require a different similarity function than a component that represents which path planning algorithm to use.

If the two evaluated behaviors, E_i and E_{max}, are sufficiently similar, based on a threshold λ (line 4), then the similarity of their time components are included in the similarity calculation (line 5). This ensures that the final similarity value only includes information from behaviors that have a highly similar counterpart in the other set. This function will return a high similarity (up to a maximum of 1.0) when similar behaviors took nearly the same time to reach the untrustworthy threshold and a low similarity (to a minimum of 0.0) when similar behaviors had a noticeable difference in the time they took to reach the untrustworthy threshold.

3.6 Evaluation

In this section, we evaluate our behavior adaptation technique in a simulated robotics environment. Two variations of trust-based behavior adaptation are used: case-based behavior adaptation and random walk behavior adaptation. While we expect both approaches to allow the robot to adapt to trustworthy behaviors, we will evaluate our claim that the case-based approach can find trustworthy behaviors more efficiently.

Algorithm 3.2: Similarity between sets of evaluated behaviors
Function: $sim(\mathcal{E}_1, \mathcal{E}_2)$ **returns** sim

```
1 totalSim ← 0; num ← 0;
2 foreach E_i ∈ 𝓔_1 do
3     E_max ← argmax_{E_j ∈ 𝓔_2} (sim(E_i.B, E_j.B));
4     if sim(E_i.B, E_max.B) > λ then
5         totalSim ← totalSim + sim(E_i.t, E_max.t);
6         num ← num + 1;
7 if num = 0 then
8     return 0;
9 return totalSim / num;
```

3.6.1 *eBotworks Simulator*

Our evaluation uses the eBotworks simulation environment (Knexus Research Corporation 2015), a multi-agent simulation engine and testbed for unmanned systems. In eBotworks, autonomous agents control simulated robotic vehicles and can receive multimodal commands from human operators. We chose to use eBotworks based on its flexibility in autonomous behavior modeling, ability to interact with agents using natural language commands, and built-in experimentation and data collection capabilities.

In our experiments, we use a single robot that is a wheeled unmanned ground vehicle (UGV). The robot uses eBotworks' built-in natural language processing (for interpreting user commands), sensing, and path-planning modules. The environment is composed of landmarks (e.g., roads, various types of terrain) and objects (e.g., houses, humans, vehicles, road barriers). The actions performed by the robot are non-deterministic and the robot also suffers from limited observability and potential sensor errors.

3.6.2 *Experimental Conditions*

Our initial study uses simulated operators to facilitate a larger-scale evaluation than if real human operators were used.[2] The simulated operators were selected to represent a subset of the control strategies used by human operators. Each simulated operator has unique preferences for how the robot should behave and these preferences will influence how the robot's performance is evaluated (i.e., when the operator allows the robot to complete a task and when it interrupts).

[2] We plan to validate these findings in a series of user studies.

Each experiment is composed of 500 *trials* and in each trial the robot interacts with a single simulated operator. At the start of a trial, the robot randomly selects (with a uniform distribution) initial values for each of its modifiable components. Throughout the trial, a series of experimental *runs* occur. Each run involves the operator issuing a single command to the robot and monitoring the robot as it performs the task. During a run the robot will complete the task, fail to complete the task, or be interrupted by the operator; it will update its trust estimate accordingly. At the end of each run the environment is reset and a new run begins. A trial concludes when the robot has either found a trustworthy behavior or evaluated all possible behaviors.

The case-based behavior adaptation approach starts each experiment with an empty case base. A case is stored at the end of a trial if the robot found a trustworthy behavior and performed at least one random walk adaptation (i.e., the robot could not find a solution in its case base so it used the *modifyBehavior* function). This case retention strategy is used to prevent adding redundant cases. An added case can be used during any of the subsequent trials in the experiment.

The robot's trustworthy threshold was set to $\tau_T = 5.0$ and its untrustworthy threshold set to $\tau_{UT} = -5.0$. These thresholds were set to allow some fluctuation between increasing and decreasing trust while still identifying trustworthy and untrustworthy behaviors quickly. When calculating the similarity between sets of evaluated behaviors, a similarity threshold of $\lambda = 0.95$ was used (i.e., behaviors must be at least 95 % similar to be matched).

3.6.3 Evaluation Scenarios

We selected two scenarios of increasing complexity: *movement* and *patrolling for threats*. While the Movement scenario is a relatively simple task, the Patrol scenario requires a more complex behavior with a larger set of modifiable components.

3.6.3.1 Movement Scenario

The initial task the robot is required to perform involves moving between two locations in the environment (Fig. 3.2). The simulated operator issues natural language commands to tell the robot where to move (e.g., "move to the flag") and the robot is responsible for navigating to that location. Three metrics are used by the operators to assess the robot's performance:

- **Task Duration**: The operator has an expectation about the amount of time the task should take to complete ($t_{complete}$). If the robot does not complete the task within that time, the operator may, with probability p_α, interrupt the robot.

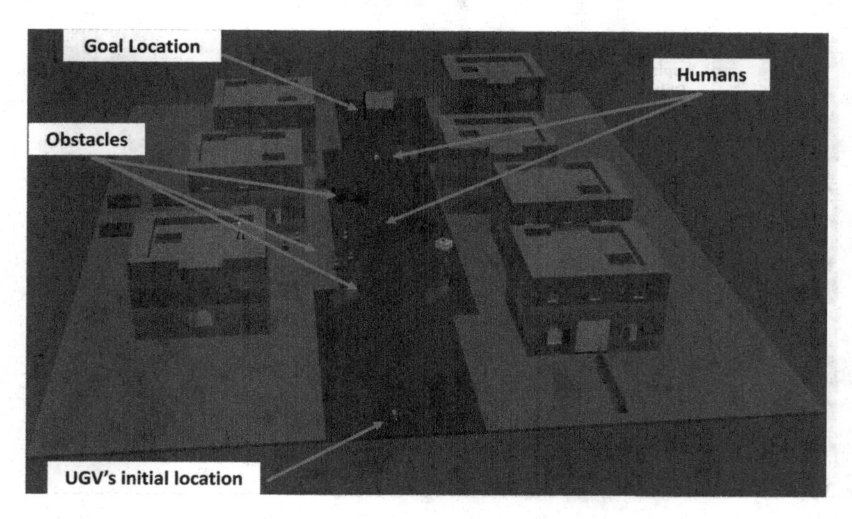

Fig. 3.2 The environment configuration for the Movement scenario

- **Task Completion**: If the operator determines that the robot has failed to complete the task (e.g., the robot is stuck or moved to the wrong location), the robot will be interrupted.
- **Safety**: The operator may interrupt the robot, with probability p_γ, if the robot collides with any obstacles.

We use three simulated operators in this scenario:

- **Speed-focused operator**: This operator prefers the robot to move to the destination quickly regardless of whether it hits any obstacles ($t_{complete}$ = 15 s, $p_\alpha = 95\,\%, p_\gamma = 5\,\%$).
- **Safety-focused operator**: This operator prefers the robot to avoid obstacles regardless of how long it takes to reach the destination ($t_{complete}$ = 15 s, $p_\alpha = 5\,\%, p_\gamma = 95\,\%$).
- **Balanced operator**: This operator prefers a balanced mixture of speed and safety ($t_{complete}$ = 15 s, $p_\alpha = 95\,\%, p_\gamma = 95\,\%$).

In this scenario, the robot has two modifiable behavior components: *speed* and *obstacle padding*. Speed, measured in meters per second, relates to how fast the robot can move. Padding, measured in meters, relates to the distance the robot will attempt to maintain from obstacles during movement. The set of possible values for each modifiable component ($\mathscr{C}_{speed}$ and $\mathscr{C}_{padding}$) are based on the robot's capabilities (i.e., minimum and maximum accepted values with fixed increments):

$$\mathscr{C}_{speed} = \{0.5,\ 1.0, \ldots,\ 10.0\}$$
$$\mathscr{C}_{padding} = \{0.1,\ 0.2, \ldots,\ 2.0\}$$

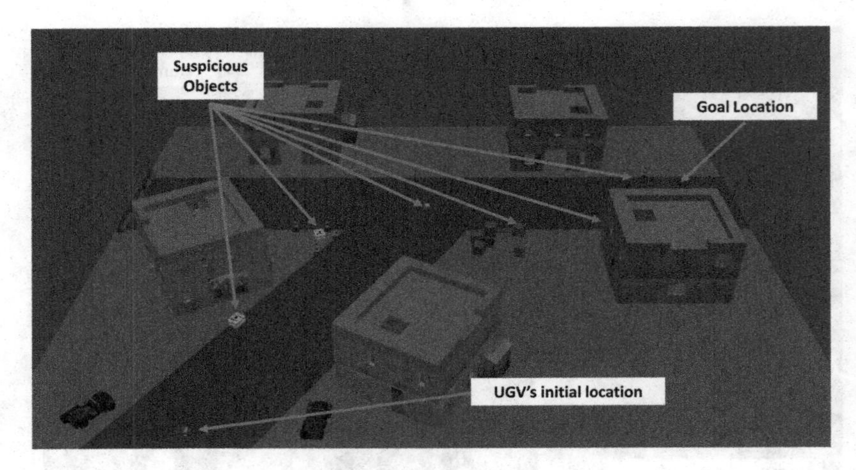

Fig. 3.3 The environment configuration for the Patrol scenario

3.6.3.2 Patrolling Scenario

In the second scenario, the robot patrols for threats as it moves between two locations in the environment (Fig. 3.3). At the start of each run, six *suspicious objects* are randomly placed in the environment. These suspicious objects represent potential threats, and between 0 and 3 (inclusive) of them are designated as hazardous explosive devices (selected randomly with a uniform distribution). The remaining suspicious objects are not hazardous to the robot or the team.

As the robot moves between the start location and the goal location (given by a natural language command from the operator), it scans for suspicious objects nearby. When a suspicious object is detected, it pauses its patrolling behavior, moves toward the object, scans it with its explosives detector, and labels the object as an *explosive* or *harmless*. The robot then resumes its patrolling behavior. The accuracy of the robot's explosives detector is a function of how long the robot spends scanning the object (long scan times result in improved accuracy) and its proximity to the object (smaller scan distances result in improved accuracy). In addition to the speed and padding, the *scan time*, measured in seconds, and *scan distance*, measured in meters, are also modifiable components of the robot's behavior. The possible values for these are:

$$\mathscr{C}_{scantime} = \{0.5,\ 1.0, \ldots,\ 5.0\}$$
$$\mathscr{C}_{scandistance} = \{0.25,\ 0.5, \ldots,\ 1.0\}$$

In addition to the task duration, task completion, and safety factors described in the Movement scenario, the simulated operators will also base their decision to interrupt the robot on its ability to successfully identify and label suspicious objects. An operator will interrupt the robot if it does not scan one or more suspicious objects (e.g., it drives by without noticing it) or incorrectly labels a harmless object as an explosive. If the robot incorrectly labels an explosive device as harmless, the object will eventually detonate and the robot will fail its task. The robot assigns higher weight to failures due to missing explosive devices (3 times higher than other failures or interruptions) because of the danger such failures cause to human teammates and bystanders.

We use two simulated operators in this scenario:

- **Speed-focused operator**: The operator prefers that the robot performs the patrol task within a fixed time limit ($t_{complete} = 120$ s, $p_\alpha = 95$ %, $p_\gamma = 5$ %).
- **Detection-focused operator**: The operator prefers the task be performed correctly regardless of time ($t_{complete} = 120$ s, $p_\alpha = 5$ %, $p_\gamma = 5$ %).

3.6.4 Trustworthy Behaviors

We found that both case-based behavior adaptation and random walk behavior adaptation resulted in similar trustworthy behaviors for each operator. This includes values falling within similar ranges of trustworthy values (e.g., for the safety-focused operator in the Movement scenario the padding never went below 0.4 m in any trial) or similar relations between values (e.g., in the Patrol scenario there was a relation between scan time and scan distance). Furthermore, the trustworthy behaviors aligned with what an outside observer would intuitively consider trustworthy for each operator (e.g., that the speed-focused operator will prefer higher speeds).

The trustworthy behaviors for each of the operators in the Movement scenario are shown in Figs. 3.4, 3.5 and 3.6. Each dot represents the trustworthy behavior found during a single trial using random walk adaptation. Although 500 trials were performed for each operator, fewer than 500 dots appear in each figure because some trials converged to the same parameter values. This is more prevalent when case-based behavior adaptation is performed since the final behaviors stored in cases occur much more frequently than other behaviors (i.e., those solutions are repeatedly reused). However, the trustworthy behaviors found by the case-based approach fall within the same regions as the random walk behaviors.

The speed-focused operator (Fig. 3.4) causes the robot to converge to higher speed values regardless of padding while the safety-focused operator (Fig. 3.5) results in higher padding values regardless of speed. For the balanced operator (Fig. 3.6), both speed and padding must be high.

In the Patrol scenario, there are similar differences in the range of values for certain behavior components. The speed-focused Patrol operator causes the robot to converge to higher speed values ($speed \geq 2.0$) whereas the detection-focused operator has no such restriction. However, unlike in the Movement scenario there are also interdependencies among behavior components. For example, none of the trustworthy behaviors for the speed-focused Patrol operator have both a medium

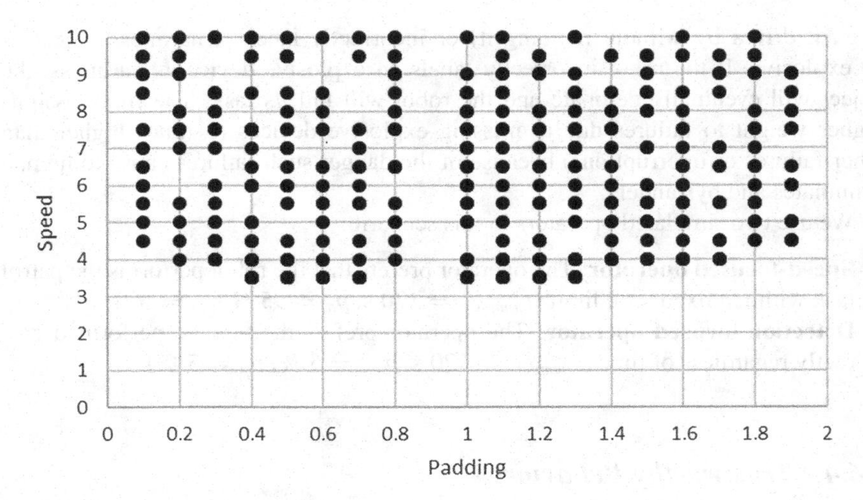

Fig. 3.4 Trustworthy behaviors for the speed-focused operator in the Movement scenario. The robot converged to behaviors with higher speed values regardless of padding

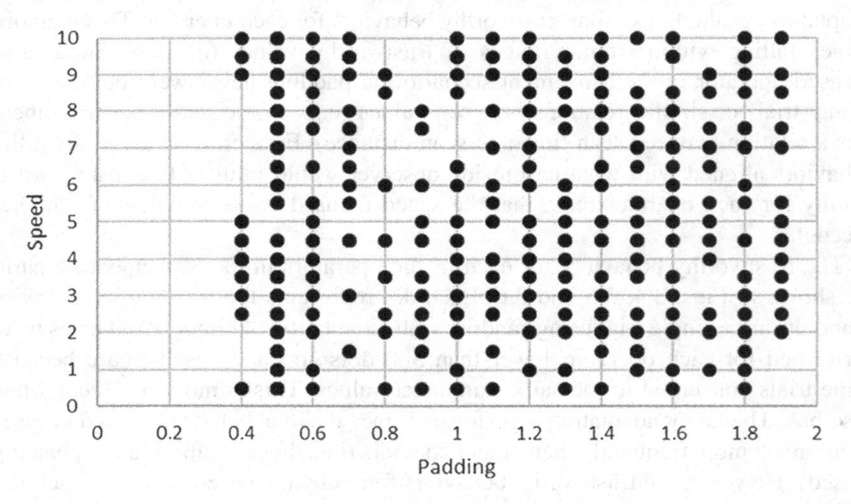

Fig. 3.5 Trustworthy behaviors for the safety-focused operator in the Movement scenario. The robot converged to behaviors with higher padding values regardless of speed

speed ($2.0 \leq speed \leq 4.0$) and high scan time. This is because the robot needs to account for longer scan times by driving faster. Similarly, there is an interdependence between scan time and scan distance. The robot only selects a poor value for one of the modifiable components (low scan time or high scan distance) if it selects a good value for the other (high scan time or low scan distance).

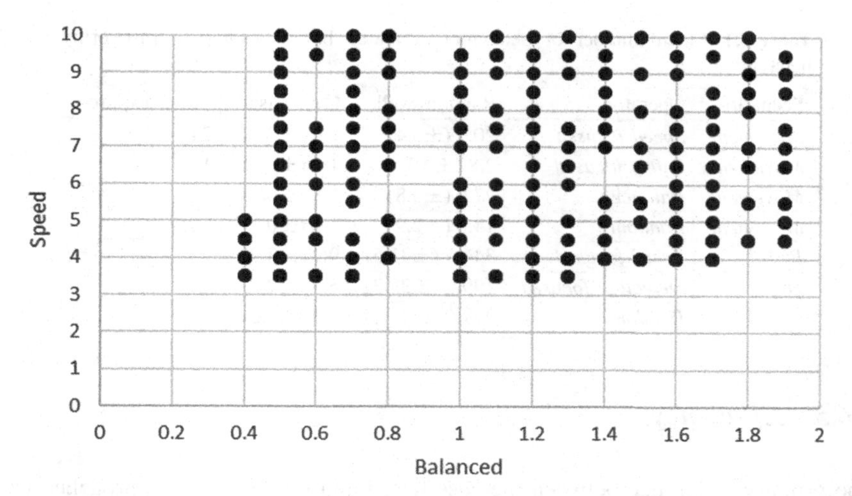

Fig. 3.6 Trustworthy behaviors for the balanced operator in the Movement scenario. The robot converged to behaviors with both high speed and padding

These results fit with the definitions of the simulated operators and our intuition on the behaviors they would find trustworthy. However, one noticeable exception occurred in the Movement scenario where behaviors that appear to be trustworthy are actually not. For both the speed-focused and balanced operators (Figs. 3.4 and 3.6), no trustworthy behaviors were found when *padding* = 0.9. For both of these operators, larger (*padding* = 1.0) and smaller (*padding* = 0.8) values were found to be trustworthy. This occurred in the results for both the case-based and random walk approaches, and in a follow up evaluation where the robot was forced to use behaviors with *padding* = 0.9. The reason this padding value was found to be untrustworthy was because of the environment. The padding value resulted in a direct, but narrow, path to the destination. This required the robot to slow down when navigating through the narrow path and caused it to exceed its time limit (this is why that padding value was not an issue for the safety-focused operator). However, when the padding was lowered the path became large enough that the robot could drive through without slowing down. Similarly, when the padding was increased the path was eliminated so the robot took a slightly longer but much easier path. These results show that even if we had a general idea about what behaviors would be considered trustworthy there is still the possibility of seemingly trustworthy behaviors being untrustworthy. It is beneficial for the robot to be able to adapt and overcome any issues that are not anticipated, especially if it operates in a dynamic or unknown environment.

Table 3.1 Mean number of behaviors evaluated before finding a trustworthy behavior

Scenario	Operator	Random walk	Case-based	Cases acquired
Movement	*Speed-focused*	20.3 (±3.4)	1.6 (±0.2)	24
Movement	*Safety-focused*	2.8 (±0.3)	1.3 (±0.1)	18
Movement	*Balanced*	27.0 (±3.8)	1.8 (±0.2)	33
Movement	*Random*	14.6 (±2.9)	1.6 (±0.1)	33
Patrol	*Speed-focused*	344.5 (±31.5)	9.9 (±3.9)	25
Patrol	*Detection-focused*	199.9 (±23.3)	5.5 (±2.2)	22
Patrol	*Random*	269.0 (±27.1)	9.3 (±3.2)	25

3.6.5 Efficiency

The primary difference between the case-based and random walk approaches was related to how many behaviors needed to be evaluated before a trustworthy behavior was found. Table 3.1 shows the mean number of evaluated behaviors (and 95 % confidence interval) when interacting with each operator over 500 trials. In addition to being controlled by a single operator during each experiment, we also examined a condition where the operator was selected at random (with equal probability) at the start of each trial. This represents a more realistic situation where the robot interacts with a variety of operators but does not know which particular operator it is currently interacting with. This variant is labeled as *Random* and was performed in both scenarios. The table also shows the number of cases acquired during the case-based behavior adaptation experiments (each experiment started with an empty case base).

The case-based approach required significantly fewer behaviors to be evaluated in all seven conditions (using a paired t-test with $p < 0.01$). This is because the case-based approach learns from previous adaptations and uses that information to quickly find trustworthy behaviors. At the beginning of an experiment, when the robot's case base is empty, the case-based approach relies on performing random walk adaptation. As the case base grows, the number of random walk adaptations decreases until the agent generally performs a single case-based adaptation before finding a trustworthy behavior. Even in the random operator experiments when the case base contains cases from several different operators (three in the Movement scenario and two in Patrol), the case-based approach can quickly differentiate between operators and select a trustworthy behavior. Operators with fewer restrictions on trustworthy behaviors (i.e., a higher percentage of the behavior space is considered trustworthy), like the safety-focused and detection-focused operators, had the lowest mean number of adaptations to find a trustworthy behavior.

3.6.6 *Discussion*

The primary limitation of the case-based approach is that it relies on random walk search when it does not have any suitable cases to use. This is especially prevalent early on when the robot has a small or empty case base. For example, if we consider only the final 250 trials for each Patrol scenario operator, the mean number of behaviors evaluated is lower than the overall mean (4.2 for the speed-focused, 2.8 for the detection-focused, and 3.3 for the random). This is because the robot performs the expensive random walk adaptation more often in the early trials, so it performs more efficiently on the later trials. These expensive adaptations occur infrequently (only in trials where a case is stored) but increase the mean number of behaviors that are evaluated.

Two primary solutions exist to reduce the number of behaviors examined during case-based behavior adaptation: improving search and seeding the case base. Random walk search is used because it requires no explicit knowledge about the domain, task, or operator. However, a more intelligent search that could identify relations between interruptions and modifiable components would likely improve adaptation time (e.g., an interruption when the robot is close to objects may require a change to the padding value). This could reduce the cost of each search, whereas seeding the case base would attempt to minimize the number of searches required. A set of initial cases could be provided to the robot so that it would not need to acquire as many on its own. However, these two solutions introduce their own potential limitations. A more informed search requires introducing domain knowledge, which may be difficult or expensive to obtain, and seeding the case base requires an expert to manually author cases (or another method of automatic case acquisition). The specific requirements of the application domain will ultimately influence whether faster behavior adaptation or lower domain knowledge are more important.

3.7 Conclusions

In this chapter we have described our approach for inverse trust estimation and how a robot can use it to adapt its behavior. Rather than traditional trust metrics that directly measure how much trust an agent has in another agent, our inverse trust estimate attempts to infer how much trust another agent has in it. As such, it cannot be thought of as an explicit measurement of trust but rather a best-guess estimate based on observable indicators of trust (e.g., the operator's response to the robot's performance). Our approach relies more on the general trends in its trustworthiness (increasing, decreasing, or constant) rather than requiring a precise numerical value.

The primary benefit of this behavior adaptation approach is that it does not require any background knowledge about the tasks, environment, context, or operators. Each time the robot successfully finds a trustworthy behavior, it stores information about the adaptation process and uses that to improve the efficiency of future adaptations. This allows it to constantly learn behavior adaptation knowledge with each trial.

We evaluated our trust-guided behavior adaptation algorithm in a simulated robotics environment by comparing it to a variation that does not learn. In the two scenarios, Movement and Patrol, both approaches converged to trustworthy behaviors but the case-based algorithm required significantly fewer behaviors to be evaluated. This is advantageous because the operator is more likely to stop using the robot the longer the robot behaves in an untrustworthy manner.

Although we have shown the benefits of trust-guided behavior adaptation, several areas of future work exist. Although much of our work is based on studies in human-robot interaction, our initial evaluation has been limited to simulation studies. An ongoing area of our research is to validate our findings in a series of user studies. Next, our robot is only concerned with undertrust. In longer scenarios, the robot should also evaluate situations of overtrust where the operator trusts the robot too much and allows the robot to behave autonomously even when its performance is poor. We also plan to expand our inverse trust estimate by incorporating other trust factors and adding mechanisms that promote transparency (Kim and Hinds 2006) between the robot and the operator. Transparency would allow information exchange between the robot and the operator, and allow the robot to verify or refine assumptions it has been using (e.g., which goals the team is currently trying to achieve). Many of these areas for future work revolve around taking our existing approach, which requires minimal domain knowledge, and allowing for extra knowledge to be incorporated if it ever becomes available.

Acknowledgment Thanks to the United States Naval Research Laboratory and the Office of Naval Research for supporting this research.

References

Baier JA, McIlraith SA (2008) Planning with preferences. AI Mag 29(4):25–36

Berlin M, Gray J, Thomaz AL, Breazeal C (2006) Perspective taking: an organizing principle for learning in human-robot interaction. In: Proceedings of the 21st National conference on artificial intelligence, pp 1444–1450

Biros DP, Daly M, Gunsch G (2004) The influence of task load and automation trust on deception detection. Group Decis Negot 13(2):173–189

Breazeal C, Gray J, Berlin M (2009) An embodied cognition approach to mindreading skills for socially intelligent robots. Int J Robot Res 28(5):656–680

Carlson MS, Desai M, Drury JL, Kwak H, Yanco HA (2014) Identifying factors that influence trust in automated cars and medical diagnosis systems. In Proceedings of the AAAI symposium on the intersection of robust intelligence and trust in autonomous systems, pp 20–27

Chen K, Zhang Y, Zheng Z, Zha H, Sun G (2008) Adapting ranking functions to user preference. In: Proceedings of the 24th International conference on data engineering workshops, pp 580–587

Desai M, Kaniarasu P, Medvedev M, Steinfeld A, Yanco H (2013) Impact of robot failures and feedback on real-time trust. In: Proceedings of the 8th International conference on human-robot interaction, pp 251–258

Esfandiari B, Chandrasekharan S (2001) On how agents make friends: mechanisms for trust acquisition. In: Proceedings of the 4th Workshop on deception, fraud and trust in agent societies, pp. 27–34

Hancock PA, Billings DR, Schaefer KE, Chen JY, De Visser EJ, Parasuraman R (2011) A meta-analysis of factors affecting trust in human-robot interaction. Hum Factors 53(5):517–527

Horvitz E (1999) Principles of mixed-initiative user interfaces. In: Proceedings of the 18th Conference on human factors in computing systems, pp 159–166

Jian J-Y, Bisantz AM, Drury CG (2000) Foundations for an empirically determined scale of trust in automated systems. Int J Cogn Ergon 4(1):53–71

Kaniarasu P, Steinfeld A, Desai M, Yanco HA (2012) Potential measures for detecting trust changes. Proceedings of the 7th International conference on human-robot interaction, pp 241–242

Kaniarasu P, Steinfeld A, Desai M, Yanco HA (2013) Robot confidence and trust alignment. In: Proceedings of the 8th International conference on human-robot interaction, pp 155–156

Kiesler S, Powers A, Fussell SR, Torrey C (2008) Anthropomorphic interactions with a robot and robot-like agent. Soc Cogn 26(2):169–181

Kim T, Hinds P (2006) Who should I blame? Effects of autonomy and transparency on attributions in human-robot interaction. In: Proceedings of the 15th IEEE International symposium on robot and human interactive communication, pp 80–85

Knexus Research Corporation (2015) eBotworks. http://www.knexusresearch.com/products/ebotworks.php

Li N, Kambhampati S, Yoon SW (2009) Learning probabilistic hierarchical task networks to capture user preferences. In: Proceedings of the 21st International joint conference on artificial intelligence, pp 1754–1759

Li D, Rau PP, Li Y (2010) A cross-cultural study: effect of robot appearance and task. Int J Soc Robot 2(2):175–186

Maes P, Kozierok R (1993) Learning interface agents. In: Proceedings of the 11th National conference on artificial intelligence, pp 459–465

Mahmood T, Ricci F (2009) Improving recommender systems with adaptive conversational strategies. In: Proceedings of the 20th ACM Conference on Hypertext and Hypermedia, pp. 73–82

McGinty L, Smyth B (2003) On the role of diversity in conversational recommender systems. In: Proceedings of the 5th International conference on case-based reasoning, pp 276–290

Muir BM (1987) Trust between humans and machines, and the design of decision aids. Int J Man Mach Stud 27(5–6):527–539

Oleson KE, Billings DR, Kocsis V, Chen JY, Hancock PA (2011) Antecedents of trust in human-robot collaborations. In: Proceedings of the 1st International multi-disciplinary conference on cognitive methods in situation awareness and decision support, pp 175–178

Richter MM, Weber RO (2013) Case-based reasoning—a textbook. Springer, Berlin

Sabater J, Sierra C (2005) Review on computational trust and reputation models. Artif Intell Rev 24(1):33–60

Saleh JA, Karray F, Morckos M (2012) Modelling of robot attention demand in human-robot interaction using finite fuzzy state automata. In: Proceedings of the International conference on fuzzy systems, pp 1–8

Schlimmer JC, Hermens LA (1993) Software agents: completing patterns and constructing user interfaces. J Artif Intell Res 1:61–89

Shapiro D, Shachter R (2002) User-agent value alignment. In: Proceedings of the Stanford Spring Symposium—Workshop on safe learning agents

Chapter 4
The "Trust V": Building and Measuring Trust in Autonomous Systems

Gari Palmer, Anne Selwyn, and Dan Zwillinger

4.1 Introduction

Autonomous systems are becoming ever more prevalent (Rosen 2012), attracting increasing numbers of researchers, papers, and symposia (IHMC 2013). The importance of autonomous capabilities has led the DoD (Department of Defense) to make autonomy one of its seven Science and Technology priorities for FY 2013–2017 (Gates 2011).

The need for machine autonomy arises from both mission priorities and system complexity. While un-manned missions drive some autonomous designs, it is principally in the area of managing complexity that issues of trust arise. Autonomous systems considered in this work are intended to include human users. Complexity can distance human users from the functional operation of a system, while autonomy in that system can bridge the gap, making operation more intuitive, responsive, and better integrated into the overall mission environment. The design and operation of these systems needs to be addressed in terms of human-system collaboration. Ideally, an autonomous system extends human capabilities and works as a team member, integrating into the mission at hand. Successful pedigree and operational

G. Palmer (✉) • A. Selwyn • D. Zwillinger
Raytheon Company, 50 Apple Hill Drive, Tewksbury, MA 01876, USA
e-mail: Gari_B_Palmer@raytheon.com

© Springer Science+Business Media (outside the USA) 2016

R. Mittu et al. (eds.), *Robust Intelligence and Trust in Autonomous Systems*,
DOI 10.1007/978-1-4899-7668-0_4

integration will engender trust in the system[1,2] and lead to sustained operator reliance on those autonomous capabilities. Development of trust in this context is an entirely human activity, and the methodologies for building trust are necessarily evolving with each new generation of autonomous capability.

While operator reliance is an end goal for an autonomous system, autonomy itself can be a challenge for system definition, and verification and validation. The design trade space contains new dimensions: optimality vs. resilience, efficiency vs. thoroughness, centralized vs. distributed, and so on (DSB 2012). Tools for verification and validation must be integrated into the system design providing high fidelity insight into the architecture as it unfolds during development. Trust cannot be established in an autonomous system unless trust has also been established for the automation attributes that provide the underlying foundation for the autonomous attributes (Stone 2011), or to put it more simply, a complex system that works is invariably found to have evolved from a simple system that works (Gall 1977).

There are many development paradigms for implementing large complex systems. System engineering development models are generally characterized by a dominant topology chosen to emphasize some overarching theme. These topologies include the waterfall, spiral, and "V". The System Engineering V is predominantly the model of choice for systems that must meet a high standard of performance assurance upon delivery, as it emphasizes verification and validation of a requirements-driven design (Forsberg and Mooz 1991, FHWA SE Guidebook 2009).

Perhaps due to its acceptance as a basic INCOSE (International Council on Systems Engineering) standard, or perhaps to its widespread use, there are many variants of the Systems Engineering V. As an Internet search reveals. These variants usually arise from domain specific concerns. In this chapter we introduce the "Trust V" framework, which provides guidance on how to build qualities that engender trust into systems with autonomous capabilities. This twist to the Systems Engineering V is a novel application of autonomous development methods to the Systems Engineering development model. While the context for our approach is the architecture of large-scale Aerospace and Defense systems that follow a rigorous Systems Engineering methodology. The general ideas may be applied to any complex system design. The framework presented here consists of a "toolbox" of reusable and adaptive trust building techniques. This toolbox is intended to exploit the commonality across system architectures, whereby each technique will be evaluated for applicability and evolve as the autonomous capabilities of the overall system design are specified.

[1]Note that systems cannot, *per se,* "have trust". Rather, a system may be trustable. This might occur through system usage or via transference of trust from one operator to another. We adopt the common convention of referring to systems as "having trust" when the system supports the creation of trust, perhaps by having built-in assessment-enabling capabilities.

[2]And the converse is also true; as when evaluating humans, an accident or mishap could greatly reduce the likelihood of obtaining trust.

Many of the tools in the Trust V toolbox directly target the validation and verification process, with the systems engineer as the target human operator. In this way an evolving autonomy framework can support both the ultimate system users and the design team. Investment in Trust V methods, which support Test, Evaluation, Verification and Validation and system operations, will reduce life cycle cost while increasing operator acceptance and system viability. This approach to robust design is a novel means of leveraging autonomous capability. It both contrasts with, and compliments other approaches to robust design such as AADL (Architecture Analysis & Design Language) (Feiler et al. 2006) and LTL (Linear Temporal Logic) (Fainekos and Pappas 2006).

4.2 Autonomy, Automation, and Trust

Complete automation and complete autonomy are often described as endpoints along a spectrum of system behavior, as depicted in Fig. 4.1. The Trust V methodology, described subsequently, can be used at any point along this spectrum. As systems transition towards dynamic and adaptive behavior, humans become increasingly uncertain of the system behavior and are less willing to trust. Hence, the value of investments in trust engendering methods will increase.

It is useful to recognize that autonomy is built upon automation. The automobile industry provides an excellent example of incremental transition from automation towards autonomy for consumers, resulting in significant autonomous capabilities. In the recent past, decreasing an automobile's speed was a manual operation performed by a human stepping on the car's brake. Today many cars are able to automatically and autonomously decrease speed based on distance between cars. In the future, automobiles will autonomously adjust speed to ensure safe distance between cars and other objects in all directions. Speed control is just one of the many automated capabilities that were needed before cars could become autonomously self-driving. In many fields, including the automotive field, the process of increasing automation, possibly trending towards autonomy, takes the form of a layered control. For speed control some of the improvements in this layering are: self-adjusting calipers, followed by the Anti-lock Braking System (ABS), followed by basic cruise control, followed by the current more sophisticated cruise control mechanisms. Each improvement builds on the last.

We study trust because we wish to develop automated and autonomous system that will be used. Perhaps unsurprisingly, an operator's use of an autonomous system is profoundly influenced by their trust in that system (Muir 1987; Dzindolet et al. 2003). From the US Department of Defense (DSB 2012)

> *"For commanders and operators in particular, these challenges can collectively be characterized as a lack of trust that the autonomous functions of a given system will operate as intended in all situations."*

While in general discussion the term "trust" is used to establish a particular quality in human interaction, trust is just a component in a behavioral pattern of

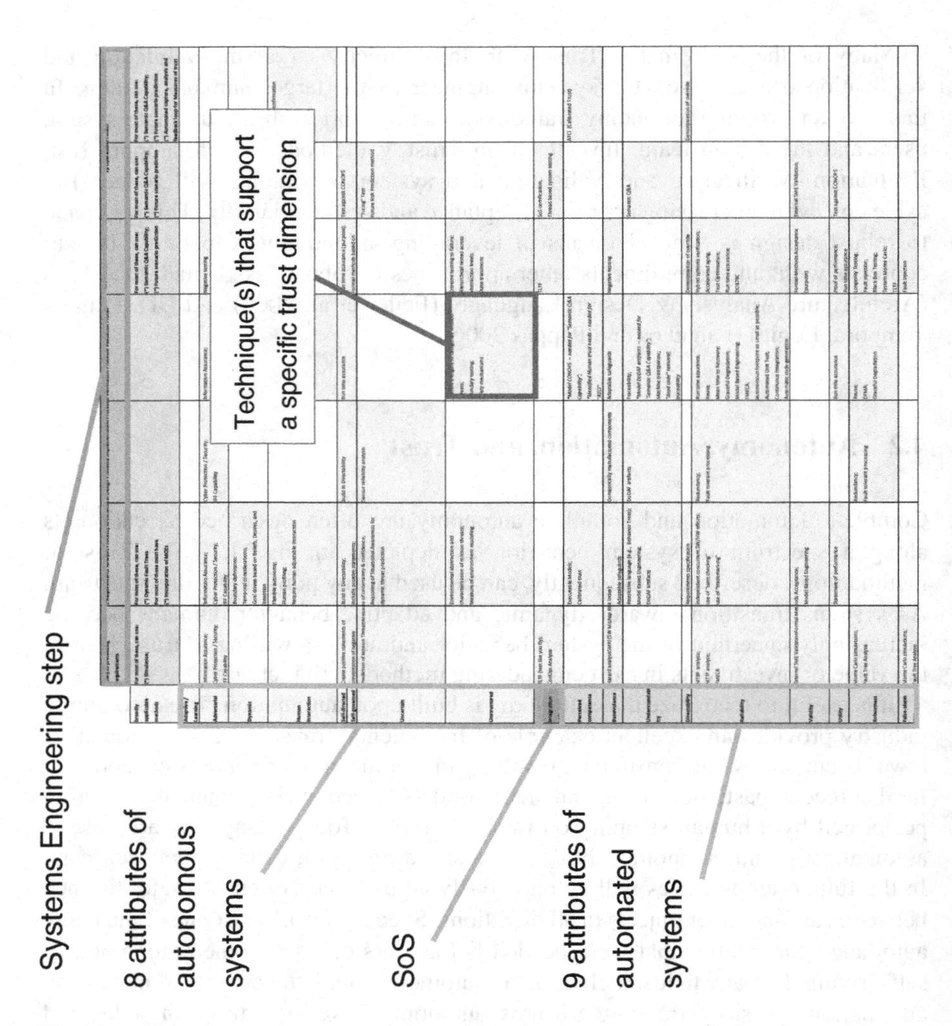

Fig. 4.A.1 The array representation of the Trust V

interaction. Trust alone is *an attitude that an agent will help achieve an individual's goals in a situation characterized by uncertainty and vulnerability* (Lee and See 2004). Given a context and a body of information, trust grants the user confidence that the agent will successfully and correctly act on the user's intention, leading the user to rely on the agent. The outcome of the agent's action feeds back into the context and body of information, closing the loop (see Fig. 4.2). By definition, trust is established in the context of the mission environment and the body of information supplied to the user. Success in establishing trust is objectively observed in the user's patterns of reliance.

	broadly applicable methods	Uncertain
Analysis preceeding Requirements	For most of these, can use: (*) Simulation	Prediction of Mission Threats
2 Requirements Analysis Process	For most of these, can use: (*) Dynamic Fault Trees For most of these must have (*) Incorporation of MOEs	Measure of uncertainty & confidence; Reasoning of Situational Awareness for dynamic threat prediction
3 Architectural Design Process		Sensor reliabiity analysis
6 Verification Process	For most of these, can use: (*) Semantic Q&A Capability; (*) Future scenario prediction	
8 Validation Process	For most of these, can use: (*) Semantic Q&A Capability; (*) Future scenario prediction	"Turing" Test Future Risk Prediction method
9 Operation Process	For most of these, can use: (*) Semantic Q&A Capability; (*) Future scenario prediction (*) Automated capture, analysis and feedback loop for Measures of trust	

Fig. 4.A.2 Trust V information showing techniques for the "Uncertain" autonomy dimension. (Note that the array has been rotated, relative to the representation in Fig. 4.A.1)

From a user's point of view, trust is generally learned through experiencing the system's patterns of behavior/execution. System pedigree may accelerate development of system trust. For example, if Operator B trusts Operator A, and Operator A has communicated their trust of Autonomous System X to Operator B, then one would expect that Operator B would be inclined to be more trusting of System X. Depending on the relationship between A and B, this communication might reduce B's need to evaluate System X.

We note that trust is an abstraction that cannot be precisely measured. We know of no absolute trust measurement. Since trust is a relative measurement, we are restricted to measuring changes in trust. That being said, when a person trusts a system it is not that a trust threshold has been met but because the person has determined that the system can adequately perform a specific purpose.

In many systems, autonomous designs have access to low-level system resources. During operational use of these systems, the software can analyze the totality of system information, thereby presenting a multi-dimensional, holistic expert viewpoint to the human operators. As discussed later, a broadly applicable trust-building technique is for the system to present the end user with commentary and rationale for system operations within the context of the mission. Sometimes, designers of

18 Trust Dimensions

MOE type: Passive, Active, Historical

MOE(s) supporting a specific trust dimension

	ype	passive	active	historical
Autonomy characteristics	Adaptive / Learning			Determine if system response to similar inputs has changed over time
	Adversarial		Evaluate results of system operation applied to operator/SME supplied what-if examples	Number of adversaries identified, and percentage of those addressed
	Dynamic	If operator must accept tainted results before they can be used then number of tainted results accepted by operator		*(if has occurred)* Assessment of quality of graceful degradation
	Human Interaction		SUS score	
	Self-directed			
	Self-governed			
	Uncertain			
	Unstructured			
	SoS integration safe			
Automation - trust attributes	Perceived competence	Percentage of time operator chooses to not override system (decrease);	Evaluate results of system operation applied to operator/SME supplied what-if examples	Number of "functions" executed per unit time by computer (vs humans) (desire increased); Ratio for completion of "Mission-Critical Objectives" vs. "Secondary Objectives" (increase); Time to respond to critical events (decrease); Time to respond to non-critical events (decrease).
	Benevolence		"Lead it into temptation and see if it delivers evil" ==> Create "tempting" scenarios and assess if malevolence results	Percentage of time that system operates outside of "safeguarded" regions
	Understandability	Number operator queries per unit time; Percentage of queries which are followed immediately by another query	Evaluate results of query usefulness by operator/SME; SUS score	Time required for operator training (desire decrease)
	Directability	Response time when operator overrides system; Time from when operator takes control until time when operator releases control back to system		
	Reliability			Measured false positive rate (desire decrease); MTBF (desire decrease); MTTF (desire decrease); MTTR (desire decrease); number of HW faults per unit time (desire decrease); number of SW exceptions per unit time (desire decrease); Inventory (spares) (desire decrease); Maintenance man-hours (desire decrease); Time to Failure (TTF) (desire decrease); Time to Support (TTS) (desire decrease); Time to Maintain (TTM) (desire decrease); *(if so configured)* If system returns multiple results with confidence (think IBM's Watson) then analysis of results with confidences
	Validity	Accuracy of task / mission completion		Key performance parameters in CONOPS are met; Number of times per unit time operator assesses system supplied trust diagnostics (desire increase?)
	Utility	Percentage of time that operator is not actively engaging the system in any way		Speed at which active goals are achieved; Decrease in manpower and personnel requirements without impact to mission effectiveness
	Robustness			Number of times that HW was radically & unexpectedly changed and system continued operation (perhaps in a limited capacity); Increase in the time the system runs autonomously (assumes effective work is being performed)
	False-alarm rate	% of alarms not dismissed by operator and acted upon.		Measured false alarm rate

Fig. 4.A.3 Measures of Effectiveness (MOEs) for Trust in three categories

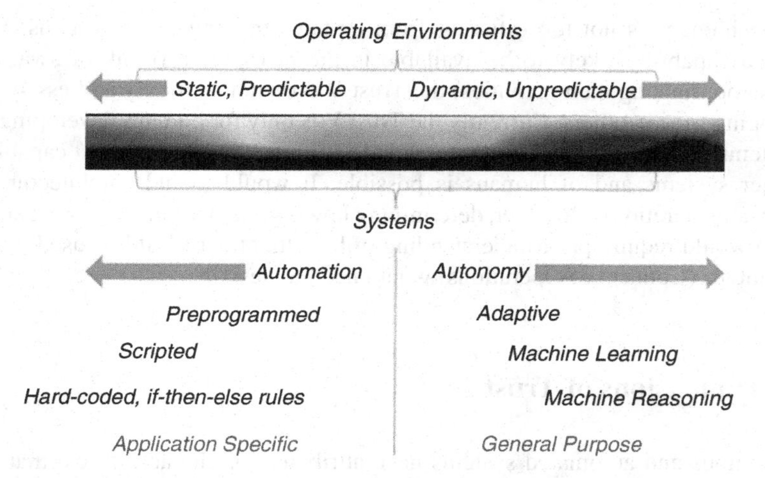

Fig. 4.1 Each operating environment is on the continuum between automated and autonomous

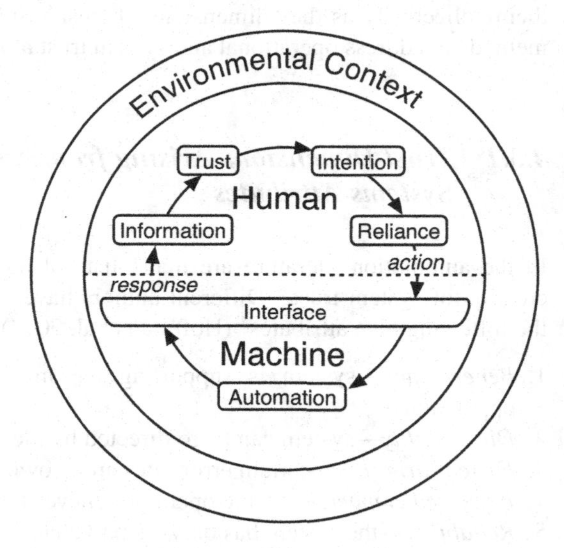

Fig. 4.2 The trust feedback loop (adapted from Lee and See 2004)

complex systems reduce human facing capability in order to manage complexity. However, it is possible for an autonomous system to demonstrate trust by reaching into a system to present a deep understanding to the human operator in a mission appropriate fashion, distilling complex data into readily usable information and knowledge.

In a utopian world, humans and systems (or machines) would seamlessly pass control back and forth as the situation required. The degree to which human-machine collaboration is successful is a measure of human trust in the system (DSB 2012). However, complete human-machine collaboration requires that the system trusts the human partner, an inversion of the usual relationship. Of course, systems trusting humans is not too different from systems trusting other systems. While this is a capability likely to be available in the near future (think of swarms of drones coordinating their activities) the Trust V does not explicitly address systems developing trust in others. Currently the Trust V is only for humans developing trust in systems. Expanding the Trust V to include systems' developing trust capabilities of other systems and of humans is possible. It would include architecture and design considerations. However, determining how a system develops trust of another system would require prior understanding of how that trust would be used—which the ConOps (Concept of Operations) would need to describe.

4.3 Dimensions of Trust

Autonomous and automated systems have attributes (or characteristics) that other systems do not have. This section identifies these special attributes and refers to them collectively as the "dimensions of trust." Subsequently, we suggest specific methods to address operational and system trust along each of these dimensions.

4.3.1 Trust Dimensions Arising from Automated Systems Attributes

In the automation literature are many lists of attributes that are needed for the creation of system trust.[3] Different authors have different lists. We choose to use the following nine attributes[4] (Hoffman et al. 2013):

1. *Benevolence*—system is supporting the mission and operator (and not in opposition)
2. *Directability*—system can be re-directed by the operator
3. *False-alarm rate*—certain error rates are known and acceptable
4. *Perceived competence*—the operator believes the system can perform a task
5. *Reliability*—the system has only a small chance of failing during a mission
6. *Robustness*—the system can appropriately handle perturbations
7. *Understandability*—the conclusions a system reaches can be understood
8. *Utility*—the system adds value
9. *Validity*—the system is solving the correct problems

[3]A synopsis of 14 previous studies in this area is in (Lee and See 2004)

[4]Definitions of the following terms, and others in this paper, are in a "Trust V" spreadsheet, available from the authors. In this paper we give the gist of each term.

4.3.2 *Trust Dimensions Arising from Autonomous Systems Attributes*

The Department of Defense (DoD) established the Autonomy Research Pilot Initiative (ARPI) in 2012, and defined autonomy as (ARPI 2012)

> *Systems which have a set of intelligence-based capabilities that allow it to respond within a bounded domain to situations that were not pre-programmed or anticipated in the design (i.e., decision-based responses) for operations in unstructured, dynamic, uncertain, and adversarial environments. Autonomous systems have a degree of self-governance and self-directed behavior and must be adaptive to and/or learn from an ever-changing environment (with the human's proxy for decisions).*

From this definition, we extract eight autonomy attributes to be used in our analysis of how to develop trust. These are attributes of a system (1,5,6), its environment (2,3,7,8), or its mission (4):

1. *Adaptive/Learning*—system can acquire information and then beneficially leverage that information
2. *Adversarial*—system can complete its mission when subject to opposing efforts
3. *Dynamic*—system can complete its mission in a changing environment
4. *Human interaction*—system can interact with humans, and humans with the system, for mutual benefit
5. *Self-directed*—system can direct itself (e.g., can determine what to do)
6. *Self-governed*—system can control itself (e.g., can do what needs to be done)
7. *Uncertain*—system can complete its mission in environments that are different from expectations
8. *Unstructured*—system can complete its mission in environments that are difficult to describe

Each of these presents their own challenges. A learning system may give different responses to the same input which makes testing a challenge. Just detecting when a clever adversarial agent is present can be a challenge, and this may be a necessary precursor to appropriately responding. In dynamic environments the time scale of the change is important, yet often unknown. When dealing with humans a system may have to compensate for errors made by the humans. Self-directed and self-governed systems will need to incorporate risk into their decision-making. Uncertain environments require a system understand what it does know, and what it does not know. In an unstructured environment the system must focus on good-enough solutions, not (impossible to obtain) perfect solutions.

4.3.3 *Another Trust Dimension: SoS*

In addition to the above trust dimensions, we add the additional attribute of "Systems of Systems (SoS) integration safe". That is, when multiple systems (autonomous or not) are combined, undesirable emergent behaviors may occur. This is of particular concern when autonomous systems are combined since operational activities become more difficult to predict. A practical example of a large scale complex system (Osmundson et al. 2008) is the North American power grid. The collapse of the Canadian power grid in Quebec province in 1989 and the power outages affecting about 50 million people in the eastern US in 1993 are examples of negative emergent behavior. Clearly, the North American power grid is automated, not autonomous. However, we are unaware of any large-scale autonomous systems, or systems of systems, for which emergent behavior (negative or positive) can be predicted. Current research indicates that for autonomous systems of systems, emergent behavior is even more unpredictable and potentially disruptive than automated systems, which tend to follow a known path.

Haskins (2011) defines Systems Engineering to be "an interdisciplinary approach and means to enable the realization of successful systems." We believe that good system engineering requires that in addition to capturing the ConOps (what a system will do) one should also capture what a system will *not* do (we can call this a "Negative ConOps"). Having a Negative ConOps analysis is especially valuable for an autonomous system since such systems are less predictable than non-autonomous system. More importantly, the information in a "Negative ConOps" can contribute to system trust. For example, if a driverless car had its speed limited to 5 miles per hour (mph), then someone's trust in going for a test ride could be increased. Fortunately, Negative ConOps do not create an additional trust dimension since they can be incorporated as a "positive statement of prevention of negative consequences". For the driverless car example, a requirement stating, "The car shall travel at speeds of less than 5 mph" would meet the need.

4.4 Creating Trust

We combine the nine automation trust attributes, the eight autonomous attributes, and SoS to obtain the "18 trust dimensions." The paradigm introduced below shows how trust may be created for (almost all) of these dimensions.

Note that autonomous systems only require trust to be demonstrated in relevant dimensions. For example, a system that is not in an adversarial environment does not need to demonstrate trust in handling adversaries. Note also that a system's relevant trust dimensions may change during system operations. For example, the need for self-direction of a system could change as communications connectivity and operator availability changes.

4.4.1 Building Trust In

An old adage is that "you cannot test quality in, you must build quality in." This is also true for many (perhaps all) of the non-functional requirements (also known as "qualities") of concern to systems engineering. Non-functional requirements are described as follows (Long 2012):

- "Functional requirements define **what** a system is supposed to do e.g. Performance
- Non-functional requirements define **how** a system is supposed to be."

That is, non-functional requirements are usually characterized by adjectives and include either: overall product properties, the character of a product's output, or the experience of the user using the product (Willis and Dam 2011). One of our key insights is that system trust can be understood as a non-functional requirement you must build "trustability" into a system. We have not seen trustability on anyone's list of qualities.[5]

The underlying concept of our Trust V approach is that trust, like quality, must be built into the system and not "bolted on" after the fact. Starting in the concept of operations (ConOps) phase and continuing through the entire Systems Engineering lifecycle, activities to engender trust should be built in. The Trust V approach identifies a selection of trust methods across all phases of the systems lifecycle to instill confidence that the system will perform correctly. Of course, the appropriate trust method(s) used for any specific system must be negotiated between customers, end users, and contractors.

Another primary driver for "building trust in" is that most of the techniques that will engender operational and system trust require trust techniques to be deeply embedded in the fabric of the system. An additional benefit of "building trust in" is the ability to generate and collect objective evidence for system certification and trust.

For example, if the operator is to be able to query the rationale and decision making of an autonomous system (what we call "Semantic Question & Answer (Q&A)"[6]) then the concept of operations must ensure that the system has the appropriate knowledge to answer an operator's query. It is likely to be cost prohibitive to add this capability after a system has been developed—this capability should be part of or an enhancement to the original requirements and design. An additional advantage to having a built-in Semantic Q&A capability is that it could substantially benefit the Test and Evaluation (T&E), and Verification and Validation (V&V) activities for all system components, not just the autonomous components.

---

[5] Willis and Dam (2011) lists 56 non-functional requirements; Long (2012) lists 65.

[6] We use "Semantic Q&A" to indicate a system response capability in which queries on system performance are answered in the language of the system operator and the ConOps.

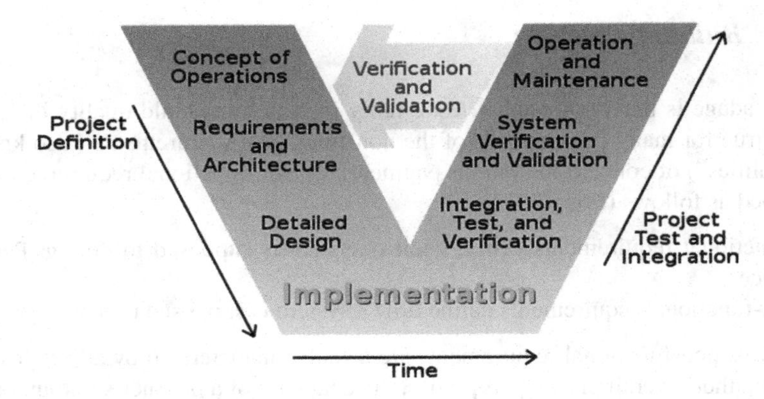

Fig. 4.3 The Systems Engineering V model (from FHA 2005)

4.5 The Systems Engineering V-Model

Large and complex systems are often managed via the Vee- or V-Model (Haskins 2011, Sect. 3.3):

> Various life-cycle models, such as the waterfall, spiral, Vee, and Agile Development models, are useful in defining the start, stop, and process activities appropriate to the life-cycle stages. The Vee model (...) is used to visualize the system engineering focus, particularly during the Concept and Development Stages. The Vee highlights the need to define verification plans during requirements development, the need for continuous validation with the stakeholders, and the importance of continuous risk and opportunity assessment.

The V-model is so-called due to the associated picture that shows the thinking behind the model (FHA 2005), see Fig. 4.3. In this figure, time/activity is going from left to right following the arrows.

The path going down the "left leg" of the V is a decomposition process, turning specifications for assemblies into specifications for sub-assemblies, those for sub-assemblies into those for components, etc. The path going up the "right leg" is a progression of system verifications against the requirements and then a validation showing that the delivered system operates correctly in the intended environment. When the top of the V is reached:

- The system has been verified to show compliance with requirements
- The system objectives and ConOps have been validated
- There is system acceptance—the system meets its intended use

A key attribute of the Trust V representation is that artifacts/capabilities needed for test on the way up (e.g., on the right leg) are put in place on the way down (e.g., on the left leg). For example, if it has been determined that the final system needs to be simulated as part of system test, then a simulation capability needs to be specified by the requirements (e.g., on the way down), built into the system, and then tested (e.g., on the way up).

4.6 The Trust V-Model

The paradigm proposed in this paper to ensure trustability of a system is the creation of a Trust V, which is aligned with the Systems Engineering V. The Trust V ensures that the activities on the way down create the necessary artifacts/capabilities for trust that can be confirmed on the way up using Test and Evaluation (T&E) and Verification and Validation (V&V). Hence, when the top of the right leg of the Trust V is reached the system should be capable of being trusted after suitable user experience.

As mentioned earlier, many of the trust dimensions/attributes can be demonstrated by a "Semantic Q&A" capability. This may be thought of as the system responding to a query of the form "Why did you do XYZ?" with an answer in the language of the operator using the concepts from the ConOps. Queries of systems/robots must be tested for reliability and validity, as with humans. Creating a Semantic Q&A capability requires capturing the rationale for all the requirements, from high level, customer-imposed requirements to lower level derived requirements. While these rationales will likely be important when having a system explain its actions, few programs capture this information in a re-usable form. Having this capability requires that the necessary meta information about the system be made available on the left leg of the Trust V. This information will be used on the right leg of the Trust V as the system is verified and validated to create system trust.

As an example, consider a self-driving car. While the car's occupants may be alarmed if the car suddenly changes lanes, they are likely to be relieved if the car can explain that its actions were based on a collision occurring up ahead. While the car's motion control system would likely involve hundreds of inputs, extracting the key single piece of information (collision up ahead) and explaining it in the operator's language (e.g., saying "An accident occurred up ahead" and not saying "line 40 of subroutine XYZ said to change lanes") would support trust generation for some of the 18 dimensions. This type of information extraction would likely be tightly coupled into the car's control system.

In addition to system trust, autonomous systems will also need operational trust; trust from the operator while the system is being used. Operational trust can be obtained by leveraging the artifacts/capabilities that were introduced on the way down the Trust V. For example, if a capability needed for operational trust is the ability to understand the system's rationale for future activities (e.g., via what-if type queries such as "If you were to observe XYZ next, what action would you take?"), then this capability must be built into the system as it is being developed—even if it is not needed to create system trust.

This is a key point. While a system is fully specified in the requirements phase, the requirements are usually derived only from the system ConOps. To ensure that both system and operational trust are created and measured, additional requirements, which may not be needed for the original mission ConOps, need to be added. This has similarities with creating a superior human-machine interface. While a system ConOps may specify an interface, ConOps rarely specify a superior interface since

Fig. 4.4 The Trust V model

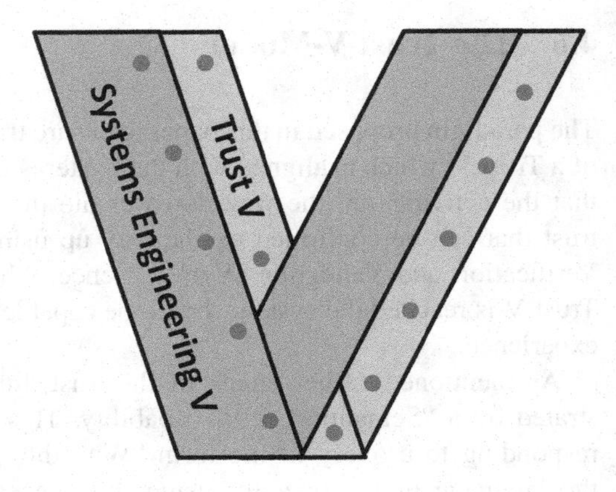

the needed specification language is not available. Yet the requirements derived from the ConOps must be properly augmented by interface requirements, as the interface can be the single most important aspect of the system to an operator. As autonomous systems become more prevalent, and need for trust increases, trust issues may become a fundamental component of a system's ConOps.

Another key component for trust is the capture and communication of system capability metrics (these are described in later sections). These capabilities are not likely to be available in an autonomous system unless they have been implemented on the way down the Systems Engineering V.

4.6.1 The Trust V Representation: Graphic

Graphically, we show the Trust V (thin part to the right in Fig. 4.4) as being aligned with the Systems Engineering V (thicker part to the left). The dots embedded within the Trust V are specific items that are needed to create system and operational trust:

- On the "left leg" of the Trust V the dots represent specific artifacts/capabilities that are added during system development for the purposes of trust.
- On the "right leg" the dots represent specific Test and Evaluation (T&E) and Verification and Validation (V&V) tests performed using those capabilities to create trust.

For example, consider the Reliability attribute of automation. Trust in a system's Reliability can be addressed (i.e., increased) in many ways. For example: introducing redundancy, use of fault tolerate computing, designing in failover capabilities, reuse of trusted components, etc. While none of these concepts are novel they all require that the system be designed to handle them. They change the specification of what and how the system is developed going down the left leg of the V.

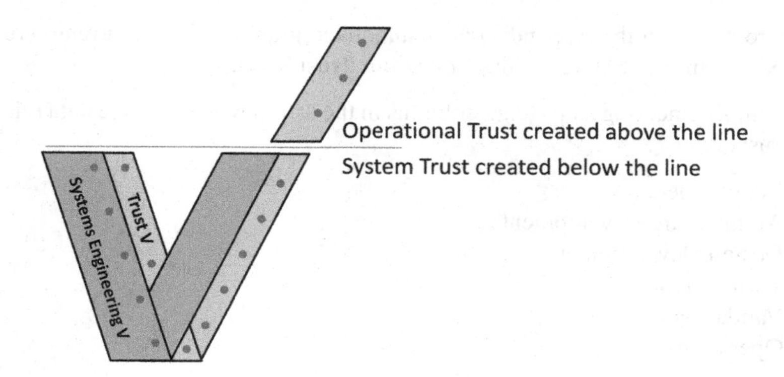

Fig. 4.5 The extended Trust V model

Note that there are also dots within the "left leg" of the Systems Engineering V. These represent artifacts/capabilities used by systems engineering to meet its non-trust requirements that can be leveraged to create trust. Considering reliability again, any of the methods mentioned may already be necessitated by the system ConOps. If so, they can be reused for trust purposes at no cost.

Operational trust is created by the operator once they start using and interacting with the system in such a way that their confidence in the system is boosted. This trust is enabled by leveraging the artifacts/capabilities that were placed in the system during the development process. To represent this we extend the Trust V on the right as shown in Fig. 4.5. Extensions to the left could also be considered, they would amount to creating trust paradigms in advance of starting system development.

The dots above the line represent trust-specific processes, such as a Run Time Assurance (RTA) (Hinchman et al. 2012) or Future Scenario Prediction (useful for the Uncertain environment attribute).

4.6.2 *The Trust V Representation: Array*

Earlier in this chapter the Trust V has only been an abstraction. We make it real by conflating the two ideas presented above: the automated/autonomous system attributes and the System Engineering V.

The information content of the Trust V can be represented by an array (e.g., a spreadsheet) in which the rows represent the 18 trust dimensions and the columns represent the Systems Engineering lifecycle steps. We use the 11 steps from ISO/IEC 15288 (Roedler 2002), which is a Systems Engineering standard. These steps correspond to a "flattened" Systems Engineering V. At the intersection of each trust dimension and each Systems Engineering step should be methods that can be used to build trust—of the specific type needed for the system attribute, at the specific phase of system development. Each method would appear as a dot on the Trust V (Figs. 4.4 and 4.5).

Figure 4.A.1 (in the Appendix) is an annotated graphic of our (current) Trust V array. Some information about the data in our Trust V array:

- System Engineering steps (e.g., columns in the array) which include data relevant to trust are:
  - Requirements
  - Architecture Development
  - Design Development
  - Verification
  - Validation
  - Operations

  Interestingly, we have not identified methods within the other system engineering steps (e.g., Integration Process or Transition Process) which would enable operational trust. Discovering these might be a side effect as the industry gains experience with Autonomous Systems.
- Verification and Validation (V&V), which is sometimes considered a single activity appears in two columns since: ISO/IEC 15288 requires it, and different trust methods apply to each activity.
- An additional column has been added entitled "Analysis Preceding Requirements". This represents an understanding of how trust and measurement of trust fit within system operations.

The rows of the Trust V array enumerate the Trust Dimensions consisting of automated, Systems of Systems and autonomous system, environment and mission attributes.

4.6.3 *Trust V "Toolbox"*

We view the individual methods in the Trust V array as a continuously evolving toolbox of trust methods. These trust methods encompass a wide range of activities from system design and architecture approaches to the use of semantic question and answer capabilities between the operator and the system. The trust methods include well known and often utilized techniques such as the proper use of boundary definitions in the requirements phase and system simulation. To accommodate the unique trust requirements imposed by autonomous systems, additional techniques have been created and added to the Trust V toolbox. These include "Chatter," "Future Scenario Prediction," and many other methods.

As an illustration, Fig. 4.A.2 shows the current entries for the autonomous attribute of "Uncertain". Many of the techniques listed are well known in systems engineering. Here are descriptions of some techniques that are not as well known:

- Semantic Q&A Capability—this capability was described earlier. It represents a paradigm in which an operator can query the system with questions such as "Why did the system do XYZ?" A simplified capability, which has been prototyped by the authors is called Chatter and applies to a layered control model. More details are in the next section.

- Future scenario prediction—in this paradigm an operator can query the system with questions such as "What will the system do next?" This is related to the usual Systems Engineering capability referred to as "simulation over live", where a simulation is built on top of a current state which uses live operational data.
  - One way this paradigm could be implemented to address trust concerns is to have the system create and run a collection of relevant simulations on top of the current state, summarize the results, and present probabilistic conclusions to the operator in a language the operator can understand. For example, a system may state, *"If the external temperature decreases more than 5° (30 % likelihood in the next 10 min) then there is an 80 % chance the battery power will be too low to complete the last mission phase"*. Proposing appropriate remedial actions could also increase trust, especially for users who engage a system at a higher level of abstraction.
  - Another way this paradigm could address trust concerns is to consider the system's layered control structure and determine when an active constraint stops being active. It is the transition of these constraints that changes the system state and influences an operator's trust in the system. For example, a system may state *"If the temperature decreases more than 10° then the auxiliary heater will start up and the mission duration will be reduced by 30 min."* Once again, recommended actions could be incorporated.
- "Turing Test"—this represent a paradigm which validates a system by comparing its performance to an expert user. We usually consider humans, especially subject matter experts (SMEs), to be the gold standard in terms of performing the right tasks (although not always at the speed of a computer). If an operator cannot discern the difference between the actual system and a human SME performing the same task as the system using the same inputs—then the system is validated.[7]

Another way in which to increase system trust is to introduce "calibrated trust", which is being used by Perceptronics (2013).[8] In "calibrated trust" an operator's decision making is enhanced by having the system provide insight into the estimated trustworthiness of the system. While the logic may appear circular it is as valid as any system that reports errors. How do we know that a system reporting an error is working well enough to report that error? The goal is for an operator to trust the system neither too little (i.e., "distrust"), nor too much (i.e., "overtrust"), but just the right amount (i.e., "calibrated trust").

[7] "The purpose of the Validation Process is to provide objective evidence that the services provided by a system when in use comply with stakeholders' requirements, achieving its intended use in its intended operational environment." (INCOSE, 4.8.1.1).

[8] Raytheon is working with Perceptronics (Perceptronics), on their Phase II SBIR involving the ATCI ("Adaptive Trustworthiness Calibration Interface").

4.7 Specific Trust Example: Chatter

Each Trust V cell, defined by one of 18 system attributes and a Systems Engineering stage, contains methods contributing to trust. When populating these cells, we recognized that some methods are broadly applicable to many automated/autonomous attributes (see Fig. 4.A.2). Chatter is one of these methods.

In Chatter a system with a layered control "chatters away" at the operator,[9] informing the operator of

- Information they would like to know, if they knew to ask
- Information on state and mode transitions in the system

To prevent information overload the operator can control:

- The frequency of communication. (How often does the system communicate with the operator: every event?, every 10th event?, a summary every 10 min?)
- The information content level. (That is, at what level in the layered control system does the information come from.)
- The information details. (The information content of each message answers the question "Why did this happen?" at a recursion level specified by the operator. Hence, a Chatter message in the abstract might be *"In Layer 3 state XYZ was changed to state ABC due to a state change in Layer 4. In Layer 4 state UVW was changed to state DEF due to a threshold value reached in Layer 5. In Layer 5 the input value of temperature reached the threshold of 20°."*.)

Chatter provides unsolicited information to the operator as follows:

- Several data streams within system operations were monitored and the statistics on each data stream were obtained.
- Whenever a data value on one of the data steams was statistically unlikely, the operator was alerted. Unlikeliness was measured by how many standard deviations away from the mean the data value was. An operator could set the number of standard deviations required for the system to raise an alert.

Chatter amounts to transparency. The operator learns when and why the system is doing what it does. The operator is also is alerted to unusual values. For example, knowing an input value (for example) was unusual might indicate that the concomitant output should be neglected (since the input was so unusual).

Chatter works best for systems represented by layered control models that make decisions. It helps if each layer's activities can be understood by a trained but perhaps unsophisticated operator. Chatter can be used in different contexts: during operator training to create Trust in a system, for on-going use in conjunction with a tactical system to assure operators that a system is working, or to supply situational awareness.

[9]It's possible for one system to Chatter to another system (and not an operator). This could be used in a "layered Trust" system or a system of interdependent autonomous components.

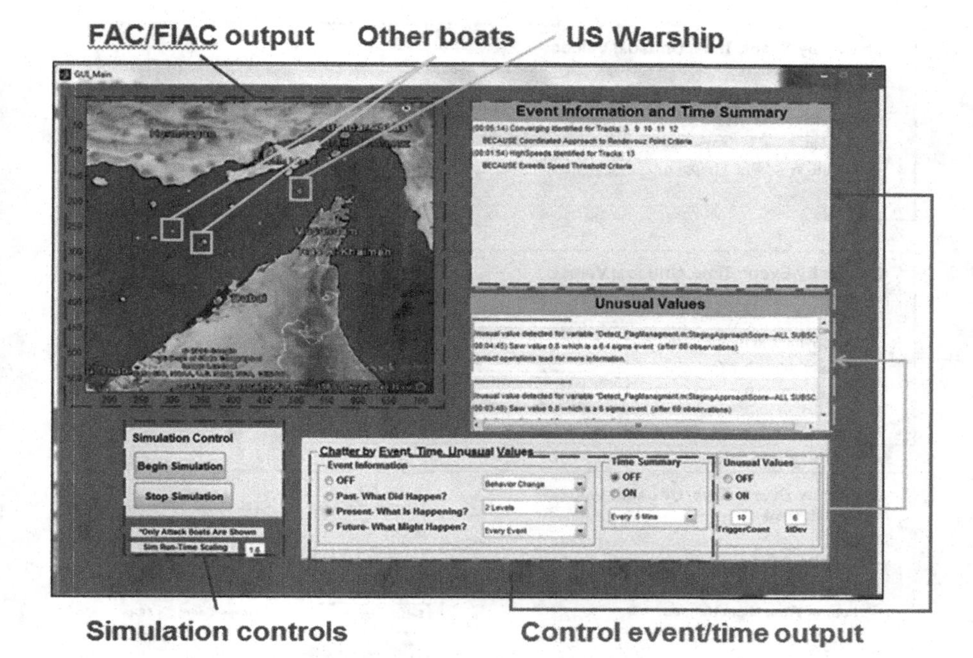

Fig. 4.6 The chatter interface for a FAC/FIAC prototype

We prototyped Chatter on a Fast Attack Craft/Fast Inshore Attack Craft (FAC/FIAC) system (FACFIAC 2014) using Matlab. The FAC/FIAC mission is to detect swarming boat attacks. Our enhancement was to layer Chatter on top of a prototype to engender Trust. The interface is shown in Fig. 4.6 and the settings available to the operator are shown in Fig. 4.7.

We performed end user testing of the FAC/FIAC tactical code both with and without the auxiliary use of "Chatter". The survey conclusions were twofold: Having Chatter added confidence in the tactical system and the Chatter capability was desired by users.

4.8 Measures of Effectiveness

Knowing that we have "built trust in" a system is not sufficient by itself. We also need ways by which to measure both system and operational trust. Since absolute measures of trust do not exist, we need ways to measure changes in trust. Like trust itself, a trust measurement and reporting capability cannot typically be bolted on to an existing system. It is best implemented as part of the system design.

An efficient way to measure trust is to have the system monitor its' use and report trends. For example, a simple but very useful metric would be how often

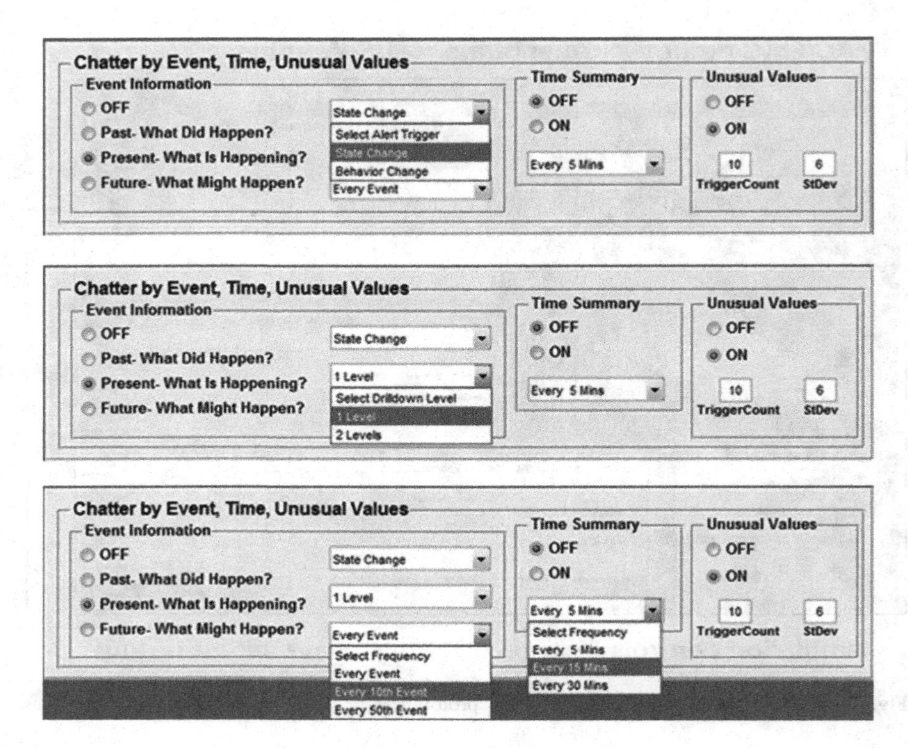

Fig. 4.7 Chatter settings available to operator on FAC/FIAC prototype

an operator overrides the system. If this measurement decreases over time, one could conclude that relative trust in the system has increased with familiarity. Another measure of effectiveness would simply be the accuracy and reliability of the system's actions/decisions.

Ideally, each of the 18 dimensions of trust would have associated measures of effectiveness (MOEs) which can assess the relative level of trust created. We have created a set of MOEs for many of the trust dimensions. They are represented in an array in which the rows represent the 18 dimensions of trust (as before) and the columns represent the "category" of MOE. We have identified three MOE categories:

- *Passive*—This MOE will be collected without special system operator effort and is available under normal system operations. This category of metric would be automatically collected, aggregated, and reported by the system. For example: an increasing value in the percentage of time that the operator chooses to *not* override an autonomous controller indicates more trust in the system.
- *Active*—For this MOE a system operator must perform some extra action, however minimal. For example: consider the frequency of "likes" that an operator gives to responses to queries of an autonomous system. An increasing value indicates more trust in the system.

- *Historical*—For this MOE very specific analyses are performed on historical operational data and may require the use of subject matter experts. For example: the percentage of times that the system "took SME approved action" without operator intervention given an appropriate opportunity. An increasing value indicates more trust in the system. Additionally, the system's historical record could be used for forensic analysis similar to the use of "black box" recorders in aircraft accidents.

Similar to methods in the Trust V, the MOEs are undergoing frequent updating. Figure 4.A.3 has an annotated graphic of the current MOE array.

4.9 Conclusions and Next Steps

Trust capabilities should be built into the system, not created after the fact. The Trust V framework adds value when used in the design and development of an autonomous system. It increases system trust for the system acquisition team, may streamline the Verification and Validation process, and will support the creation of operational trust by a system operator. These goals are achieved by identifying the specific methods required to engender trust based on system, environmental, and mission attributes. The additional trust created by using the framework is beneficial to any system, not just systems with autonomous or automated components.

The claims of usefulness of the Trust V process have been validated. As indicated earlier the Chatter technique was successfully prototyped and well received by potential operators.

To continue demonstrating the value of the Trust V framework the authors will:

- Mature the Trust V and associated MOEs
  - Incorporate relevant industry and DoD trust elements
- Mature selected ubiquitous trust methods
  - Determine the trust methods and MOEs that are most ubiquitous
  - Perform "Make/Buy" trade studies for selected trust methods. That is, evaluate existing Trust methods developed by academia and industry for incorporation into the Trust V.
- Pilot the Trust V and associated MOEs on additional internal and customer programs to measure and document specific benefits of the Trust V methodology and associated trust methods
- Explore the relationship between system complexity, autonomous attributes and thresholds for establishing trust. Since building trust into a system is time consuming and expensive, we need to determine which strategies should be adopted to ensure that we have sufficient but not excessive trust built into a system.

- Evaluate how well trust can be engendered in one part of a system while not impacting the overall operational end user experience. If an autonomous component is isolatable from the rest of the system, perhaps the autonomous component is where the primary trust methods and execution can be focused.
- Understand if the Trust V approach can be utilized for machines trusting humans or perhaps even for machines trusting machines.
- Evaluate instances of system mistrust and analyze where trust methods and approaches might have been used to rectify the mistrust

A.1 Appendix

References[10]

ARPI (2012) Autonomy Research Pilot Initiative (ARPI): invitation for proposals. www.defenseinnovationmarketplace.mil/resources/AutonomyResearchPilotInitiative.pdf. Accessed 11 Nov 2013

DSB (2012) Defense Science Board, The Role of Autonomy in DoD Systems, July 2012. https://www.fas.org/irp/agency/dod/dsb/autonomy.pdf. Accesses 11 Nov 2013

Dzindolet MT, Peterson SA, Pomranky RA, Pierce LG, Beck HP (2003) The role of trust in automation reliance. International Journal of Human-Computer Studies 58:697–718

FACFIAC (2014) U.S. Navy Ships Conduct FAC/FIAC Exercises. www.navy.mil/submit/display.asp?story_id=29911. Accessed 30 Oct 2014

Fainekos GE, Pappas GJ (2006) Robustness of temporal logic specifications. In: Havelund K, Núñez M, Roşu G, Wolff BAC (eds) Proceedings of International the Workshop on formal approaches to software testing and runtime verification. Lecture notes in computer science, vol 4262. Springer, Berlin, pp 178–192

Feiler P, Gluch D, Hudak J (2006) The architecture analysis & design language (AADL): an introduction (Technical Report CMU/SEI-2006-TN-011). Software Engineering Institute, Carnegie Mellon University, Pittsburgh

FHA (2005) Clarus concept of operations, Publication No. FHWA-JPO-05-072, Federal Highway Administration (FHWA). ntl.bts.gov/lib/jpodocs/repts_te/14158_files/14158.pdf,20. Accessed 19 Nov 2013

Forsberg K, Mooz H (1991) The relationship of systems engineering to the project cycle. First Annual Symposium of the National Council On Systems Engineering (NCOSE)

Gall J (1977) Systemantics: how systems work and especially how they fail. Quadrangle/New York Times Book, New York

Gates R (2011) Science and Technology(S&T) Priorities for Fiscal Years 2013–17 Planning. https://www.acq.osd.mil/chieftechnologist/publications/docs/OSD%2002073-11.pdf. Accessed 21 Nov 2015

FHWA SE Guidebook (2009) Systems engineering guidebook for intelligent transportation systems, ver. 3.0, U.S. Dept. of Transportation, Federal Highway Administration, California Division

Haskins C (ed) (2011) INCOSE Systems engineering handbook v. 3.2.1, INCOSE-TP-2003-002-03.2.1

[10] A spreadsheet containing the latest Trust V data and the latest MOEs is available from the authors.

Hinchman J. et al. (2012) Towards safety assurance of trusted autonomy in air force flight critical systems. https://www.acsac.org/2012/workshops/law/AFRL.pdf. Accessed 30 Oct 2014

Hoffman RR, Johnson M, Bradshaw JM, Underbrink A (2013) Trust in automation, human centered computing. IEEE InTeLLIGenT SySTeMS. www.jeffreymbradshaw.net/publications/50.%20Trust%20in%20Automation.pdf. Accessed 11 Nov 2013

IHMC (2013) Executive summary—Workshop on human-machine trust for robust autonomous systems. Florida Institute for Human and Machine Cognition (IHMC). www.ihmc.us/groups/hmtras/wiki/9d7ea/Executive_Summary.html. Accessed 11 Nov 2013

Lee JD, See KA (2004) Trust in automation: designing for appropriate reliance. Human Factors 46(1):50–80. www.engineering.uiowa.edu/~csl/publications/pdf/leesee04.pdf. Accessed 11 Nov 2013

Long A (2012) Proposed unified "ility" definition framework. www.dtic.mil/ndia/2012system/track714832.pdf. Accessed 04 May 2014

Muir B (1987) Trust between humans and machines, and the design of decision aids. Int J Man Mach Stud 27(5–6):527–539

Osmundson JS, Huynh TV, Langford GO (2008) Emergent behavior in systems of systems. INCOSE. faculty.nps.edu/thuynh/Conference%20Proceedings%20Papers/Paper_14_Emergent%20Behavior%20in%20Systems%20of%20Systems.pdf. Accessed 29 Oct 2014

Perceptronics (2013) Perceptronics company website. www.percsolutions.com in Sherman Oaks (CA), Encino (CA), and Falls Church (VA). Accessed 11 Nov 2013

Roedler G (2002) What is ISO/IEC 15288 and why should I care? www.incose.org/delvalley/iso_iec_15288.pdf. Accessed 04 May 2014

Rosen RJ (2012) Google's self-driving cars: 300,000 miles logged, not a single accident under computer control. The Atlantic, 9 August 2012. www.theatlantic.com/technology/archive/2012/08/googles-self-driving-cars-300-000-miles-logged-not-a-single-accident-under-computer-control/260926/. Accessed 04 May 2014

Stone M (2011) DoD Priorities for autonomy research and development, 21 October 2011, NDIA Disruptive Technologies Conference. www.acq.osd.mil/chieftechnologist/publications/docs/2011%2011%207%20Autonomy%20PSC%20Roadmap.pdf. Accessed 11 Nov 2013

Willis JD, Dam S (2011) The forgotten "-ilities". www.dtic.mil/ndia/2011system/13166_WillisWednesday.pdf. Accessed 16 Nov 2013

Chapter 5
Big Data Analytic Paradigms: From Principle Component Analysis to Deep Learning

Mo Jamshidi, Barney Tannahill, and Arezou Moussavi

5.1 Introduction

System of Systems (SoS) are integrated, independently operating systems working in a cooperative mode to achieve a higher performance. A detailed literature survey on definitions to applications of SoS can be found in recent texts by Jamshidi (2008, 2009). Application areas of SoS are vast indeed. They are software systems like the Internet, cloud computing, health care, and cyber-physical systems all the way to such hardware dominated cases like military, energy, transportation, etc. SoS's are among main sources of big data, e.g. social networks, smart grid, healthcare data, traffic, military, etc. Data analytics and its statistical and intelligent tools including clustering, fuzzy logic, neuro-computing, data mining, pattern recognition and post-processing such as evolutionary computations have their own applications in forecasting, marketing, politics, and all domains of SoS.

A typical example of SoS is the future Smart Grid, destined to replace the conventional electric grid. A small-scale version of this SoS is a micro-grid designed to provide electric power to a local community. A micro-grid is an aggregation of multiple distributed generators (DGs) such as renewable energy sources, conventional generators, in association with energy storage units which work together as a power supply networked in order to provide both electric power and thermal energy for small communities; these may vary from one common building to a smart house or even a set of complicated loads consisting of a mixture of different structures such as buildings, factories, etc. (NREL 2014). Typically, a

M. Jamshidi (✉) • A. Moussavi
ACE Laboratories, The University of Texas, San Antonio, TX, USA
e-mail: moj@wacong.org; arezou.moussavikhalkhali@utsa.edu

B. Tannahill
Southwest Research Institute (SwRI), San Antonio, TX, USA
e-mail: tannahill@swri.org

© Springer Science+Business Media (outside the USA) 2016

R. Mittu et al. (eds.), *Robust Intelligence and Trust in Autonomous Systems*,
DOI 10.1007/978-1-4899-7668-0_5

micro-grid operates synchronously in parallel with the main grid. However, there are cases in which a Micro-Grid operates in islanded mode, or in a disconnected state (Jamshidi 2009). Accurate predictions of received solar power can reduce operating costs by influencing decisions regarding buying or selling power from the main grid or utilizing non-renewable energy generation sources. The object of this chapter is to use big data on energy to forecast wind energy availability and traffic jams in an attempt to derive an unconventional model, paving the way towards robust intelligence in big data analytics.

Section 5.2 first describes the wind energy data used for a micro-grid that will be used as the SoS of interest for this chapter. Section 5.3 then discusses the application and effectiveness of different data analytics tools in the generation of wind speed models. Section 5.4 provides an introduction to deep architectures, leading to deep learning. Section 5.5 presents conclusions.

5.2 Wind Data Description

Predicting wind power availability can be useful in energy trading or control algorithms for power companies. In order to make accurate predictions, a significant amount of environmental data was gathered (NREL 2014; IEM 2014) in order to be used as a training data set for artificial neural networks (ANN), a key tool in data analytics. It was decided that the output of the data analytics for this work should be predicted values of wind speed at three different altitudes (19 ft, 22 ft, and 42 ft). These values were shifted by 60 min so that they would serve as output datasets for the training of the neural networks investigated. Figures 5.1, 5.2 and 5.3 show the 19 ft, 22 ft, and 42 ft wind speed data throughout the October 2012 evaluation period.

Two different input data sets were used for this investigation. The first was merely reduced to a dimension of 21 using PCA (Smith 2002; Shlens 2009). The second was first expanded to include derived, preprocessed values including slope and average values from the past parameter values. Then, this second dataset was also reduced to a dimension of 21 using PCA.

5.3 Wind Power Forecasting Via Nonparametric Models

This section describes application of big data analytic to forecast wind energy availability. Section 5.3.1 discusses the high level results of the different ANN architectures investigated, and Sect. 5.3.2 shows the detailed results of the best performing architecture.

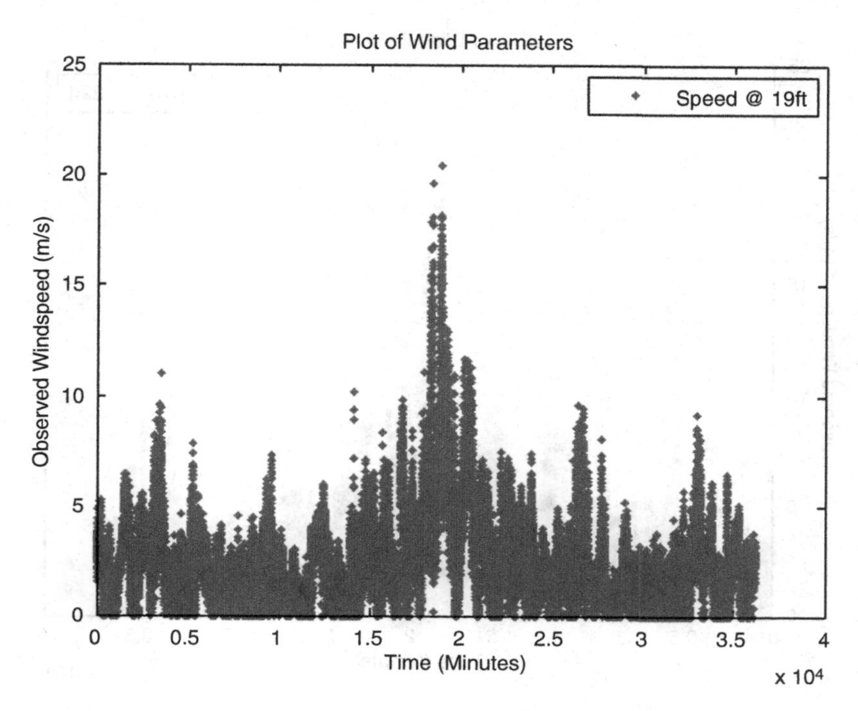

Fig. 5.1 Actual wind speed data @ 19 ft

5.3.1 *Advanced Neural Network Architectures Application*

During this effort, the following network architectures included in the MATLAB Neural Network Toolbox (Beale et al. 2014) were investigated:

- Standard Feed-Forward Neural Network
- Time Delay Network
- Nonlinear Autoregressive Network with Exogenous Inputs (NARXNET) Neural Network
- Layer Recurrent Neural Network

The first architecture used was the standard feedforward neural network. This is the default architecture used by MATLAB. When using this architecture, the training tool uses the Levenberg-Marquardt backpropagation method by default to train the network to minimize its mean squared error performance. This feed forward ANN had 21 inputs variables and one hidden layer comprised of 10 neurons.

The different configurations tested for this exercise are listed in Table 5.1, where results are in RMSE (Root Mean Square Error).

Note that when out-of-memory errors occurred during training using the default *trainlm* training function (Levenberg-Marquardt backpropagation), the lower

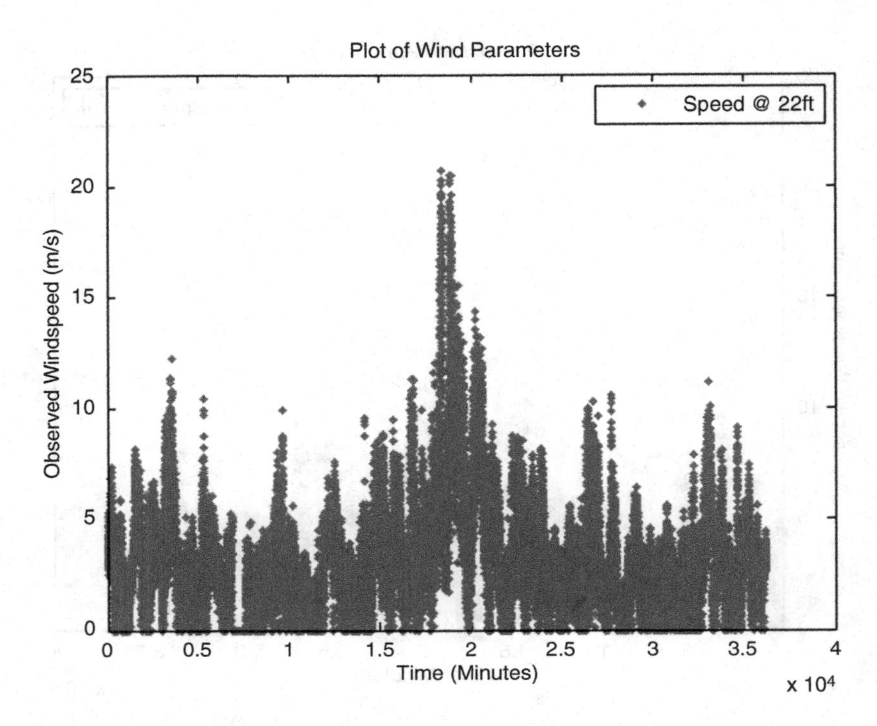

Fig. 5.2 Actual wind speed data @ 22 ft

memory requiring scaled conjugate gradient backpropagation (trainscg) training function was utilized instead. From the results in Table 5.1 the number of neurons can have an initial improvement in the error (RMSE) measure, but after it will not improve the convergence of ANN any more.

The second architecture used was the time delay neural network. This time-series sequential architecture allows for past values of the input parameters to be fed into the neural network. Note that the major difference between the Feed Forward Neural Network and this is the case of 0–10 delay block at the input to the network. This block represents that the actual input to the neural network is [inputs(t) inputs(t-1) inputs(t-2) ... inputs(t-10)]. As opposed to the Feed Forward Neural Network, this architecture allows for the network to have memory without including it during the preprocessing stage as described in Sect. 5.2.

One configuration was tested for this architecture, as listed below:

Time Delay Neural Network Configuration 1

- Number of Neurons in Hidden Layer = 10
- Delay Vector = [0:10]
- Using raw PCA-reduced data set
- Trained using trainscg
- RMSE Error: 2.1430

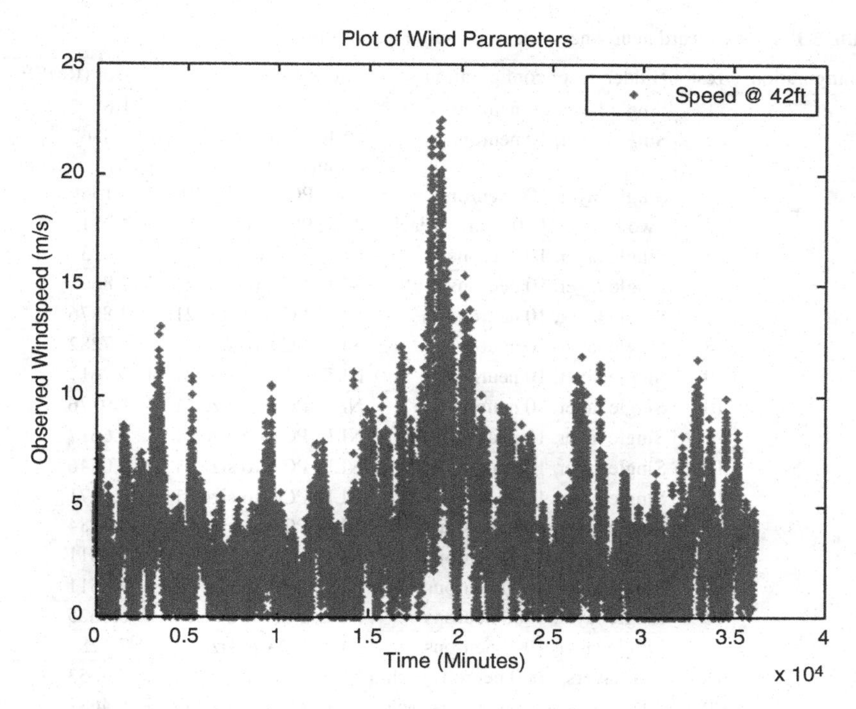

Fig. 5.3 Actual wind speed data @ 42 ft

Note that since this is a sequential neural network, the training vs. verification data sets were separated manually. Seventy percent of the data was used for training, and the entire data set was used for verification. Other architectures such as the NARXNET ANN (open and closed-loop variety) and Layered Recurrent Neural Network were similarly used, and more details are in (Tannahill 2014).

The NARXNET configurations tested for this simulation are listed in Table 5.2.

The layered recurrent neural network configurations tested for this study are listed in Table 5.3.

5.3.2 *Wind Speed Results*

Considering all the configurations tested, surprisingly the best performing neural networks were those using a pre-expanded (via NLE: nonlinear expansion) data set fed into a conventional feed forward neural network with ten neurons in the hidden layer. Figures 5.4, 5.5, 5.6, 5.7, 5.8 and 5.9 show the results and error generated using this network to predict wind speed an hour in advance.

As discussed in Tannahill and Jamshidi (2014), predicted wind power availability was then calculated assuming the wind speed across the entire wind turbine blade

Table 5.1 Feed forward neural network architecture test results

Configuration	Test	Hidden layer configuration	Data set type	Error (RMSE)
1	1	Single layer, 10 neurons	Raw	1.6187
2	2	Single layer, 10 neurons	NLE (nonlinear expansion)	1.3863
3	3	Single layer, 100 neurons	Raw, PCA to size 21	2.1556
4	4	Two layers, 1000 neurons each	Raw, PCA to size 21	4.2116
5	5	Single layer, 10 neurons	Raw, PCA to size 21	9.823
6	6	Single layer, 10 neurons	NLE, PCA to size 21	5.867
6	7	Single layer, 10 neurons	NLE, PCA to size 21	2.8876
6	8	Single layer, 10 neurons	NLE, PCA to size 21	2.7282
6	9	Single layer, 10 neurons	NLE, PCA to size 21	2.7617
6	10	Single layer, 10 neurons	NLE, PCA to size 21	1.9776
6	11	Single layer, 10 neurons	NLE, PCA to size 21	2.6617
6	12	Single layer, 10 neurons	NLE, PCA to size 21	4.0546
6	13	Single layer, 10 neurons	NLE, PCA to size 21	5.4171
7	14	Single layer, 1000 neurons	NLE, PCA to size 21	10.3084
7	15	Single layer, 1000 neurons	NLE, PCA to size 21	9.4811
7	16	Single layer, 1000 neurons	NLE, PCA to size 21	7.3511
7	17	Single layer, 1000 neurons	NLE, PCA to size 21	6.8625
7	18	Single layer, 1000 neurons	NLE, PCA to size 21	7.128
8	19	Two layers, 1000 neurons each	NLE, PCA to size 21	7.3653
8	20	Two layers, 1000 neurons each	NLE, PCA to size 21	8.4681
8	21	Two layers, 1000 neurons each	NLE, PCA to size 21	9.0292
8	22	Two layers, 1000 neurons each	NLE, PCA to size 21	8.9808

area was the same. It was also assumed that the air density was 1.23 kg/m^3 throughout the year, and that the power coefficient Cp was 0.4. These assumptions were used in conjunction with the wind turbine equation found to calculate power density availability (Watts/m^2). This quantity can be multiplied by the cumulative sweep area of a wind turbine farm to calculate total available wind power; however, this step was left out of this exercise to keep the results more generalized. The statistics of the resulting data from the year-long wind power prediction are included in Tannahill (2014).

5.4 Introduction to Deep Architectures

Neural networks are identified by their ability to capture the nonlinearity of data without being exposed to the dynamics of a model or the bounding function of input-output variables. Multi-layer perceptron (MLPs), also known as feedforward neural networks (FFNNs), contribute to many achievements in classification and regression problems in different science fields. In classification tasks, the focus

Table 5.2 NARXNET neural network architecture test results

Configuration	Test	Hidden layer configuration	Data set type	Input delays	Feedback delays	Closed loop error (RMSE)
1	1	Single layer, 10 neurons	Raw, PCA to size 21	[1:3]	[1:3]	3.6389
2	2	Two layers, 5 neurons each	Raw, PCA to size 21	[1:3]	[1:3]	3.5218
3	3	Single layer, 20 neurons	Raw, PCA to size 21	[1:3]	[1:3]	2.9939
4	4	Two layers, 10 neurons each	Raw, PCA to size 21	[1:3]	[1:3]	2.9461
5	5	Single layer, 100 neurons	Raw, PCA to size 21	[1:3]	[1:3]	36.2431
6	6	Single layer, 10 neurons	Raw, PCA to size 21	[1:10]	[1:10]	4.1676
7	7	Two layers, 100 neurons each	Raw, PCA to size 21	[1:3]	[1:3]	5.7898
8	8	Two layers, 100 neurons each	Raw, PCA to size 21	[1:10]	[1:10]	3.478
9	9	Single layer, 1000 neurons	Raw, PCA to size 21	[0:4]	[1:4]	59.1439
10	10	Single layer, 10 neurons	NLE, PCA to size 21	[0:4]	[1:4]	3.9389
10	11	Single layer, 10 neurons	NLE, PCA to size 21	[0:4]	[1:4]	4.4305
10	12	Single layer, 10 neurons	NLE, PCA to size 21	[0:4]	[1:4]	6.1877
10	13	Single layer, 10 neurons	NLE, PCA to size 21	[0:4]	[1:4]	10.7736
11	14	Two layers, 100 neurons each	NLE, PCA to size 21	[1:10]	[1:10]	6.8203

Table 5.3 Layered recurrent neural network architecture test results

Configuration	Hidden layer configuration	Data set type	Error (RMSE)
1	Single layer, 10 neurons	Raw, PCA to size 21	2.5788
2	Two layers, 100 neurons each	Raw, PCA to size 21	2.3513

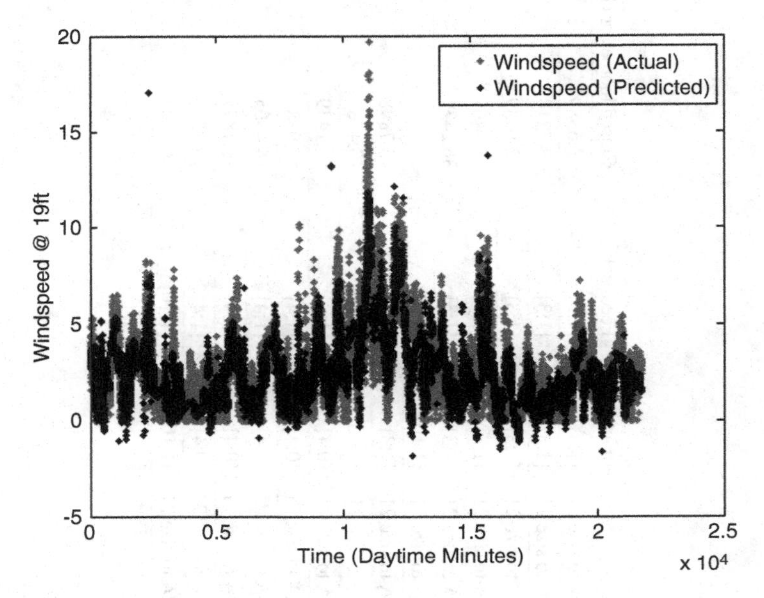

Fig. 5.4 19 ft Wind speed predicted results

is on classifying the inputs into one of the nominal outputs (i.e., predicting the class or category of the input). On the other hand, regression problems deal with forecasting real-valued output corresponding to the inputs. Despite their ability to perform well in both classification and regression tasks, shallow FFNNs consisting of less than two or three hidden layers are not capable of handling high-dimensional input data. Alternatively, increasing the layers does not lead to a better performance. The reasons are yet to be discovered, but there are a couple of theories that tried to address this problem: vanishing the gradient effect in a backpropagation algorithm through the layers and getting trapped in local minima, which is the result of random initialization of the parameters. Figure 5.10 shows a multi-layer neural network, where the number of layers equal to 3 (for notational simplicity), counting the output as the last layer. As the depth (number of hidden layers) of the architecture increases, the fewer computational elements, which are neurons in neural networks, are needed, and thus fewer examples are required to tune the neurons. Indeed, more computational elements are required to be employed in architectures with fewer layers to capture the complex relationships of the data (Bengio 2009). Based on different applications and training sets of deep architectures, the depth of the architecture varies.

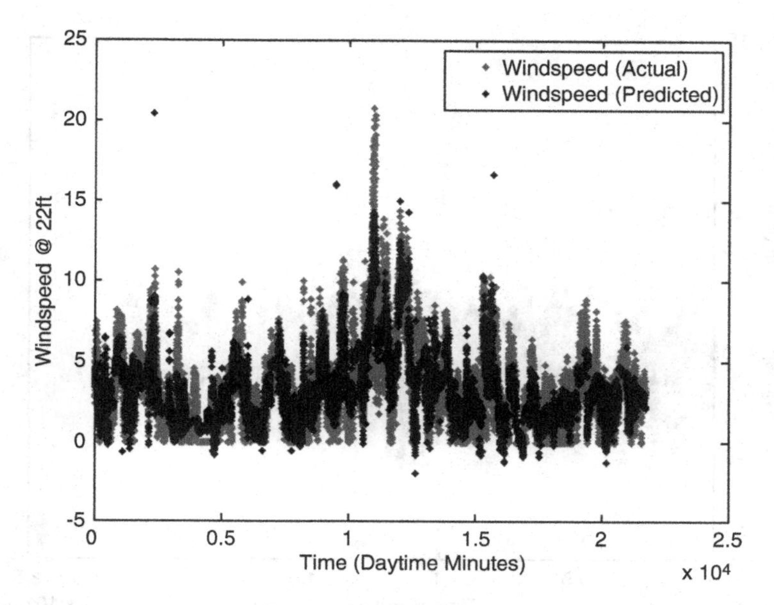

Fig. 5.5 22 ft Wind speed predicted results

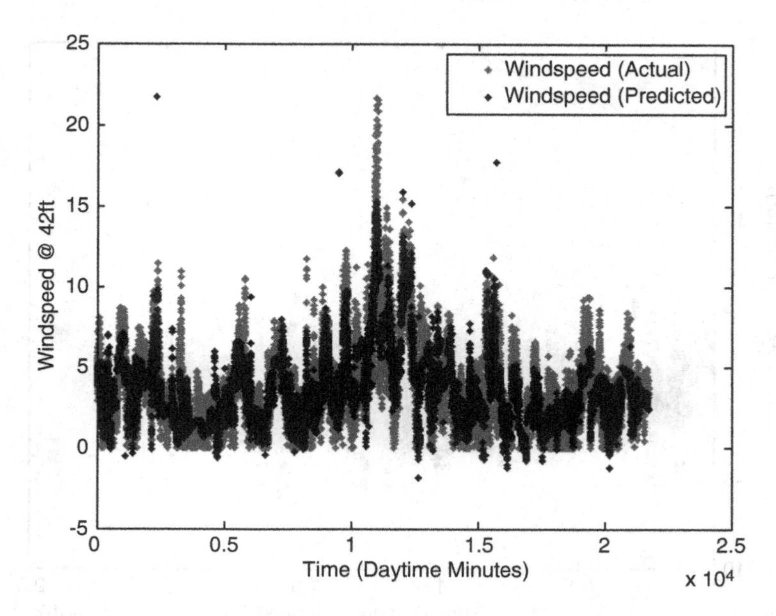

Fig. 5.6 42 ft Wind speed predicted results

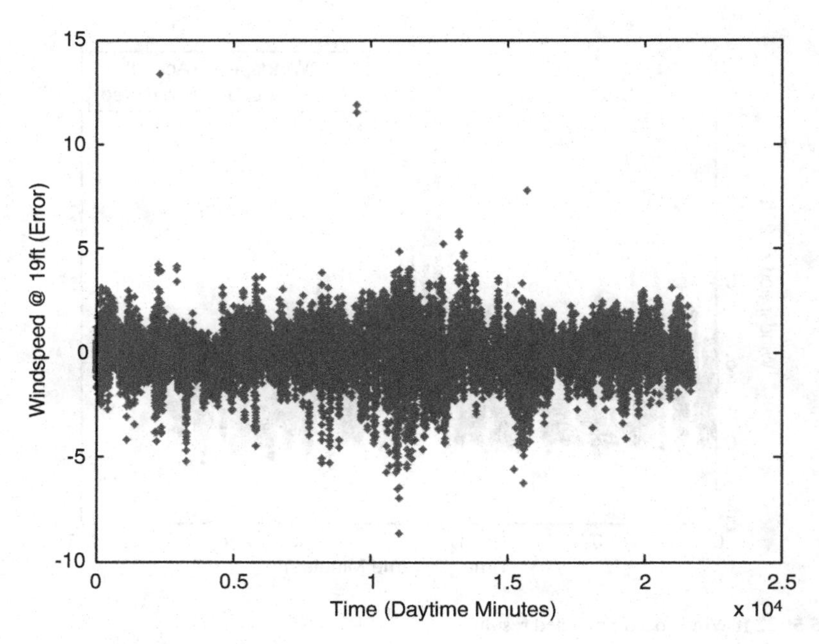

Fig. 5.7 19 ft Wind speed prediction error

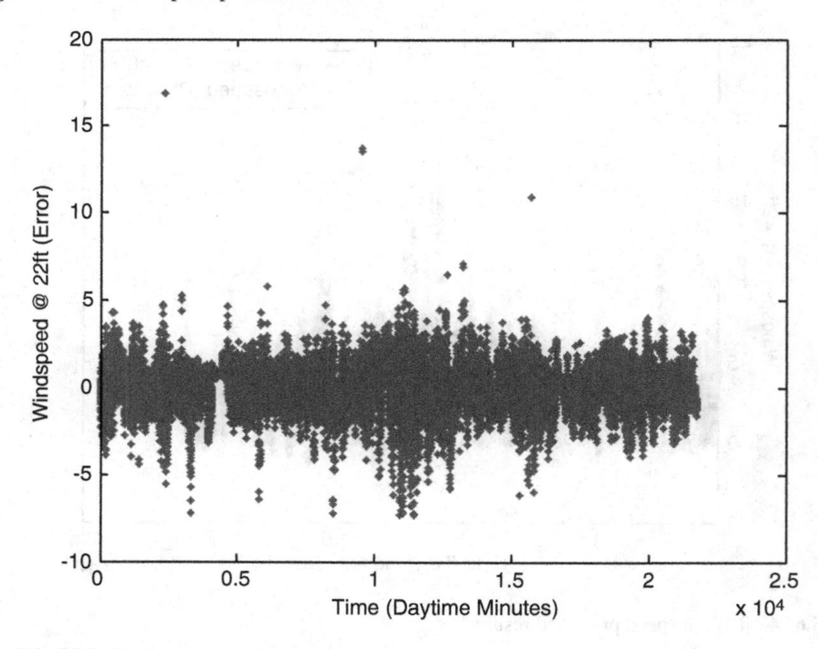

Fig. 5.8 22 ft Wind speed prediction error

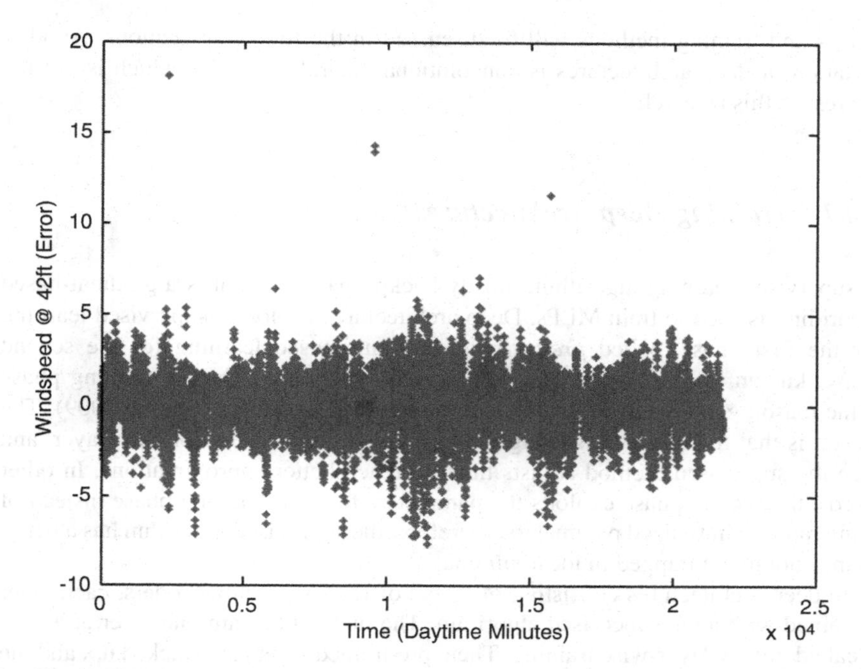

Fig. 5.9 42 ft Wind speed prediction error

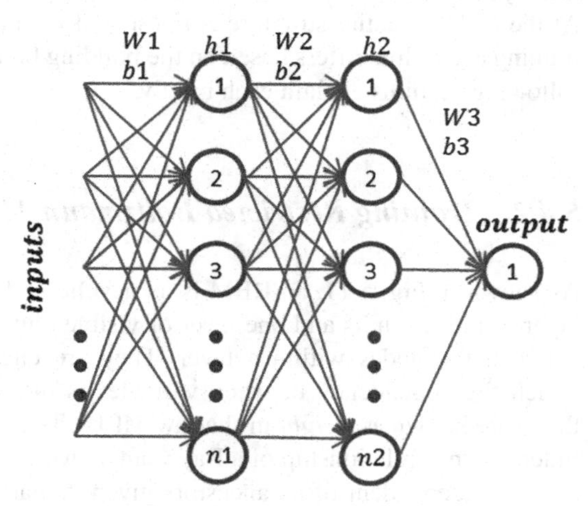

Fig. 5.10 A typical architecture of a multi-layer perceptron with three layers consisting hidden layer h1 with n1 nodes, hidden layer h2 with n2 nodes, and the output layer

In addition to dimensionality reduction, deep architectures are able to find complex relationships in data through different training principles than MLPs. For this study, deep neural networks built from Restricted Boltzmann Machines (RBMs) and autoencoders are considered [Autoencoders are also capable of performing non-linear PCA for feature extraction; therefore, applying autoencoders to data with high non-linearity is superior to linear PCA]. Each of the above mentioned building blocks and training methods will be discussed in the following sections. Another variation of deep architectures is convolutional neural networks, which is not the interest of this research.

5.4.1 Training Deep Architectures

A supervised learning algorithm, mostly backpropagation that is a gradient-based algorithm, is used to train MLPs. Deep architectures exploit unsupervised learning for the first phase, called *pre-training,* and supervised learning for the second phase, known as *fine-tuning*. Due to research experiences, the pre-training phase is the reason good results derive from training deep structures (Bengio 2009). The reason is that the unsupervised algorithm finds the parameters for each layer, and then the supervised method adjusts them to reach better approximations. In other words, the second phase exploits the parameters found in the first phase instead of using random initialized parameters; therefore, the supervised algorithm has a better chance not to get trapped in local minima.

In deep architectures consisting of layers of RBMs or autoencoders, each layer is trained with an unsupervised algorithm. The method to train one layer at a time is called greedy layer-wise training. Then, pre-trained layers are stacked up, and the last layer is added to the entire stack to perform the regression or classification task. At the end, the entire structure is fine-tuned with a supervised algorithm. The pre-training algorithm differs based on the building block of a deep architecture, and the following sections explain each briefly.

5.4.2 Training Restricted Boltzmann Machines

As shown in Fig. 5.11, an RBM is an acyclic undirected graph that comprises one layer of hidden units and one layer of visible units, where there are no connections between the nodes within a layer. They are energy-based generative models in which the dependency between variables is measured by *energy* (Murphy 2012) that were known as *weight* in shallow MLPs. Their special structure and conditional independence relationship of a Bayesian network (or Markov property), where each node is independent of its ancestors given its parents, make the training of RBMs tractable. "Contrastive divergence" (CD) learning or other variations of the CD is usually used to train each layer of an RBM. The stacked RBMs construct a Deep Belief Network (DBN).

Energy functions, rooted in statistical physics, which have relations to bidirectional models or RBMs, are used to show the probability distribution in RBMs (see (5.1)). Low energies are corresponding to high probabilities (Hinton 2010).

In Fig. 5.11, the dashed circles show the visible variables (inputs), undashed circles relate to hidden variables, and the relations between two layers are bidirectional

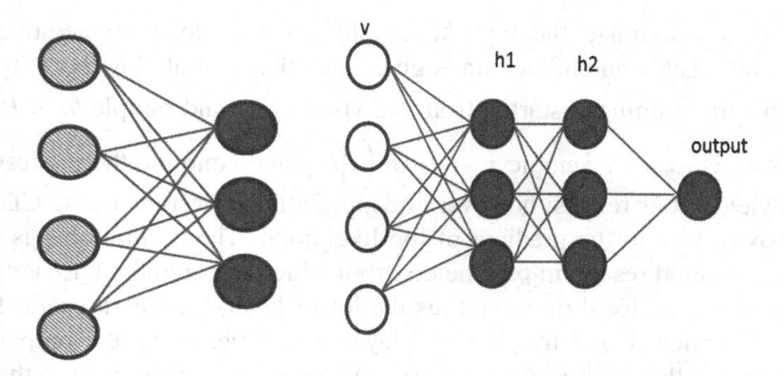

Fig. 5.11 An RBM is an acyclic undirected graph

(left). A deep neural network consisted of two RBMS as its hidden layers each with 3 units (right) as follows:

$$P(x,h)=\frac{e^{-Energy(x,h)}}{\sum_x e^{-Energy(x,h)}} \tag{5.1}$$

where data, also called visible units, are denoted by x, and hidden layers are shown by h. The likelihood is

$$p(x)=\sum_h \frac{e^{-Energy(x,h)}}{\sum_x e^{-Energy(x,h)}} \tag{5.2}$$

Then the Energy of a joint configuration is:

$$Energy\,(x,h)=-(bx+ch+hwx) \tag{5.3}$$

Where b and c show the bias matrices for visible and hidden units, and w is a matrix containing the weights between visible and hidden units.

For Bernoulli RBM, where units are binary, the conditional probability on each unit is given by:

$$P\left(x_i=1\middle|h\right)=sigmoid\left(\sum_j w_{ij}h_j+b_i\right) \tag{5.4}$$

$$P\left(h_i=1\middle|x\right)=sigmoid\left(\sum_j w_{ij}x_i+c_i\right) \tag{5.5}$$

The goal is to maximize the log-likelihood, which is done by iterative Gibbs sampling from visible and hidden units given the other. Initializing the weights and biases with zero, training is started from the visible unit and sample $h_i \sim P\left(h \middle| x_i\right)$. The next step is to get a sample $x_{i+1} \sim P\left(x \middle| h_i\right)$ and continue the process. More steps will yield better results; however, the more efficient way is to use CD, which is an approximation to the gradient of log-likelihood. The training that is done in a layer-wise method results in parameters from which the gradient based algorithm benefit to escape the local minima, thus the lower layers can learn useful features in data. It is believed that training one layer at a time, using the output of the previous layer as the input to the next layer, helps each layer to capture the lower-level features, as well as the statistical characteristics of data [9]. Therefore, the pre-training phase is a key to successful implementation of deep architectures. Explaining the detail of training RBMs and its requirements are beyond the scope of this study. Hence, for more information on how to train RBMs refer to Bengio, 2009; Murphy 2012; Hinton 2010.

5.4.3 *Training Autoencoders*

Training RBMs were first carried out by Hinton et al. (2006, 2010), and shortly after that stacked autoencoders were implemented by Bengio et al. (2006, 2009). Training autoencoders, also known as autoassociators, or Diabolo networks, are similar to training RBMs. However, training autoencoders is easier than training RBMs since there is no need for Gibbs sampling. Figure 5.12 shows a single-layer and a stacked autoencoder with 3 hidden units. Output is equal to the input, since each of the autoencoders in a deep architecture aims to reconstruct its input.

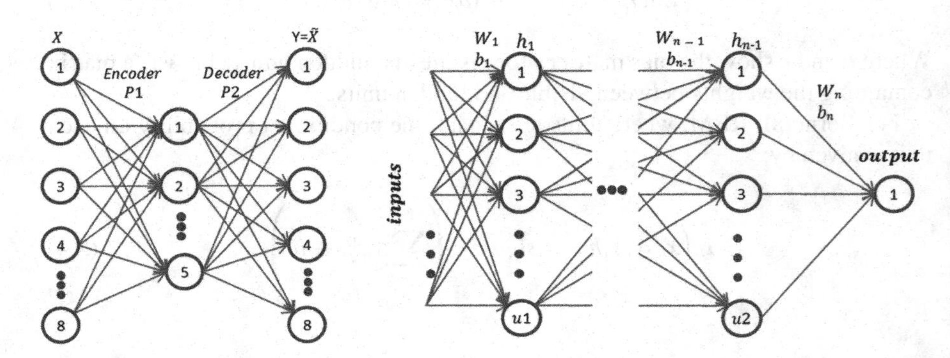

Fig. 5.12 A single-layer autoencoder with 8 input units and 5 hidden units (*left*). A stacked autoencoder with *n–1* hidden units (*right*)

The reconstruction cost is the squared differences between the actual inputs (x) and the reconstructed inputs ($\tilde{x}$), which is computing $(x - \tilde{x})^2$ over all of the training set.

As the input and the output of the autoencoder are the same, the autoencoder learns nothing more than the identity function *f*, where x = f(x). Therefore, some constraints are applied to autoencoders to make them capable of learning the useful features of the input data. One of the techniques is to make a bottleneck in the autoencoder by limiting the number of hidden units to a number less than the input units. This way, the hidden layer is forced to learn the meaningful features of the data. There are also methods for working with larger hidden units; i.e., when the number of hidden units is greater than the number of input units. Autoencoders implemented by using the above mentioned techniques are called *sparse autoencoders* and *denoising autoencoders*, which benefit from the sparse distribution of the hidden units and the training set respectively.

Sparse autoencoders exploit deactivating some of the latent neurons to cause sparsity in the hidden layer. In other words, restraining the hidden units to remain deactivated most of the time is a desired effect (Ng 2011) which resembles the way that the human brain works (Bengio 2009). Hence, the sparsity parameter is defined as the average activations of each neuron over the training set. To panelize the activations with far more or far less than a certain value of the sparsity parameter, a penalty term is added to the cost function.

Denoising autoencoders are another variation of autoencoders, where the autoencoder tries to reconstruct the input from the randomly corrupted version of input. The corruption is done by equating a percentage (usually 50 %) of the input data to zero (Bengio 2009) that is also known as the dropout method. This way, a denoising autoencoder will learn the statistical distribution of the training set; therefore, denoising autoencoders resemble the functionality of RBMs although training autoencoders is less complex than training RBMs.

Algorithm 5.1 shows how to train an autoencoder and includes comments to train a sparse autoencoder using the guidelines given by (Ng 2011).

Algorithm 5.1: Training a (Sparse) Autoencoder

AE stands for autoencoder; parameters are weights and biases

```
For i = 1 to number of AEs
    Initialize weights ~ U (-a, a), biases ← 0, y
                              ← input data
   // a is a real number where 0 < a < 1; y is the
  output
   // for the sparse AE, initialize the sparsity
   //parameter, and the weight of sparsity penalty
    While stopping criterion not met
          train the AE using backpropagation
          algorithm
         //the cost function has an additional
        sparsity term
```

```
    End while
    Read parameters of input-hidden layer
   (p1)
   //ignore the parameters of the hidden-output layer
   (p2)
    Compute the activation of hidden layer unit using
   (p1)
   //activation ← f (input data*p1); f is the
    activation
   //function
   input data ← activation
 End for
```

As the algorithm shows, training autoencoders does not need to perform Gibbs sampling, because the gradient of the log-likelihood has a closed solution in autoencoders.

Training several autoencoders and stacking them up, then adding a classification layer or a regression method, will construct a deep classifier or a deep regression model. When the entire structure is built up, the last phase of training, also known as fine-tuning, may be applied to adjust the parameters more.

For future work of this study we plan to apply the stacked sparse autoencoders to the traffic and biological data. Since deep architectures have gained success in applications dealing with big data, more accurate predictions are expected by employing deep regression models. The last layer, which is responsible to perform the regression analysis, exploits stochastic gradient descent to minimize the cost function between the actual measurements and the predicted values. An application of online stochastic gradient descent for traffic flow forecast is performed by Moussavi-Khalkhali and Jamshidi (2014)

5.5 Conclusions

This chapter presents a high level look at some of the data analytic tools available that enable the user to extract information from "Big Data" sources in order to draw useful conclusions. As described in Sect. 5.3, one of the specific applications discussed in this chapter is the prediction of the amount of wind power generated by a micro-grid through estimation of wind speed. Section 5.3 then discusses the data that was gathered to support this exercise. Section 5.4 discussed the deep architecture leading to advanced artificial neural networks or deep learning. Applying the same data to a deep architecture and the related comparison analysis are left for future studies. Although deep architecture and deep learning have, thus far, produced mixed results, much more work is needed. Initial experience with this research has shown that deep architectures may have a better chance for robust intelligence with machine learning for image data which will be the subject of a future publication.

Acknowledgements This work has been supported, in part, by University of Texas System's Lutcher Brown Endowment, University of Texas at San Antonio. Original version of this chapter, is in parts, based on a keynote by first author at AAAI Workshop on Big Data, Stanford, CA, April 24, 2014.

References

Beale M, Hagan M, Demuth H (2014) MATLAB neural network toolbox user's guide. http://www.mathworks.com/help/pdf_doc/nnet/nnet_ug.pdf

Bengio Y (2009) Learning deep architectures for AI. Found trends Mach Learn 2(1):1–127

Bengio Y, LeCun Y, Yann S, Chopra R, Hadsell M, Ranzato HF (2006) A tutorial on energy-based learning. In: Bakir G, Hofman T, Schölkopf B, Smola A, Taskar B (eds) Predicting structured data. MIT Press, Cambridge, pp 1–59

Hinton G (2010) A practical guide to training restricted Boltzmann machines. Momentum 9(1):926

Hinton G, Osindero S, The Y-W (2006) A fast learning algorithm for deep belief nets. Neural Comput 18(7):1527–1554

IEM—Iowa Environmental Mesonet (2014) ASOS/AWOS data download. http://mesonet.agron.iastate.edu/request/download.phtml

Jamshidi M (ed) (2008) Systems of systems engineering—principles and applications. CRC—Taylor & Francis Publishers, Oxford

Jamshidi M (ed) (2009) System of systems engineering—innovations for the 21st century. John Wiley & Sons, New York

Moussavi-Khalkhali A, Jamshidi M (2014) Leveraging machine learning algorithms to perform online and offline highway traffic flow predictions. In: Machine Learning and Applications (ICMLA), 2014 13th International Conference on, IEEE, Dec 2014, pp 419–423

Murphy KP (2012) Machine learning: a probabilistic perspective. MIT Press, Cambridge

NREL—National Renewable Energy Laboratory (2014) MIDC SOLPOS Calculator. http://www.nrel.gov/midc/solpos/solpos.html

Ng A (2011) Sparse autoencoder. CS294A Lecture notes 72

Shlens J (2009) A tutorial on principal component analysis. http://www.snl.salk.edu/~shlens/pca.pdf Accessed 22 April 2009

Smith LI (2002) A tutorial on principal components analysis. http://www.cs.otago.ac.nz/cosc453/student_tutorials/principal_components.pdf. Accessed 26 Feb 2002

Tannahill B (2014) Big data analytic techniques: Predicting renewable energy. MS Thesis, ACE Laboratory, University of Texas, San Antonio, April, 2014

Tannahill B, Jamshidi M (2014) System of systems and big data analytics—bridging the gap. Int J Comput Electr Eng 40:2–15, Elsevier

Chapter 6
Artificial Brain Systems Based on Neural Network Discrete Chaotic Dynamics. Toward the Development of Conscious and Rational Robots

Vladimir Gontar

6.1 Introduction

It seems reasonable to suppose that the next step in developing artificial intelligent systems, having human thinking abilities, should be based on a better understanding of existing and new laws of nature responsible for the dynamics of thinking systems. One should remember how the physical sciences succeeded in opening and exploiting the physical principles and laws of nature for the creation of the special theories and mathematical tools necessary for breaking into the micro-world of atoms and molecules, constructing new processes and machines along the way.

We desire to combine the structural complexity of the brain's neural networks with the mathematical models derived from the laws of nature responsible for complex heterogeneous biochemical reaction dynamics, accompanied with storage, processing and exchanges of information. Biochemical reactions and processes, which are taking place within and between the brain's neurons, combining to compose neural networks, are responsible for specific brain functions. We hardly expect serious progress in the improvement of modern *AI* systems along the way to the creation of "artificial brain" systems without a detailed understanding of the internal mechanisms of biochemical processes in the brain, which should include the physicochemical meaning and roles of "information" and "information exchange" currently not presented elsewhere. We also need to accept a lack of complete fundamental physical principles and mathematical models for living and especially the thinking systems responsible for the origin and functioning of human intelligence and its connection to consciousness, cognition, creativity, learning, and

V. Gontar (✉)
Biocircuits Institute, University California San Diego (UCSD), San Diego, CA, USA

Ben-Gurion University of the Negev, Beersheba, Israel
e-mail: galita@bgu.ac.il

© Springer Science+Business Media (outside the USA) 2016

R. Mittu et al. (eds.), *Robust Intelligence and Trust in Autonomous Systems*,
DOI 10.1007/978-1-4899-7668-0_6

rational decision making. In order to address these questions, we need to define the physicochemical meaning of "information" and "information exchange" in relation to regular processes such as the mass, charge and energy exchanges taking place during biochemical reaction dynamics within the brain's neurons and neural networks. The classical meaning of "information", introduced by Shannon (1948) and Brillouin (1962), was based on the physics of thermodynamics and probabilistic principles related to measures of the quantity of "information" without considering the quality of information which is important for thinking systems.

The meaning of "information exchange", which we are introducing, reflects the extreme sensitivity of chaotic states of the neurons to the infinitesimal portion of energy (which we intend to relate to "information") contained by the internal and external stimuli delivered to a neuron(s) that causes unique patterns that can be associated with the brain's mental properties. From the point of view of delivered energy, those infinitesimal stimuli drastically change the current chaotic state of an individual neuron and the whole neural network. The ability of neurons to receive and react to infinitesimal signals we associate with "information exchange" within and between neurons. For "information exchange" to occur, dynamical process within the physical or biological system of neurons and neural networks by necessity should have chaotic regimes to be able to change under infinitesimal influences (stimuli or signals) for specific patterns to emerge. "Information exchange" takes place in parallel with the regular biochemical reactions between the neuron's biochemical constituents (atoms, molecules, ions, etc.). The fundamental difference between the process of "information exchange", where an infinitesimal amount of energy produces large effects, and regular biochemical reactions is that for regular exchanges, the more energy consumed, the bigger effect that could be expected from the interaction. It is also the important that while all constituents participating in a regular biochemical reaction could be in any physical state, the "information exchange" requires that constituents are in a chaotic state, the only state that could be changed by an infinitesimal (small) portion of energy and be considered as "information".

To construct a general theoretical approach and mathematical model of neural network dynamics for "information exchange", we introduce a new extremal dynamical principle for multicomponent biochemical reaction dynamics. This new principle leads to a system of non-linear difference equations for the numerous embedded chaotic regimes in the mathematical modeling of "information exchange" within and between neurons. This proposed principle results from the extension of the maximum entropy principle, the $\pi-$ theorem of the theory of dimensionality and stoichiometry for the multicomponents of the chemical reactions occurring (Gontar 1993, 2004)

As will be shown, the equations derived from the dynamical principle enable the simulation of specific natural neural network features, namely "self-organization" and "self-synchronization". These features lead to the emergence of a new "phenomenological" state (s) within the "artificial brain" in the form of the specific discrete time and space patterns which we intend to correlate with human consciousness, cognition, creativity (which is the ability of the system, natural or artificial, to generate innovative results in the form of art, music, poetry, technical inventions, etc.) and intelligence, which should correspond to general rational behavior and decision making. We are presenting here results of numerical simulations, performed by the proposed approach, to demonstrate the 2D patterns generated in the form of ornaments and mandalas (Figs. 6.2 and 6.3) to support the idea that artificial neural networks, when constructed from a first physical principle, necessarily lead to the variety of dynamical artistic "patterns" traditionally considered to be the prerogative of human creative abilities.

Formulated here, the first physicochemical dynamical principle could serve as a possible explanation of the origin of the "driving force" for thinking system dynamics, thereby opening a new perspective to simulate the brain's cognitive functions with the goal of eventually developing artificial brain systems.

6.2 Background

The idea of translating the properties of the interconnected neurons of the human brain into mathematical models gave impetus to the development of an *ANN* functioning as interconnected individual neurons simulated by step, linear and sigmoid functions (Haykin 1998). Even this pure mathematical approach applied to the complexity of neural networks has demonstrated the ability of *ANNs* to perform numerous and "intelligent" operations, including image and signal recognition, assisted decision making, and control and navigation among many other applications associated with human intelligence. At the same time it should be clear that an *ANN* based on pure mathematics includes a variety of solutions that may not be relevant to real processes. Therefore, the use of mentioned above *ANN* makes it problematic for autonomous and intelligent systems when the time or data for training and learning is limited and when innovative and rational solutions are required.

When we are talking about the scientific understanding of intelligence, we should realize that its origin and explanation could be found only within the understanding of living cells (neurons) and their biochemical processes. The way to the creation of artificial intelligence lays in understanding the physicochemical laws responsible for brain functioning. In spite of the fact that living cells are composed of the same atoms and molecules as non-living matter, they do not appear to obey the physical laws of quantum mechanics and statistical physics formulated for non-living matter. It seems that on the scale of their operations, living and thinking cells and systems, such as the brain, may not obey the laws of thermodynamics and the second law of thermodynamics in particular. Numerous attempts to depict the existing laws of physics for the dynamics of living cells have not allowed biologists to understand any better what are thoughts, consciousness and cognition, and the many other specific features of living and thinking matter. The extreme complexity of the structural and behavioral properties of brain neurons and networks does not manifest dynamics similar to that observed and simulated in inert matter physics. Living cells, such as neurons, present behavior comparable to that of a well-organized factory under optimal control and synchronization, with "information" and biochemical exchanges between constituents taking part in living and thinking cycles and processes that "rationally" and "creatively" respond to internal and external stimuli. Self-reproduction, "information exchange", memory, aging and emerged and self-organizing mechanisms make a living as thinking systems. They form an extremely complex theoretical object of research that requires new fundamental principles and laws which should reflect a specificity for living and thinking matter in contrast to non-living systems. Classical physics, initially focused on inert matter dynamical processes, traditionally exploits as a mathematical tool continuous time and space with differential equations known also as the calculus of the infinitesimal. We think that "living and thinking systems" require the introduction of a new calculus, which we call the "calculus of iterations" and leading to systems of difference equations. These equations should be directly derived from first principles, reflecting a specificity of living systems, for further use in the mathematical modeling of the dynamics of living and thinking systems (Gontar 1995). Under some assumptions, these two calculi intersect when $\Delta t \rightarrow 0$, but we intend to benefit from using difference equations independently from differential equations for a source of mathematical models. Difference equations, by their very nature, have numerous embedded chaotic regimes which could be applied for mathematically modeling one of the basic concepts of thinking system dynamics: the "information exchange" based on chaotic regimes (Gontar 1995, 2004). To emphasize our preference of difference equations for mathematical modeling of living and thinking systems, we need to mention that differential equations have a limited list of equations with chaotic regimes which exist within a narrow range of parameters. Numerical integration of systems of differential equations are always accompanied by the contradiction between the continuous variables and discrete computer calculations that complicate the identification of the computational artifacts and real chaotic regimes of the simulated physical system. These are the reasons why difference equations, with their clear physicochemical meanings for the variables and parameters derived from first physicochemical principles and laws of nature, are preferable to differential equations for modeling living and thinking systems (Gontar 2000a, b).

As already mentioned, the brain consists of neurons interconnected to form complex neural networks. Another empirical fact is that each neuron operates as a "biochemical reactor" where numerous chemical, electrochemical reactions and biochemical reactions occur. Before introducing our basic hypothesis about thinking system mathematical models, let us remind the reader that chemical reactions between the original "simple and non-living" elements (atoms, molecules, etc.) can lead to the creation of more complex systems such as bacteria with the manifest new properties in the emergence of "life". By analogy, the brain's specific properties, such as consciousness, cognition and creativity could result from the biochemical reactions and the information exchange within and between the neurons composing a neural network. All kinds of brain activity, including cognitive properties, are fully defined by the states of neurons and their dynamics

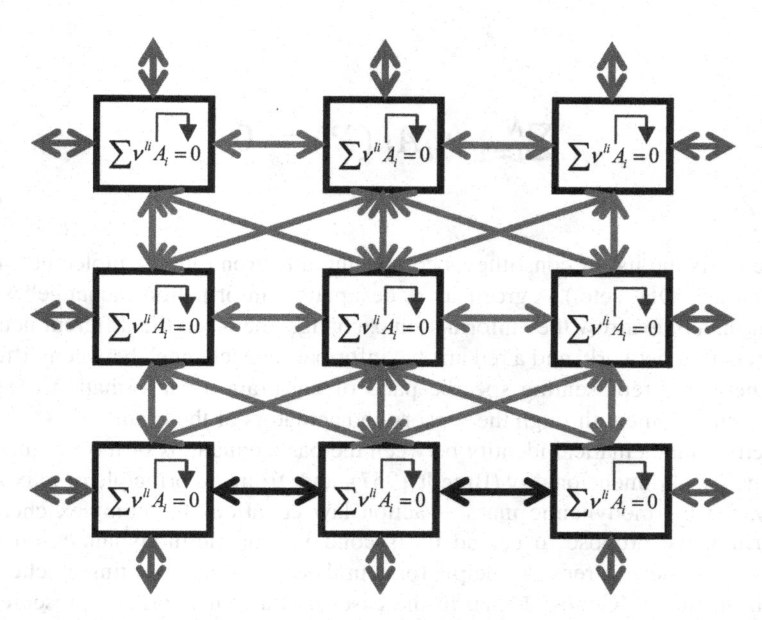

Fig. 6.1 A neural network composed by interconnecting through "information exchange" neurons (blue arrows). Each neuron is represented by the mechanism of biochemical reaction dynamics with "information exchange" between the neuron's constituents (green arrows) and between constituents composing other neural networks representing different parts of the "artificial brain" (red arrows)

that depend on a neuron's internal *i*th chemical constituent concentrations, y_i. To simulate a brain's cognitive functions, we construct a mathematical model that describes the dynamics of the chemical constituents distributed among the brain's neural networks. We hypothesize that a neuron's chemical constituent concentrations distributed among the neural network associate with the "phenomenological states" manifesting as consciousness, cognition and creativity among the brain's other properties. These phenomenological states are represented by the calculated concentrations y_i (t_q, $\boldsymbol{R}$) of the *i*th chemical constituents distributed on the brain's neural networks for any network state, t_q, within the discrete space $\boldsymbol{R}$. Structurally, the human brain is composed of frontal, parietal, occipital and temporal lobes, cerebellum, etc. Each brain's part, for the purpose of mathematical modeling, could be represented by a specific form of 2D or 3D neural network interconnected with other parts of the brain to promote information exchanges. In Fig. 6.1, one can see nine "mathematical neurons" with the discrete coordinates (r_x, r_y) interconnected via information exchanges (designated by arrows) within the 2D artificial neural network $R\left(r_x, r_y\right)$; $r_x, r_y = 1, 2 \ldots N'$. Each neuron is represented through the mechanism of its biochemical reactions by the matrix of stoichiometric coefficients $|v_{li}|$ (Gontar 1997):

$$\sum_{i=1}^{N} \nu_{li} A_i(\boldsymbol{R}) = 0 \tag{6.1}$$

Here, A_i is the list of constituents composing a neuron (atoms, molecules, ions: H, H_2O, Ca^+, OH^-, etc.). A green arrow designates "information exchange" within the neuron, a blue arrow the "information exchange" between the different neurons within a neural network, and a red arrow "information exchange" between different neural networks representing specific parts of the brain, or information received from the environment through the sensors and actuators of the brain.

Based on mathematical identity between the basic equations of the π– theorem of the theory of dimensionality (Brandt 1957), and, from the principle of maximum entropy, the thermodynamic mass – action law equations for complex chemical equilibrium, we propose to extend the second law of thermodynamics on open systems with a new extremal principle for neural networks representing biochemical reaction dynamics (Gontar 2004). In the case of neural networks, represented by the neurons with internal biochemical reactions and information exchange within and between the neurons, as well as with neurons from other networks, the new extremal principle can be formulated as follows: the evolution of neural networks proceed in such a way that at any discrete time t at state q, t_q, $q = 1, 2 \ldots Q$, each neuron within the network with discrete coordinates $\boldsymbol{R}$ (r_x, r_y) is fully defined by its chemical constituent concentrations y_i $(t_q, \boldsymbol{R})$, a minimizing function (6.2) for the space of constituent concentrations $0 < y_i\left(t_q, \boldsymbol{R}\right) < 1$, and under the constraint of the mass conservation law (6.3):

$$min_{y_i\,(t_q,\,\boldsymbol{R})}\ \Phi\left(y_i\ (t_q, \boldsymbol{R})\right) = \sum_{i=1}^{N} y_i\ (t_q, \boldsymbol{R})\ \left(\ln\left(y_i\ (t_q, \boldsymbol{R})\right) + f_i\left(\pi_l^d,\ w_l',\ \rho_{li}',\ \beta_{li}^{\otimes},\ t_{q-s}, \Im_{l,g}\right)\right) \tag{6.2}$$

$$\sum_{i=1}^{N} \alpha_{ij}^T y_i\left(t_q, \boldsymbol{R}\right) = b_j^0 \tag{6.3}$$

$$f_i = \sum_{l=1}^{N-M}\left(-v_{li}^T \ln\left(\pi_l^d\ \exp - \left(w_l + \sum_{i=1}^{N} \rho_{li} y_i\ (t_{q-s}, \boldsymbol{R}) + \sum_{i=1}^{N'} \beta_{li}^{\otimes} y_i(t_{q-s},\ \boldsymbol{R}^{\otimes}) + \Im_{l,g}\right)\right)\right) \tag{6.4}$$

Here $\boldsymbol{R}^{\otimes}$ are coordinates of the neighboring neurons participating in an information exchange with the currently considered neuron $\boldsymbol{R}(r_x, r_y)$; π_l^d and w_l are empirical parameters characterizing the rate of the lth biochemical reaction; ρ_{li} and $\beta_{li}^{\otimes}$ are empirical parameters characterizing the intensity of information exchange within and between the neurons; the α_{ij}^T are the elements of transposed molecular matrix $|\alpha_{li}|$ to indicate the number of system basic constituents of type j ($j = 1,2 \ldots M$) in the constituent of type i ($i = 1,2 \ldots N$); the b_j^0 reflect the total concentration of the jth constituent in a neural network; and $s = 1, 2 \ldots$ is the index characterizing "system memory" and indicates the state prior to the currently considered state, t_q, (in this work we are considering only the previous state, t_q or $s = 1$). $\mathfrak{I}_{l,g}\left(y_{il'}^g\left(t_{q-1}, \boldsymbol{R}^g\right)\right)$ is the function characterizing information exchange between the lth reaction within neural network with coordinates $\boldsymbol{R}$ and the other $g^{th}(1, 2 \ldots G)$ neural networks with coordinates $\boldsymbol{R}^g$ (for example, frontal lob coordinates denoted as $\boldsymbol{R}$, occipital lob $\boldsymbol{R}^1$, sensors $\boldsymbol{R}^2$, actuators $\boldsymbol{R}^3$, etc.). As an initial approximation to the explicit form of the unknown function $\mathfrak{I}_{l,g}$, we approximate it with a linear regression corresponding to the neural network constituent concentrations $y_i^g\,(t_{q-1})$ and empirical parameters ξ_i:

$$\mathfrak{I}_{l,g}\left(y_{il'}^g\left(t_{q-1}, \boldsymbol{R}^g\right)\right) = \sum_{l'}^{L'} \sum_{i=1}^{N''} \xi_i\, y_{il'}^g\left(t_{q-1}, \boldsymbol{R}^g\right) \tag{6.5}$$

N'' is the number of constituents within the gth neural network, l' is $\left.1, 2 \ldots L'\right)$ for the number of reactions in the neurons representing the gth neural network.

In the case when an interaction between the neurons from different neural networks is not limited by "information exchange", the exchange of chemical constituents could be introduced into (6.1) through the extension of the $|v_{li}|$ matrix by adding the corresponding chemical reactions between the neurons.

The formulated dynamical extremal principle (6.2) and (6.3) equivalent to the solution of the following system of N non-linear difference equations has a unique solution for all $y_i\,\left(t_{q-s}, \boldsymbol{R}\right) > 0$ (Gontar 1993, 2004):

$$\prod_{i=1}^{N} y_i^{v_{li}}\left(t_q, \boldsymbol{R}\right) = \pi_l^d \exp\left(-w_l + \sum_{i=1}^{N} \rho_{li} y_i\left(t_{q-s}, \boldsymbol{R}\right) + \sum_{i=1}^{N'} \beta_{li}^{\otimes} y_i\left(t_{q-s}, \boldsymbol{R}, \boldsymbol{R}^{\otimes}\right) + \mathfrak{I}_{l,g}\right) \tag{6.6}$$

$$\sum_{i=1}^{N} \alpha^T{}_{ij}\, y_i\left(t_q, \boldsymbol{R}\right) = b_j^0;\ s = 1, 2 \ldots \tag{6.7}$$

Mathematical model (6.6) and (6.7) could simulate brain dynamics, since according to our assumptions, it is fully defined by the evolution of a neuron's constituent concentrations y_i (t_q , $\boldsymbol{R}^g$) distributed on the neural network $\boldsymbol{R}^g$. Specific cognitive brain functions could be interrelated with a neuron's constituent concentration distributions y_i (t_q, $\boldsymbol{R}^g$) which, as it will be shown, represent complex patterns that could be related to specific cognitive functions of the brain such as the creation of a work of art like a mandala.

The formal meaning of "information exchange" introduced here reflects a special type of interaction between complex and living systems, unlike an energy exchange, and has specific features. The energy can be delivered or transmitted from its source to any receiver to change its state without any requirements for the receiver to be under predefined conditions. However, "information exchange" in our view could take place only if the receiver is "ready" for that type of interaction. For the receiver "to be ready" means that it should react ("perceive") to infinitely small transmitted signals, since information conveyed usually contains small amounts of energy that nevertheless could drastically change the state of the receiver. As we know from deterministic chaos, any physical system (a network of neurons operating as a receiver in our case) could accept infinitely small signals only when it is in a chaotic state. This type of interaction, which we have named "information exchange" in comparison to regular energy exchanges, complements the living and thinking system dynamics which, as it is now well known, contained chaotic regimes. The meaning of the "information exchange" presented here and being applied to the interaction between humans could be illustrated by the fact that even "one word" (bad or good) exchanged between humans could cause a strong emotional reaction. This is supporting the idea about the use of mathematical models with embedded chaotic regimes to simulate the basic thinking system properties by information exchange. Information exchange could exist on the level of individual neurons, neural networks and between the interconnected neural networks of a whole brain.

Equations (6.6) and (6.7) written for the initial hypothesis about a mechanism of biochemical transformation and a scheme of information exchange within the neuron for any given parameters π_l^d, $w_l, \rho_{li}, \beta_{li}^{\otimes}$ and b_j^0 enable us to compute the unique distribution of each ith constituent's concentration y_i (t_q, $\boldsymbol{R}^g$) with a neural network at state t_q. The obtained distributions represent a visual dynamical pattern (e.g., a mandala) where equal values of y_i (t_q, $\boldsymbol{R}^g$) can be marked by the same color taken from an arbitrary palette (Gontar and Grechko 2006, 2007).

As shown in Fig. 6.2, the extremal principle denoted by (6.2) and (6.3) followed by (6.6) and (6.7) enable the generation of various dynamical patterns related to those observed or produced by complex, living and thinking systems: spirals, rings, waves and artistic patterns in a form of creative ornaments and mandalas. These results support the idea that this proposed principle could be applied to the mathematical modeling of any physicochemical system with chemical reactions like the human brain because its functioning is defined by the biochemical reactions occurring within neurons and neural networks. By this process, the different brain functions such as consciousness, cognition, creativity, and decision making could be directly related to the biochemical reaction mechanisms and dynamics within the neural networks and then associated with the specific patterns that emerge in a form that composes the neuron chemical constituent distributions and their dynamics.

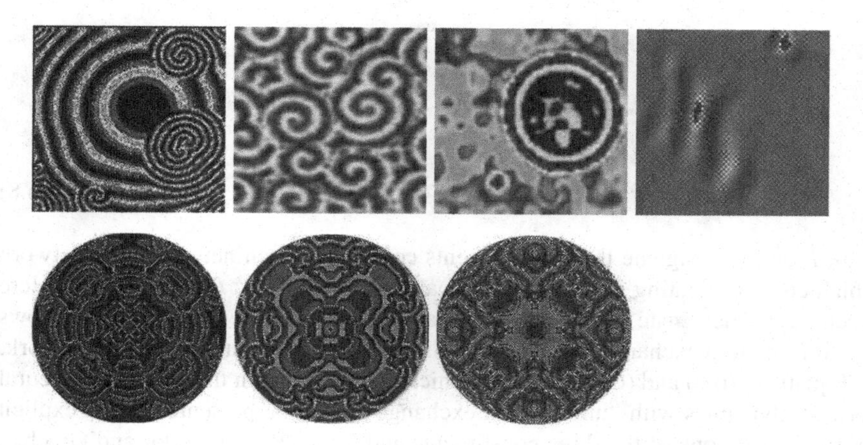

Fig. 6.2 Selected examples of the 2D patterns generated by the (6.9) sequences y_A $(t_q\ ,\boldsymbol{R})$ corresponding to the arbitrarily chosen $t_q, q = 1.2 \ldots Q$ for different sets of parameters $\pi_1^d, \pi_2^d, b^0, \rho_1, \rho_2, \rho_3, \beta_1^{\otimes}, \beta_2^{\otimes}, \beta_3^{\otimes}$; network $\boldsymbol{R}$ (100 × 100)

The proposed extremal principle denoted by (6.2) and (6.3) can be considered as a "driving force" for brain functioning by consuming and exchanging energy and information and used as mathematical tool for the creation of autonomous "artificial brain" systems. Mentioned above, the brain's cognitive functions should be interrelated with the specific complex patterns emerging from the "artificial brain" and controlled by the internal and external stimuli and by special training and learning of the *ANN*. This supervised and unsupervised training could provide a rational interaction of the artificial brain systems with the environment, artificial agents and humans. The proposed mathematical model has demonstrated its ability to generate an almost unlimited variety of complex and creative 1, 2 and 3D dynamical patterns (Gontar 1997, 2000a, b). The problem of the creation of autonomous conscious artificial brain systems then becomes the technical problem of how to provide training and learning for such a system by finding the concrete mechanism of biochemical reactions and parameters of the mathematical model (6.6) and (6.7) that correspond to the desired "intelligent" or rational behavior.

6.3 Numerical Simulations

As an example of using the proposed paradigm to simulate brain creativity in a form of 2D images such as ornaments and mandalas, we developed a system for the automatic finding of the model parameters that correspond to desired patterns. The general mechanism of the biochemical reactions expressed by (6.1) and written for the two reactions between three constituents with information exchange looks as follows:

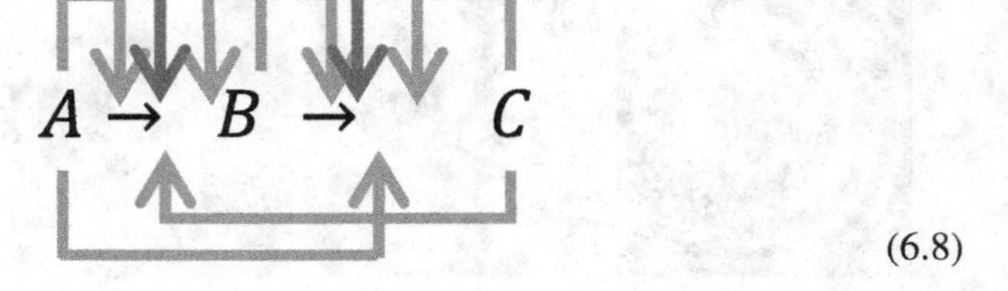

(6.8)

A, *B* and *C* designate three constituents composing each neuron in the network which are participating in two biochemical reactions: $A \rightarrow\ B$ and $B\ \rightarrow C$. Here green arrows designate "information exchange" within the neuron, and blue arrows the "information exchange" between the different neurons within a neural network.

Equations (6.6) and (6.7) for this chemical reaction within the neuron and neural network dynamics with "information exchange" could be presented in the explicit form for every one of the three constituents and for $y_1(\boldsymbol{R}, t_q)$ (Gontar and Grechko 2006):

$$y_1\left(\boldsymbol{R}, t_q\right) = \frac{b_A^0}{1 + \pi_1^d \mathscr{D}_1 + \pi_2^d \mathscr{D}_2} \tag{6.9}$$

$$\mathscr{D}_1 = exp\left(-\left(w_1 + \sum_{i=1}^{3} \rho_{1i} y_i\ \left(\boldsymbol{R}, t_{q-1}\right) + \sum_{i'=1}^{N'} \beta_{1i}^{\otimes} y_1\left(t_{q-1},\ \boldsymbol{R}, \boldsymbol{R}^{\otimes}\right)\right)\right)$$

$$\mathscr{D}_2 = exp\left(-\left(w_2 + \sum_{i=1}^{3} \rho_{2i} y_i\ \left(\boldsymbol{R}, t_{q-1}\right) + \sum_{i'=1}^{N'} \beta_{2i}^{\otimes} y_1\left(t_{q-1},\ \boldsymbol{R}, \boldsymbol{R}^{\otimes}\right)\right)\right),$$

with the initial and boundary conditions:

$$y_i\ \left(t_{q=0},\ r_x, r_y\right) = \begin{cases} b_j^0,\ i = 1, 2 \ldots M \\ 0,\ i = M+1, M+2, \ldots N \end{cases} \tag{6.10}$$

$$y_i\ \left(t_q,\ r_x, r_y\right) = \begin{cases} y_i\ \left(t_q,\ r_x, r_y\right),\ 0 < r_x, r_y < |\boldsymbol{R}|\ (inside\ the\ network) \\ 0,\ \ r_x, r_y \geq |\boldsymbol{R}|\ (outside\ the\ network) \end{cases}.$$

$$M = 1 \text{ is the number of main constituents (components)};$$

$$N = 3 \text{ is the total number of consituents.}$$

We also put constraints on the empirical parameters in equation (6.9):

$$w_1 = w_2 = 0;$$

$$\rho_{li} = \rho_{l'i};\ l \neq l';, l, l' = 1, 2 \tag{6.11}$$

$$\beta_{li}^{\otimes} = \beta_{l'i}^{\otimes} \begin{cases} l \neq l',\ l, l' = 1, 2 \\ r_x \neq r_y\ \ r_x, r_y = 1, 2 \ldots 9 \end{cases}$$

By making these assumptions, we reduce the number of controlled parameters to 9: π_1^d, π_2^d, b^0, $\rho_1, \rho_2, \rho_3, \beta_1^{\otimes}, \beta_2^{\otimes}, \beta_3^{\otimes}$. The number of parameters can be extended if we need to generate more complex patterns to better correspond to the experimental data. In our examples, each neuron for the neural network considered has coordinates (r_x, r_y) fully characterized by the concentrations of its $N = 3$ constituents $y_i\ (t_q, r_x, r_y)$ at any state t_q.

The values of the parameters ρ_{li}, $\beta_{li}^{\otimes}$ could be used as a quantitative characteristic of the level of information exchange. Qualitative conclusions about information exchange could be made from the obtained results: if the desired output (a specific pattern) has not appeared for a given set of parameters, it means that the scheme used and the level of information exchange should be changed.

For the purpose of a visualization, generated by the (6.9) array of data, we chose one of the three constituents, for example $y_1\ (t_q, r_x, r_y)$. Selected results of the patterns generated are presented in Fig. 6.2. As can be seen, even simple mechanisms of biochemical reactions for (6.8) with its reduced number of parameters in (6.10) reflect the simplified scheme of information exchange within and between neurons; (6.9) possesses different solutions, which could be observed both in reality (sand, spirals and ring waves) and in the form of mandalas produced by artists. By varying the neuron's biochemical reaction internal mechanism, as a scheme of information exchange for the model parameters, we can use (6.6) and (6.7) to generate an unlimited source of complex patterns including symmetrical images in the form of mandalas. It also should be clear that for each chosen biochemical reaction, any type of pattern exists in a limited domain of the model's parameters, found by inverse problem solutions. For that purpose, we need an automatic search of the parameters to generate the desired pattern. Such an automatic system, based on a specially constructed genetic algorithm, has been developed and has demonstrated a high level of performance in finding desired symmetrical patterns such as specific mandalas (Gontar and Grechko 2006). Each mechanism of the neural network biochemical reaction should be considered as an initial hypothesis for finding the desired solution as a specific pattern by performing a search in parameter space. If the initial hypothesis does not result in the desired pattern, it should be changed and repeated with a new search of parameter space until the pattern corresponding to the formalized criteria, such as a desired shape, symmetry, etc., has been found.

Obtained by (6.6) and (6.7), the symmetrical patterns are similar to the analogous patterns produced by human artists in the form of mandalas, as shown on Figs. 6.2 and 6.3 and similar to those mandalas presented by Jung (1973). Thus, the proposed *ANN* could be expanded to different areas of human mental and cognitive activity. One can suppose that, in general, human brain cognitive functions could be connected with the specific patterns emerging in the brain from a neural network's constituent concentrations, demonstrating real mental activity as in the case of an art painting that results in a desired mandala. If so, by using search methods, such as a genetic or simulated annealing algorithm for given fitness function (desired artistic pattern, optimal robot's trajectory, etc.), an *ANN* architecture (6.6) and (6.7) and its parameters could be obtained for its further use as an artificial brain system with cognitive properties.

The dynamical patterns shown in Fig. 6.3 are usually accompanied by a series of discrete chaotic states $y_i\ (t_q)$, representing each neuron of a network with coordinates $\boldsymbol{R}(r_x, r_y)$. Based on the results we have obtained, systems of interconnected neurons could demonstrate well-organized collective behavior by an *ANN* in a form of 2D symmetrical pattern at t_q, $q = 250$, while each individual neuron demonstrates a chaotic regime by its constituents $y_i\ \left(t_q\right)$, $q = 1, 2, 3 \ldots$ as it shown on Fig. 6.4. These symmetrical patterns demonstrate self-organization and self-synchronization within the *ANN* composed of interconnected "chaotic oscillators" (chaotic regimes provided by (6.9)). This supports the statement that "chaos is creative" in a sense that interconnected chaotic regimes are usually accompanied by a high level of collective organization in a form of specific time-space distributed patterns (Gontar 2007).

At this point we would like to discuss what is in common and what is the difference between the approach presented here as a neural network for distributed discrete biochemical reaction dynamics and 2D cellular automata (CA) (Wolfram 2002). Both approaches are operating in discrete space and time and both are using neighboring cell states updated from the previous state of a neuron (each cell of the CA lattice corresponds to the neuron in our network). The main difference between these two approaches is that the CA is not including in its algorithm any physicochemical meaning or constraints from the laws of nature as we presented here, namely the conservation laws, the second law of thermodynamics and the stoichiometry of chemical reactions. This difference makes the CA limited in the control of the patterns it generates. Another difference is that the CA rules are discrete and therefore any type of pattern, corresponding to a particular rule, cannot be transformed into another pattern smoothly, as we know usually happens with natural processes. Other well-known discrete time and space mathematical models, such as fractals (Mandelbrot's, Julia sets) and L-systems, should also be considered as purely empirical computational models; but they do not have any relation to fundamental physicochemical principles and laws of nature, and therefore their solutions hardly can be used for extrapolation; they lack a clear definition of "discrete time and space" related to continuous time and space; and they do not establish relations between model parameters and experimentally obtained data. These reasons limit these other approaches to exploit discrete mathematical

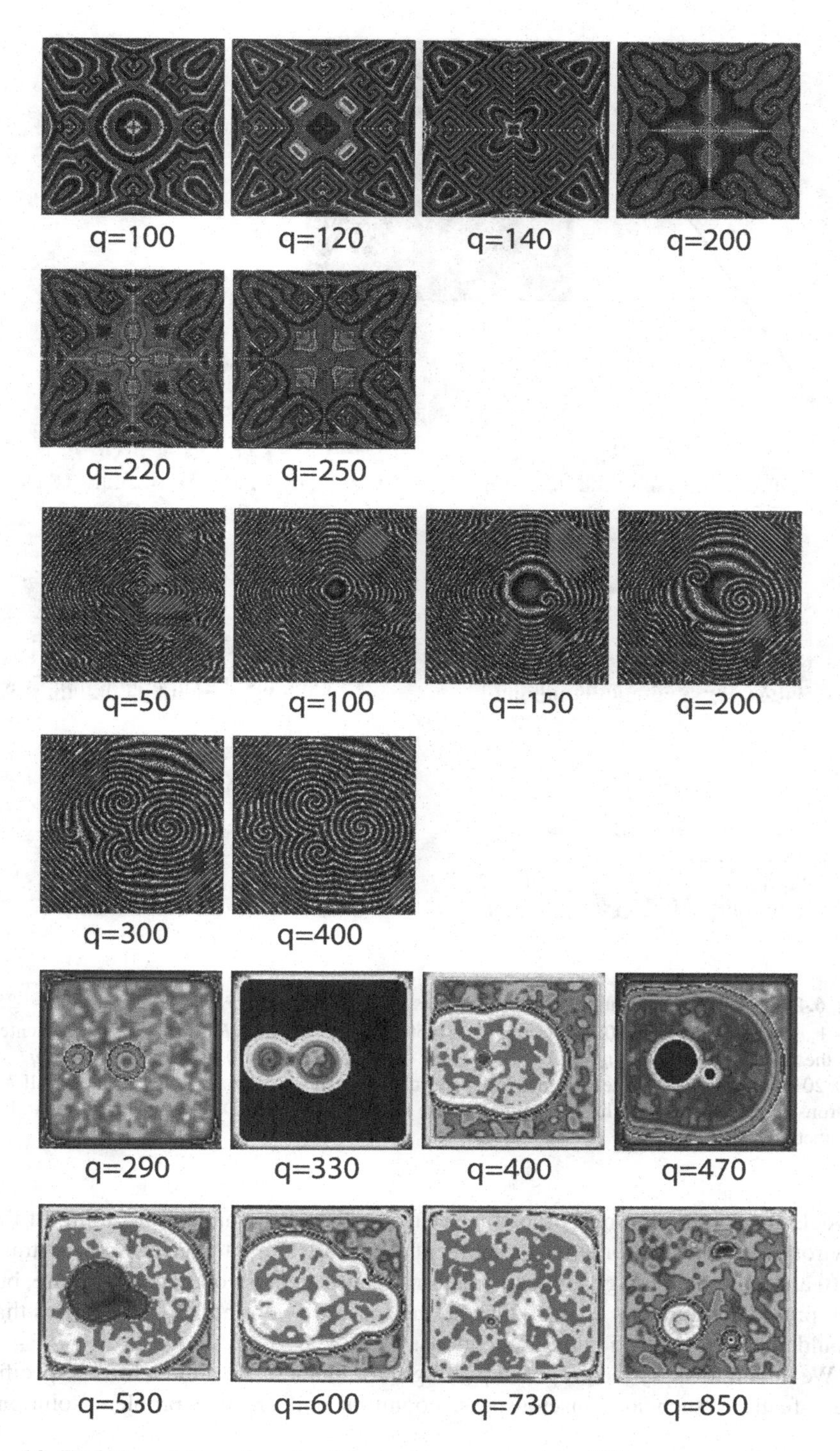

Fig. 6.3 Evolution of the pattern for discrete states generated by (6.9), represented by selected states $t_q, q = 1.2\ldots Q,\ q = 120, 130, \ldots$ correspond to the concrete state of the *ANN*

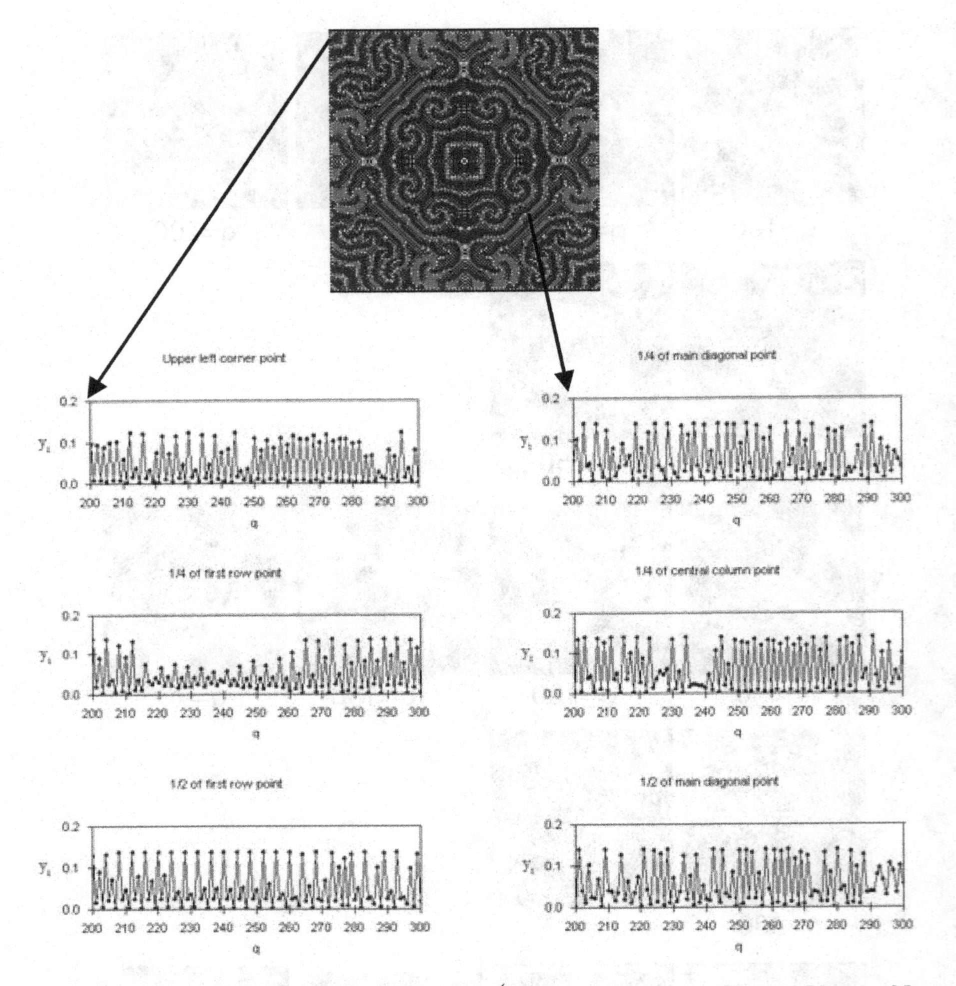

Fig. 6.4 Evolution of six neurons with coordinates $\left(r_x = 1, r_y = 100; r_x = 25, r_y = 75; r_x = 25, r_y=1; r_x=50, r_y=25; r_x=50, r_y=1; r_x=50, r_y=50\right)$ for the 2D *ANN**R*** (100 × 100) represented by the concentration y_1 (t_q, r_x, r_y) sequences that contained 100 discrete states $t_q(q = 200, 201 \ldots Q = 300)$. The symmetrical pattern presented corresponds to $q = 250$. All six neurons are demonstrating chaotic regimes, while for 100 states the 2D patterns are different, but symmetrical

models for solving the real problems related to the mathematical modeling of the environment where autonomous intelligent robots are likely to perform. In contrast, (6.6) and (6.7) not only generate the patterns related to those observed in nature, but also provide continuous control through the variation of a model's parameters that should enable future rational robot actions.

We intend to apply the proposed paradigm for mathematical modeling to specific brain features such as consciousness, cognition and creative problem solutions

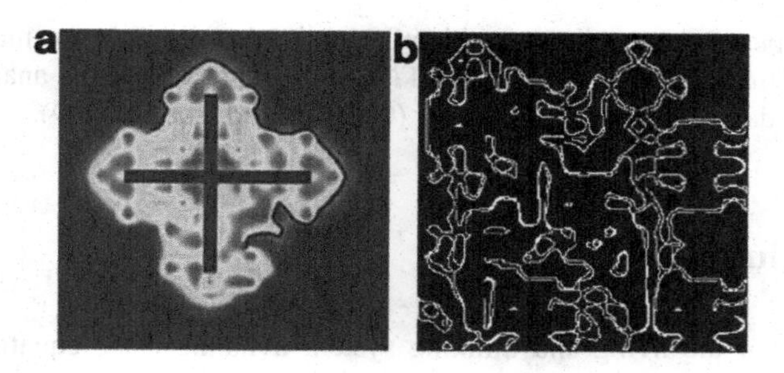

Fig. 6.5 (**a**) A trajectory for an autonomous agent (marked by black curve), composed of neurons on the edge of a pattern generated by (6.6) and (6.7) in the presence of an obstacle in the form of a cross. (**b**) Trajectories composed by the neurons with equal states $y_A\,(t_q, \boldsymbol{R})$ extracted from the generated 2D pattern

in order to construct the "artificial brain" systems with the cognitive properties for autonomous robot rational behavior that resemble human behavior. From our point of view, conscious "artificial brain" systems are those systems that possess the ability to generate the "phenomenological states" associated with the complex dynamical patterns shown in Figs. 6.2 and 6.3. These "phenomenological states" could be used to illustrate rational, innovative and cognitive actions by feeding the data collected into *ANN* for learning and forecasting. By "artificial consciousness", we plan to determine the "phenomenological states" as a form of specific dynamical patterns defined by the mathematical model parameters in the (6.6) and (6.7) that correspond to the internal (learning) and external stimuli (environmental data) that provide the desired rational actions of an intelligent agent or robot.

For example, the autonomous "conscious" robot navigation in an unknown environment could be realized by the proposed approach if we extracted its continuous trajectories from the generated patterns in a form of continuous curves connected to the pattern's internal edges. Environmental data, for example, about obstacles could be introduced into the neural network as shown in Fig. 6.5a. Neurons with coordinates occupied by the obstacle are not changing their state during the network's dynamics. Another option for extracting trajectories could be realized by connecting neurons with equal states as shown in Fig. 6.5b. The obtained trajectories could then be transferred to a robot's navigation system for movement across a real terrain.

The choice of the concrete trajectory for navigation satisfies the conditions of rationality applied to a human as conscious behavior: the minimum distance to a destination, or the avoidance of collision with an obstacle, etc. (Gontar and Tkachenko 2012). This approach could be extended to an artificial agent's intelligent functions by extracting the desired information embedded within the patterns generated by (6.6) and (6.7).

We underline the difference between the mathematical modeling of an "artificial conscious brain" and its process of "learning": the former is related to the generation of desired patterns with the embedded rational information by the (6.6) and (6.7) with known parameters ("direct problem"), while "learning" is the mathematical procedure of finding the model (6.6) and (6.7) parameters from the analysis of the experimental data about the environment (Grechko and Gontar 2009).

6.4 Conclusion

Development of the living and thinking system dynamic basic equations should be the basis for a new generation of artificial neural networks and artificial brain systems. It will require the formulation of new fundamental principles and laws of nature which would reflect the main features of living and thinking systems. Formulated within classical physics and chemistry, the known principles and physical laws of nature have been directed to explain non-living system dynamics and hardly could be applied to living and thinking systems.

Instead, we have suggested new principles and basic equations in the form of difference equations, reflecting specific living and thinking system characteristics, such as "information" and "information exchange". These equations have enabled us to describe the biochemical reaction dynamics that accompany the information exchanges that occur between neurons and neural networks. These difference equations possess numerous chaotic regimes that can simulate the emergence of collective states as complex dynamical patterns. Similar to living systems (those that have emerged from non-living elements to reproduce in and to communicate with the environment), thinking systems composed of "non-thinking" neurons and its constituents when interconnected into networks demonstrate properties similar to the emergence of life, such as emergence of "thoughts", learning, memorizing, consciousness, cognition, creativity, and communication with other "thinking systems" and with the environment.

We associated a neural network's creative dynamical patterns (an artificial brain's "phenomenological states") with consciousness, cognition and creativity involved in the artwork of a mandala. We believe that this application can be extended for future research into robotics since these patterns suggest rational solutions to the problems arising for autonomous intelligent robots during their missions. Application of the extracted solutions from the simulated "phenomenological states" (patterns) for an autonomous robot's actions will look like intelligent behavior to an observer.

We presented one possible approach to formulate the new principle for thinking system dynamics basic equations, and how to apply it to autonomous robot navigation.

Our proposed paradigm for living and thinking system dynamics opens a discussion about the physical meaning of discrete space and time versus continuous space and time, "deterministic chaos" versus probabilistic approaches, and continuous differential equations versus discrete difference equations. This discussion puts more emphasis on developing new principles and laws of nature that might be responsible for brain functioning, including consciousness, cognition and creativity among other brain functions. On that basis, we have suggested how we may be able to create a conscious and cognitive artificial brain system.

Acknowledgment I would like to express my special thanks to Prof. William Lawless for fruitful critiques and discussions about the topic and for help in editing this manuscript.

References

Brandt L (1957) The π-theorem of the theory of dimensionality. Arch Ration Mech Anal 1:35
Brillouin L (1962) Science and information theory. Academic, New York
Gontar V (1993) New theoretical approach for physicochemical reactions dynamics with chaotic behavior. In: Field RJ, Györgyi L (eds) Chaos in chemistry and biochemistry. World Scientific, Singapore
Gontar V (1995) Calculus of iterations and dynamics of physicochemical reactions. Math Comput Simul 39:603–608
Gontar V (1997) Theoretical foundation for the discrete chaotic dynamics of physicochemical systems: chaos, self-organization, time and space in complex systems. Discret Dyn Nat Soc 1(1):31–43
Gontar V (2000a) Entropy as a driving force for complex and living systems dynamics. Chaos Soliton Fract 11:231–236
Gontar V (2000b) Theoretical foundation of Jung's "Mandala Symbolism" based on discrete chaotic dynamics of interacting neurons. Discret Dyn Nat Soc 5:1
Gontar V (2004) The dynamics of living and thinking systems, biological networks, and the laws of physics. Discret Dyn Nat Soc 8:2
Gontar V (2007) Some creative properties of the 2D and 3D lattice distributed interconnected chaotic oscillators and neuronal networks. PAMM 7:2030047–2030048
Gontar V, Grechko O (2006) Generation of symmetrical colored images via solution of the inverse problem of chemical reactions discrete chaotic dynamics. Int J Bifurcation Chaos 16:5
Gontar V, Grechko O (2007) Mathematical imaging using discrete chemical reactions dynamics. Fractals 15:4
Gontar V, Tkachenko C (2012) Autonomous robot path planning algorithm based on neural network discrete chaotic dynamics. Math Model 7:1
Grechko O, Gontar V (2009) Visual stimuli generated by biochemical reactions discrete chaotic dynamics as a basis for neurofeedback. J Neurother 13:1
Haykin SO (1998) Neural networks: a comprehensive foundation. Prentice Hall, Englewood Cliffs
Jung CG (1973) Mandala symbolism. Princeton University Press, Princeton
Shannon CE (1948) A mathematical theory of communication. Bell Syst Tech J 27(3):379–423
Wolfram SA (2002) New kind of science. Wolfram Media, Champaign

Chapter 7
Modeling and Control of Trust in Human-Robot Collaborative Manufacturing

Behzad Sadrfaridpour, Hamed Saeidi, Jenny Burke, Kapil Madathil, and Yue Wang

7.1 Introduction

Traditional industrial robots have been designed for implementation inside safety peripheral equipment (Shi et al. 2012). However, the advent of collaborative robots is changing manufacturing plants by more flexible and efficient robotic automation. Human and robot collaborative manufacturing opens up a new realm of industrial mass production where humans and robots are co-workers (Charalambous 2013). In this paper, we consider hybrid manufacturing systems (Krüger et al. 2009). In such hybrid cells, the associate and a peer human-friendly robot (for example, Rethink Robotics Baxter (Robotics), KUKA LBR iiwa (Bischoff et al. 2010), and Universal Robots UR5 and UR10 (Ostergaard 2012)) collaborate with each other to fabricate customized products (Goodrich and Schultz 2007; Shi et al. 2012) in the same workspace at the same time. For instance, a skilled associate can collaborate with a lightweight, flexible, and human friendly robot to perform an assembly operation. In such applications, human's capability in performing highly skilled tasks such as assembly are combined with the advantages of robots such as precision, performance consistency in performing repetitive jobs, data processing, sensor, and actuator based assistance (Krüger et al. 2009). The resulting collaboration between human and robot in production cells (Tan et al. 2009) is

B. Sadrfaridpour (✉) • H. Saeidi • Y. Wang
Department of Mechanical Engineering, Clemson University, Clemson, SC 29634, USA
e-mail: bsadrfa@clemson.edu; hsaeidi@clemson.edu; yue6@clemson.edu

J. Burke
Boeing Research & Technology, North Charleston, SC 29418, USA
e-mail: jennifer.l.burke2@boeing.com

K. Madathil
Department of Industrial Engineering, Clemson University, Clemson, SC, USA
e-mail: kmadath@clemson.edu

© Springer Science+Business Media (outside the USA) 2016

R. Mittu et al. (eds.), *Robust Intelligence and Trust in Autonomous Systems*,
DOI 10.1007/978-1-4899-7668-0_7

expected to lead to high productivity, flexibility, and safety, as well as balanced human working experience. However, improper HRC may cause counter effects such as misuse of machine and/or safety issues and hence there arises a need for investigating HRC in advanced manufacturing (Krüger et al. 2010). There are potentially many issues worth addressing, but in this paper we focus on human-robot trust as a critical element in HRC manufacturing because trust will directly affect the degree of autonomy that a human delegates to the industrial robot, which determines the efficiency as well as quality of the manufacturing processes. We adopt the concept of trust among humans to study HRC in manufacturing automation (Lee and See 2004). Thus we investigate empirical as well as theoretical studies to utilize trust analysis (Lee and See 2004) in HRC manufacturing. There exist two types of trust related to the automation use among different individuals, i.e. dispositional trust and history-based trust (Merritt and Ilgen 2008). Dispositional trust reflects trust in other persons (or machines) upon initially encountering them, even if no interaction has yet taken place. In contrast, history-based trust is founded on interactions between the person and another person or machine. We study the history-based trust in this paper due to the dynamic nature of HRC. Several works have developed mathematical models for trust (Moray et al. 2000; Lewandowsky et al. 2000; Itoh and Tanaka 2000; Gao and Lee 2006). In our previous works, inspired by Lee and Moray's (1992) trust study for an automated juice plant (Lee and Moray 1992), we used a model for human-robot trust in HRC manufacturing tasks and showed examples of changing robot performance based on human's trust (Sadrfaridpour et al. 2014a,b). In this paper, we extend our work by proposing a time-series model of human-robot trust for real-time control allocation in HRC manufacturing tasks, a model of robot performance that ties speed to flexibility, a model of human performance that includes muscle fatigue, and a series of experimental validations to capture the impact of performance on trust within the HRC system. The proposed dynamic trust model is a function of prior trust, change of robot performance, and change of human performance, as well as fault occurrence.

The robot performance can be described in terms of reliability, flexibility, dexterity, etc. Because robot reliability is almost always guaranteed in manufacturing applications, here we will focus on understanding and improving the robot flexibility assuming the robot is reliable. Flexibility is required for factory environments with frequent changes, varying positions of transport containers, and various uses of machine tools. Flexibility is envisioned to increase productivity and humanization of the work place (Stopp et al. 2001). In fact, it is one of the advancements brought by the new generation of manufacturing robots and is achievable via instructable or adaptable robots. To model the performance of a human associate of doing a repetitive kinesthetic task, which is typical in manufacturing tasks, we adopt the muscle fatigue and recovery model (Ma et al. 2009, 2010; Liu et al. 2002; Fayazi et al. 2013). This model shows how the performance of the human associate changes as his/her muscles gradually get tired or recovered.

Artificial neural networks (ANNs) are powerful tools that can be used for realizing artificial intelligence (White 1992). They have been widely applied in aviation industry, business, financial forecasting, control systems, security systems, etc. (Widrow et al. 1994). Neural networks are capable of function approximation, pattern recognition and nonlinear mapping (Mehrotra et al. 1997). Their learning ability and adaptability also introduce robustness to a tool (Cichocki and Unbehauen 1996). In this paper, we are interested in the applications of neural networks in intelligent control such as black box model identification, adaptive inverse control and model predictive control (Hagan and Demuth 1999). More specifically, we will use neural networks to learn the desired pattern of robot speeds in order to collaborate with a specific human associate and to use the result for autonomous adjustments of the robot's speed.

Next, we design control allocation schemes to switch between manual and autonomous modes in order to increase the human-robot trust. To do so, three approaches are designed. One way is to increase or decrease the robot performance exclusively based on manual inputs. Another way is to predict the human requests and autonomously adjust the robot performance using the neural network-based intelligent control. The last way is to use a collaborative control scheme to adjust the robot performance using both autonomous and manual inputs.

To study the trust evolution and human working pattern during HRC manufacturing, we present both a numerical example and a set of experimental validations. The numerical example is simulated for a typical 9 h workday starting at 8 AM. The exclusively manual, exclusively autonomous, and collaborative control modes are compared. The experiments are designed as HRC assembly tasks where the robot picks the parts and places them in front of the participant and the participant assembles these parts. Such collaborations require the robot to keep pace with the human and can be applied in many manufacturing processes to partially automate the assembly tasks.

The major contributions of this paper are threefold: (1) We propose and experimentally validate a new dynamic, quantitative trust model specifically for HRC assembly manufacturing; (2) We develop a neural network based robust intelligent scheme for autonomous robot speed control; (3) We integrate the quantitative trust models with robust intelligence for improved performance in HRC manufacturing.

The rest of the paper is organized as follows. Section 7.2 introduces the time-series trust model, robot and human performance models. Section 7.3 develops the neural network based robust intelligence control algorithm for learning the human working pattern in controlling the robot speed. Section 7.4 discusses three control allocation schemes, i.e. exclusively manual, exclusively autonomous, and collaborative control of the robot speed. We simulate the proposed trust model and intelligent control scheme using a numerical example of a typical work day in a manufacturing plant in Sect. 7.5. A set of experimental validations on assembly tasks are performed and major results analyzed in Sect. 7.6. We conclude the paper in Sect. 7.7.

7.2 Trust Model

7.2.1 Time-Series Trust Model for Dynamic HRC Manufacturing

Based on Lee and Moray's (1992) time-series trust model and the more recent meta-analysis (Hancock et al. 2011) and survey (Hoff and Bashir 2014), a human's trust in the robot depends on the robot performance, human performance, and fault occurrences. In this section, we introduce a time-series dynamic model of human-robot trust for HRC manufacturing based on these results from human factors research. To clarify the manufacturing application, let us start with an example. Consider the case when a skilled associate collaborates with a flexible robot on a product, such as inserting screws into parts or welding, in a hybrid cell. The robot picks up a part and then holds it still in specific positions and orientations near the associate so that the associate can focus on the assembly operations. As the working speed of the associate varies during the working hours, a constant speed of the robot will cause trust degradation of the associate when he/she feels that the robot is working faster or slower than what he/she expects, i.e. the robot lacks flexibility to keep the same pace as the associate. This discrepancy indicates the robot's inflexibility. To recover trust, the robot speed should be adjustable so that the associate feels more comfortable in the collaboration. Moreover, the associate's performance has influence on his/her trust in the robot. For example, due to physical and/or mental fatigue resulting from continuous work during a day, the associate may tend to rely more on the automation and thus his/her trust in the robot increases. With this mindset, we propose the following time-series model for the dynamics of human-robot trust

$$T(k) = AT(k-1) + B_1 P_R(k) + B_2 P_R(k-1) + C_1 P_H(k) + C_2 P_H(k-1) + D_1 F(k) + D_2 F(k-1), \tag{7.1}$$

where P_R, P_H, and F are robot performance, human performance, and fault, respectively. We use k to indicate the time step. The coefficients A, B_1, B_2, C_1, C_2, D_1, and D_2 are constants to be determined through experiments. Note that we seek to obtain a computational model of a human's trust for HRC in assembly lines in general. In practice, these parameters of the trust model can be tuned for different individuals to fit their subjective trust to some extent. Moreover, similar to Lee and Moray (1992) we assume that the trust dynamics follow a lag model and there are some delays before changes of trust. As long as there is a considerable difference between the human and robot working speeds, the robot performance (P_R, flexibility) will decrease regardless of which speed is greater than the other. Therefore, the trust value decreases accordingly. In contrast, if there is no considerable decrease in robot flexibility over time, the trust will increase.

We design robust intelligent control schemes to increase human trust in a robot as described in Sect. 7.3. To obtain the trust model (7.1), we need to develop robot and human performance models as discussed in the subsequent sections.

7.2.2 Robot Performance Model

In manufacturing, machine reliability is almost always guaranteed in order to avoid huge loss under even small malfunctions. Meanwhile, for the new type of flexible manufacturing tasks, the robot needs to seamlessly collaborate with the human coworker. Hence, robot performance in this case can be evaluated by its flexibility in accommodating a human's work behavior. In our study, we consider specially the robot capability in adjusting its speed so as to keep the same pace as the associate. Hence, the difference between human and robot speed will determine the robot flexibility. We denote robot working speed, $V_R \in [0, 1]$, as the normalized speed of the robot for doing a specific task where "0" represents the situation when the robot stops working, and "1" represents the situation when the robot works at its maximum speed. We denote the human working speed, V_H, correspondingly. Note that both V_H and V_R are defined as normalized non-dimensional numbers in [0, 1]. Based on our definition of the robot flexibility, P_R, we can write:

$$P_R(k) = P_{R,max} - |V_H(k) - V_R(k)| \,. \tag{7.2}$$

Since we use the normalized values of V_H and V_R, we have $P_{R,max} = 1$ and hence P_R is always bounded between [0, 1]. In the ideal case when the robot works at its highest flexibility in adapting to the associate's speed, the speed difference is minimum and $P_R = 1$. In the worst case when the robot is fully incapable of adjusting to the associate's speed, the speed difference is maximum and $P_R = 0$.

7.2.3 Human Performance Model

A human's performance in physical tasks such as assembly manufacturing depends on his/her state of muscle fatigue or recovery. In such scenarios, an associate usually performs repetitive kinesthetic tasks. We adopt the muscle fatigue and recovery model proposed in Ma et al. (2010) and Fayazi et al. (2013) for our human performance model. This model explains how a muscle or group of muscles get fatigued or recovered during performing physical tasks and shows how the performance of an associate changes as his/her muscles gradually get tired or recovered. We assume that the higher the fatigue level is, the lower the performance would be. The maximum human performance occurs at the situation when he/she is not subjected to any fatigue, and the minimum value when he/she is experiencing the maximum level of fatigue. We first present the muscle fatigue and recovery model and then develop the human performance model based on the muscle fatigue and recovery model.

For the modeling of muscle fatigue and recovery, we introduce a model for isometric force generation, i.e. when the muscles do not move but they apply force. When a muscle applies some force for an amount of time, the maximum isometric force that one can produce, $F_{max,iso}(k)$, decreases. The dynamic model of fatigue for $F_{max,iso}(k)$ is a function of time, the initial maximum isometric force one can generate at rest, called Maximum Voluntary Contraction (MVC), and real-time applied force $F(k)$ (Ma et al. 2009). On the other hand, when the muscle does not apply any force, it gets recovered. The recovery process is also a function of the time and MVC (Ma et al. 2010). Based on Liu et al. (2002), when the muscle fibers work, some of them become fatigued and some recover. That is to say, fatigue and recovery occur simultaneously (Ma et al. 2010). We develop the discretized version of the combined fatigue and recovery model in Fayazi et al. (2013) using the first-order Euler approximation

$$\begin{aligned} F_{max,iso}(k) = F_{max,iso}(k-1) - C_f F_{max,iso}(k-1)\frac{F(k-1)}{MVC} \\ + C_r(MVC - F_{max,iso}(k-1)), \end{aligned} \tag{7.3}$$

where C_f is the fatigue constant and C_r is the recovery constant. Both C_f and C_r are individual-specific. Equation (7.3) is for isometric muscle contraction and has an equilibrium point at which the fatigue and recovery balance out. This point is the lowest limit (threshold) of the $F_{max,iso}(k)$. This threshold force, F_{th}, can be calculated by assuming that $F_{max,iso}(k) = F_{max,iso}(k-1)$ at the threshold:

$$F_{th} = MVC\frac{C_r}{2C_f}(-1 + \sqrt{1 + \frac{4C_f}{C_r}}). \tag{7.4}$$

Theoretically, at the threshold force, the fatigue and recovery occur at the same rate and one can generate this threshold force for a long time. Since the fatigue and recovery model predicts the human muscle status related to workload, this model can be used to measure the physical performance of an associate during manufacturing tasks. Hence, we propose the following performance model for human, P_H

$$P_H(k) = \frac{F_{max,iso}(k) - F_{th}}{MVC - F_{th}}. \tag{7.5}$$

Note that in Eq. (7.5), $F_{max,iso}$ varies between the minimum value F_{th} and the maximum value MVC, therefore it is a normalized value between 0 and 1. The maximum value MVC, is assumed when the associate starts the task, i.e. $F_{iso,max}(k = 0) = MVC$.

Remark 1. The threshold force, F_{th}, is the minimum value of $F_{max,iso}$. Hence, the forces below F_{th} are not theoretically achievable.

7.3 Neural Network Based Robust Intelligent Controller

The goal of using a neural network in this problem is to design a robust intelligent controller for adjusting the robot speed autonomously during the work cycle which is a black box model identification. This controller is designed so that it reduces the associate's workload for adjusting the speed of the robot manually. To do so, a neural network with a proper method of training and also some training data are required. One way of training the neural network is to mimic the behavior of the associate in adjusting the robot speed manually, which can be regarded as the desirable pattern for the robot flexibility when collaborating with the associate. We performed human-in-the-loop experiments to collect the training data. In this data set, the current robot speed, human speed, and current work-cycle time index is used as the input to the neural network and the estimation of robot speed at the next cycle is the output.

The structure of the neural network used in this paper is illustrated in Fig. 7.1. This network consists of an input layer, a hidden layer, and an output layer of neurons which form a Perceptron artificial neural network (Hagan and Demuth 1999). This type of neural network has the capability of approximating many nonlinear functions. The additional input "1" (as seen in the first and second layers of Fig. 7.1) represents the effect of bias in the neural network. Using bias increases

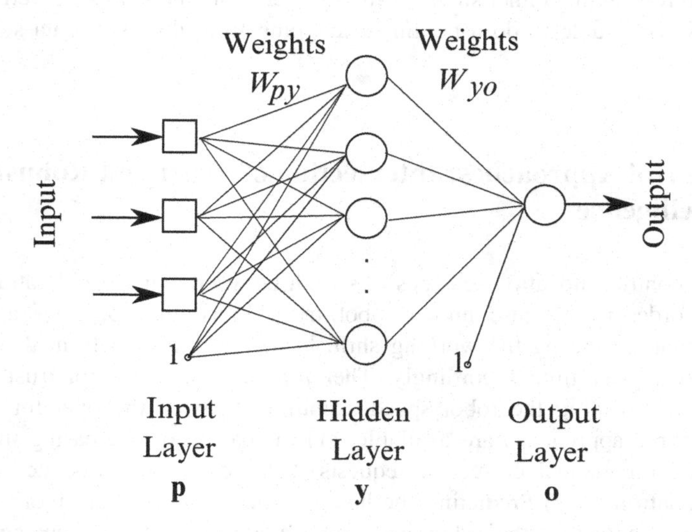

Fig. 7.1 The structure of neural network used for learning the robot speed

the learning capability of a neural network by providing an additional degree of freedom through an adjustable offset. We utilize two different activation functions for the hidden layer and the output layer, respectively. The activation functions determine the output of the neurons in each layer as a function of the weighted sum of the inputs to that layer. The activation function of the hidden layer $\mathbf{y}$ is a tangent sigmoid function as follows

$$\text{tansig}(x_{py}) = \frac{e^{x_{py}} - e^{-x_{py}}}{e^{x_{py}} + e^{-x_{py}}}, \tag{7.6}$$

where x_{py} is the input for the tangent sigmoid function. In the neural network shown in Fig. 7.1, this variable is defined as $x_{py} = W_{py} \times [\mathbf{p}\ 1]$ where W_{py} represents the weights of the neural network that connect the input layer $\mathbf{p}$ (i.e. the current robot speed, human speed, and current work-cycle time index) to the hidden layer $\mathbf{y} = \text{tansig}(x_{py})$. The output of this function is in $(-1, 1)$ region which produces the inputs to the next (output) layer. The activation function for the output layer $\mathbf{o}$ is chosen to be the linear function according to the following

$$\text{purelin}(x_{yo}) = x_{yo}, \tag{7.7}$$

where $x_{yo} = W_{yo} \times [\mathbf{y}\ 1]$ for the output layer are the weights of the neural network that connect the hidden layer to the output layer. This layer determines the robot speed at the next work cycle. Once enough data are collected, the Levenberg-Marquardt Backpropagation training algorithm (Hagan and Demuth 1999) is used to train the neural network. This algorithm is a gradient decent based optimization algorithm for minimizing the mean square estimation error of the neural network. It can be used for training either single or multi-layer neural networks. A well-trained neural network is able to do a nonlinear mapping from the input data set to the output data set.

7.4 Control Approaches: Intersection of Trust and Robust Intelligence

We design control allocation schemes to switch between manual and autonomous modes in order to increase human-robot trust. Since the speed of a human associate changes during the working shift, his/her expectation from the partner robot changes over time accordingly. Therefore, the human-robot trust can be increased by adjusting the robot speed according to what the operator desires. To do so, three approaches are available: (1) Increasing or decreasing the robot speed based on manual corrective requests that the human associate sends to the robot controller; (2) Predicting the human requests at different moments and autonomously adjusting the robot performance without sending any corrective request; or (3) Using a collaborative control scheme to adjust the robot speed using the autonomous control and manual inputs interchangeably. The prediction approach can be achieved by the robust intelligence algorithm which seeks to learn the pattern of human requests as he/she collaborates with the robot over time. Here we use the artificial neural networks as the robust intelligence algorithm as discussed in Sect. 7.3. In the collaborative mode, the robust intelligence algorithm is used to autonomously control the robot speed by default. However, the human associate can adjust the robot speed at the times when the robust intelligence fails to mimic the human pattern in adjusting the robot performance. We now explain the details of implementation of the three different approaches for adjusting the robot speed.

7.4.1 Manual Mode

For the manual mode, a human-sensitivity based approach is adopted to predict how the human coworker adjusts the robot speed. Most of the time, the robot speed does not match the human working speed exactly. However, it is only when the difference between these two speeds exceeds a certain threshold, then the associate would feel the significance and send some corrective commands to change the robot speed. Let this threshold be human sensitivity, H_S. With this setting, the robot speed at the next time step is adjusted by the associate as follows

$$V_R(k+1) = V_{RH}(k), \tag{7.8}$$

where $V_{RH}(k)$ represents the manual control input whenever the associate changes the robot speed. Other than these moments, we have $V_R(k+1) = V_R(k)$.

7.4.2 Autonomous Mode

Based on the explanations in Sect. 7.3, to train the artificial neural network, we collect data on how an associate sends commands to the robot in the manual mode for some period of time. There are different ways to construct the neural network based on the inputs and the training algorithm. For example, we can predict the pattern of the speed commands that the associate sends to the robot only based on time parameters or we can include other parameters into the network as well. Figure 7.1 shows the neural network with current time, human speed, and robot speed as inputs. The output is the robot speed at the next time step. After training the neural network, it will predict the desirable robot speed based on the inputs. With this setting we have

$$V_R(k+1) = V_{RI}(k), \tag{7.9}$$

where $V_{RI}(k)$ represents the autonomous control input calculated by the neural network for the next time step. The neural network is the only source of robot speed adjustment in this mode, and thus it is used at each time step whether it generates a new command or the similar command as the previous step.

7.4.3 Collaborative Mode

The autonomous mode reduces the human workload through the use of robust intelligence algorithms. However, the manual mode offers more accurate control over the robot speed. In the collaborative mode, we combine both advantages. The robot speed is controlled autonomously by the neural network by default and the associate can change the robot speed whenever he/she wants to. Therefore, we can describe the process of controlling the robot speed by the following equation

$$V_R(k+1) = \sigma(k)V_{RH}(k) + (1-\sigma(k))V_{RI}(k), \tag{7.10}$$

where $V_{RH}(k)$ and $V_{RI}(k)$ are as in Eqs. (7.8)–(7.9) respectively, and $\sigma(k)$ is the activation mode

$$\sigma(k) = \begin{cases} 1 & \text{manual control} \\ 0 & \text{autonomous control} \end{cases}$$

In this setting, the robot speed at the next time step is determined either directly by the human commands or the predictions of the robust intelligence algorithms. Examples of utilizing this scheme will be presented in Sects. 7.5 and 7.6.

7.5 Simulation

In this section, we present a numerical example using MATLAB R2014a software for three different control schemes described in previous sections. This example shows (1) how the human trust evolves according to the human and the robot performances; and (2) how the control workload of the human associate changes. The human performance dynamics (7.5) described in Sect. 7.2.3 are simulated for a typical 9 h workday starting at 8 AM. In the simulation we shift the time origin to 8, i.e. we use $k' = k - 8$ instead of k in all of the equations. For a fixed repetitive task we assume that the external force applied by the human associate is constant. Moreover, the associates do not need to apply their full strength (MVC) to finish the manufacturing tasks. Therefore, we use a constant value for the external force, i.e., $F(k) = \frac{MVC}{4}$. The maximum value for both human and robot performance is 1, i.e. $P_{H,max} = 1$ and $P_{R,max} = 1$. The associate is assumed to start with P_H between [0.95, 1]. The associate working speed, V_H is set to be half of his/her performance value, i.e. $V_H = \frac{1}{2} P_H$ in the simulation. The robot is set to start with half of the maximum robot working speed, $\frac{1}{2} V_{R,max}$. We also assume that initial trust of the associate is the half of its maximum value. In all the simulation modes, we assume that the associate works according to the following pattern. He/She starts to work at 8 AM and ends at 5 PM. There is an approximately 1 h lunch break around noon. There are also two short breaks (15–20 min) in mid-morning and mid-afternoon (around 10 AM and 3 PM, respectively). During such a workday, based on the Eq. (7.5) the human performance decreases from the beginning of the day through the end of the day, except for the break times and the lunch time when the human performance recovers. We simulate the three control methods in Sect. 7.4.

Based on the explanations in Sect. 7.3, to train the neural network, we simulate and collect the corresponding data for the human-robot interaction of a particular associate for a period of 4 months. According to the data, as in Fig. 7.1, we have 3 inputs to the artificial neural network, namely month, day and time of the day, and one output which is the performance of the robot. The number of hidden layer neurons is chosen to be 10 and the Error Backpropagation training algorithm is used to train the neural network. The results for each of the three control schemes are presented in the next subsections.

7.5.1 Manual Mode

According to the explanations in Sect. 7.4.1, we set the human sensitivity as $H_S = 0.05$. The results of this simulation are shown in Fig. 7.2a. As can be seen in this figure, at the start of the day both human and robot start fresh with high working speeds and consequently the robot performance is high. As time passes, the working speed of the associate decreases but the robot working speed does not change, so the difference between the human and the robot speed increases and thus the robot performance decreases. The human performance also decreases during this time. Although both robot and human performance decreases, since they have high values the trust increases before 9 AM. The trust value decreases slightly when the human performance declines after 9 AM. Therefore, when the human speed decreases during the time interval 8 AM to 10 AM, the associate sends corrective commands to decrease the robot speed. After that, the associate takes a break and his/her speed increases when going back to work again. We use the same trend for the rest of the day with breaks at 12 PM and 3 PM, respectively. The trust value does not change during the breaks.

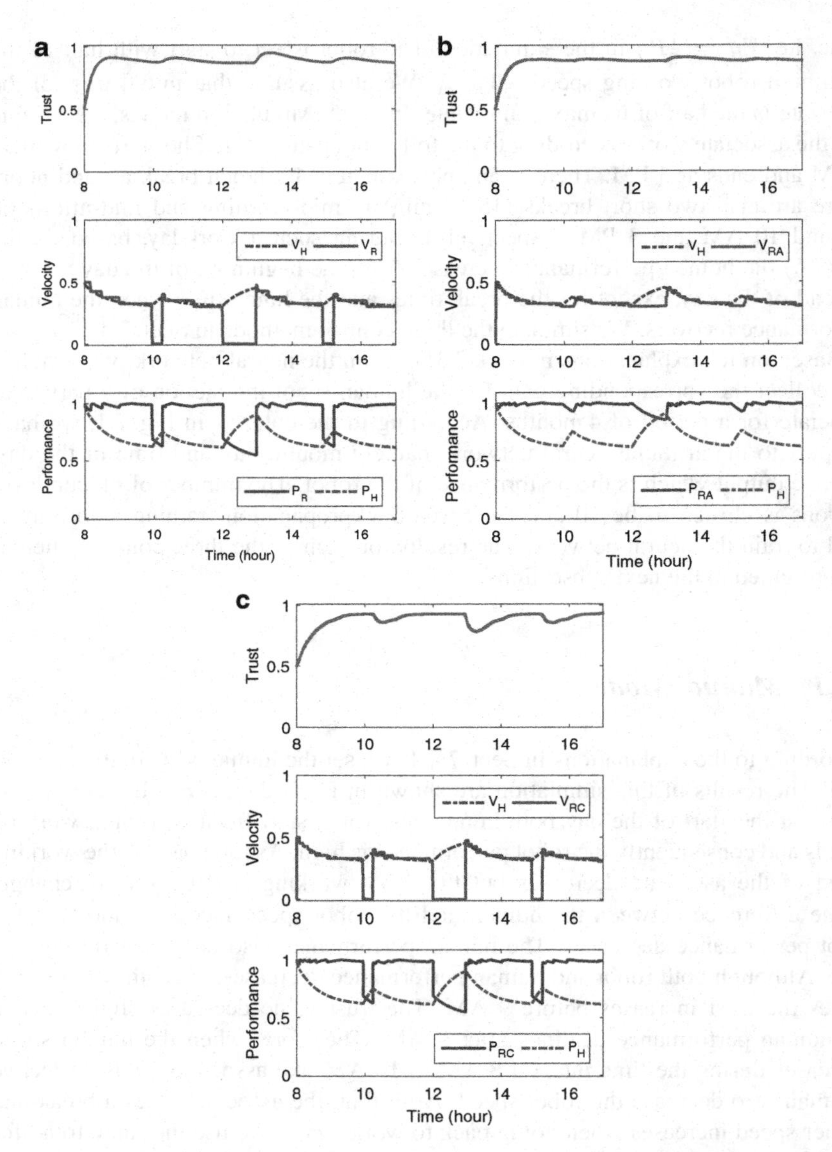

Fig. 7.2 Evolutions of human speed V_H, robot speed V_{RA}, human performance P_H, robot performance P_{RA}, and trust T in (**a**) manual mode, (**b**) autonomous mode, and (**c**) collaborative mode

7.5.2 Autonomous Mode

According to the explanations in Sect. 7.4.2, we use the neural network for adjusting the robot performance autonomously. The results of this simulation are shown in Fig. 7.2b. As shown in this figure, the autonomous mode can adjust the robot speed properly most of the times. For the autonomous mode, the trust level has a similar trend as in the manual mode except for the end of the break times, where the neural network cannot predict the desired robot speed accurately. This leads to a sudden momentary drop of trust due to a temporary difference between the human and robot speed.

7.5.3 Collaborative Mode

For simulation of this mode, we use the same configuration of the manual and autonomous control modes described in this section. We then combine them as described in Sect. 7.4.3 to simulate the collaborative mode. The results are shown in Fig. 7.2c. The team starts to work in the autonomous mode at the beginning of the workday. After some time, if the robot speed does not match the human speed, the level of trust decreases. Moreover, if the robot performance is high and the human performance declines, the level of trust increases. In contrast to the autonomous mode, except autonomous adjustment, the associate can also switch to the manual mode by sending corrective commands. Note that the associate sends commands whenever he/she feels that the autonomous adjustments are not correct. If the system switches back to the autonomous mode right after the manual correction, the adjustments might not be correct and hence the associate needs to adjust the robot speed again. This leads to frequent switches back and forth between the manual and autonomous mode. To prevent such problems, once the manual mode is activated, it will be kept for a fixed time period (5 min) before it is allowed to switch back to the autonomous mode. After that, the system switches back to the autonomous mode and remains in the autonomous mode if no corrective commands are sent.

7.5.4 Comparison of Control Schemes

We can measure the human control workload under the manual, autonomous, and collaborative mode, respectively. The control workload for the manual mode is 100 % since the human associate always changes the robot velocity by him/herself. The control workload under the autonomous mode is 0 % since the human associate does not change the robot speed at all. The amount of control workload for the collaborative mode depends on the amount of time when the manual mode is activated. In our example, this value is 61.4 %. We cam also compare the average value of trust under these three modes. In the autonomous mode, the average trust value is 0.8803 which is lower than this value in manual mode, 0.8825. The average trust value in collaborative mode is 0.8816. This shows that using the collaborative mode, we can increase the trust compared to the autonomous mode while the control workload is smaller than the manual mode.

7.6 Experimental Validation

In this section, we provide detailed description about our experiments to validate the quantitative trust model (7.1) and the effectiveness of the proposed control schemes. We will measure the overall task performance of the collaborative control scheme versus exclusively manual and autonomous control as well as the difference in human workload.

7.6.1 Experimental Test Bed

As shown in Fig. 7.3a, we employ a humanoid manufacturing research robot Baxter made by Rethink Robotics (Guizzo and Ackerman 2012) to collaborate with the participant. The robot has two arms. Each arm provides 7 degrees of freedom. The arm joints are compliant as they are built with back-drivable motors and compliant actuators. The robot has a rotary screen at its head where informative messages or affective expressions can be displayed. It has a moveable base. The robot control program is coded in Python language and is interfaced with the robot hardware through ROS software. Baxter is very suitable for light-weight material handling and intelligent assembly, testing and sorting, and especially for small batch productions. We use the Impulse X2 motion tracking system from PhaseSpace to track the human hand for speed measurement (as shown in Fig. 7.3b). The tracking system includes 8 cameras, a set of active markers and a workstation for tracking rigid bodies in a 3D environment. The workstation combines the data from the cameras, which track the active markers mounted on an object (for example, a participant's hand in this study), to calculate its 3D position. The resulting position and timing information is sent to a client machine to calculate the hand motion speed.

7.6.2 Experimental Design

The experiment resembles the task that an associate performs in the manufacturing assembly lines. In such an environment, associates are required to perform a series of assembly tasks within a fixed period of time. For making a final product, different components need to be assembled together. Each of these components need to be

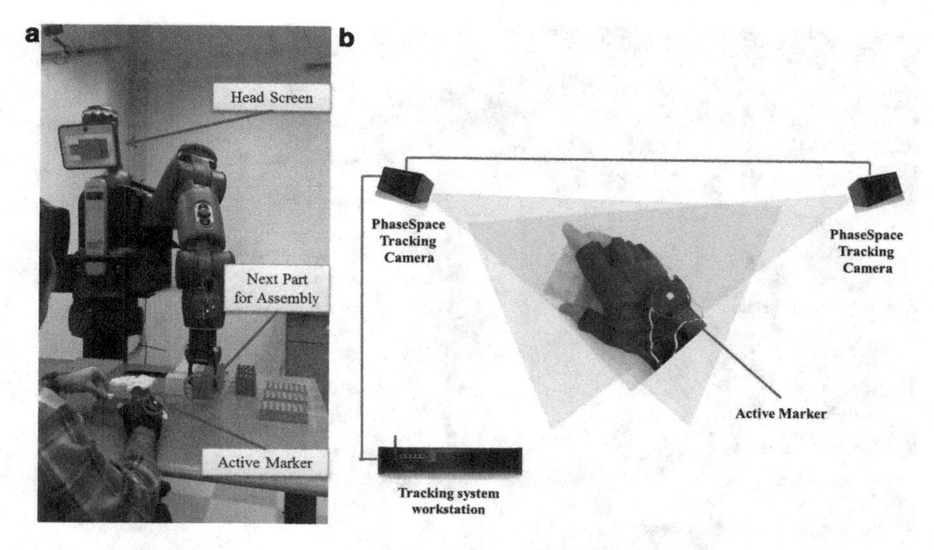

Fig. 7.3 (**a**) Collaboration of a participant and Baxter, and (**b**) PhaseSpace tracking system for tracking participants's hand motion

assembled by different parts as well. This procedure of component assembly is called subassembly, which is common in airplane and automobile assembly and usually done by the associates manually. We will consider such a subassembly task in our experiments. In such tasks, the parts need to be assembled are usually stacked near the workbench of the associate. The associate picks these parts and assembles them. If the component is customized, there will be a variation of choice for some of the parts. These customized parts can be delivered to the associate by means of automatic delivery systems such as belt feeders. Once the component is assembled, it needs to be mounted on the final product. The experimental setup of this study is very similar to these tasks in a real assembly line except that there is a humanoid robot (Rethink Robotics Baxter) that collaborates with the participant. Within this collaboration, the robot helps the participant by picking up and placing the customized parts needed for the assembly task while the associate performs tasks that robots are not capable of, e.g. assembling these parts together. The details of experiment scenario are as follows.

7.6.2.1 Experiment Scenario

The participant is asked to perform a cooperative assembly task with Baxter within a fixed period of time. For each experiment condition, the task is assembling 10 components within 17 min (102 s per task cycle). Figure 7.3a shows the collaboration of a participant and Baxter. The task is to assemble a customized component (e.g. component G in Fig. 7.4) made from different parts (Lego bricks, e.g. bricks

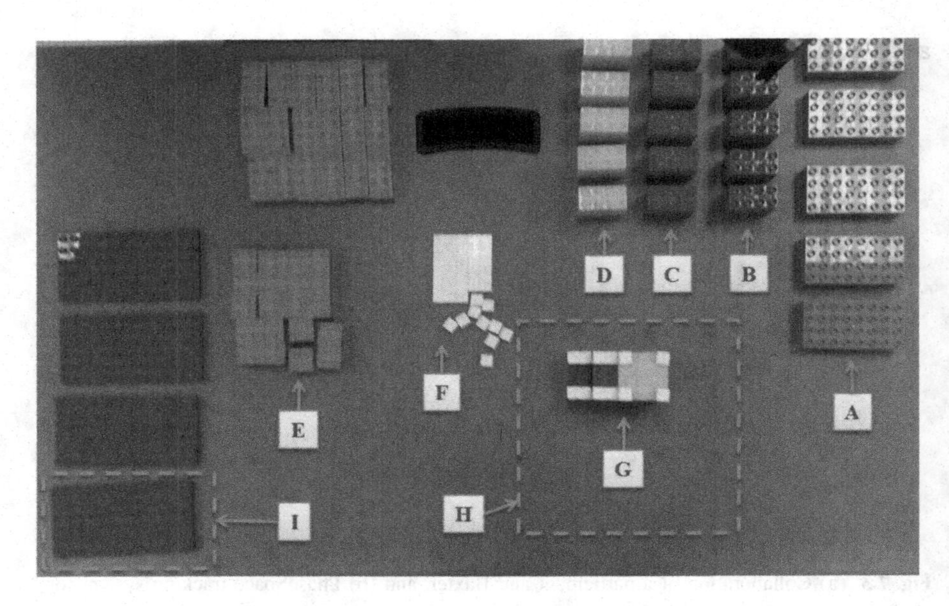

Fig. 7.4 Different assembly parts and regions on the experiment table

A, B, C, D, E, F in Fig. 7.4) and mount it to another component (here is another larger Lego brick, e.g. component I in Fig. 7.4). The example assembly task we consider here can be found commonly in automobile and airplane assembly, e.g. center console subassembly and airplane wing spar assembly. There are 10 trials in total in each trial. The participant and Baxter share meme workspace on a table and the assembly parts are placed at different regions on the table as shown in Fig. 7.4. In this figure, the Lego bricks that need to be assembled together are A, B, C, and D. At the beginning of each task cycle, Baxter picks up a required part (brick A) and places it in front of the participant (region H) and displays a picture of the assembled part via its head screen (Fig. 7.3a). The participant is required to look at Baxter's screen and assemble the part exactly as appeared on it. The participant is also required to add fitting parts (bricks E and F in Fig. 7.4) on top of the assembled Lego bricks similar to tightening screws or bolts in real manufacturing. When the participant finishes assembling the last part, he/she is required to pick and mount the whole component to another Lego brick located at the other side of the table (component I in Fig. 7.4). Meanwhile, Baxter picks and places the next part in front of the participant and displays the next picture of the assembled part. The similar process is repeated until Baxter picks and places the last required part in front of the participant. Figure 7.5 shows the instruction pictures that Baxter shows to the participant in each cycle. Each of these pictures show the correct assembly of current Lego bricks and corresponding fitting parts (needed to be mounted on the top of the Lego bricks). F and E are the fitting parts for assembling and mounting, respectively. Figure 7.6 provides a flow chart to summarize the required actions for both Baxter and the participant and their collaboration in every task cycle.

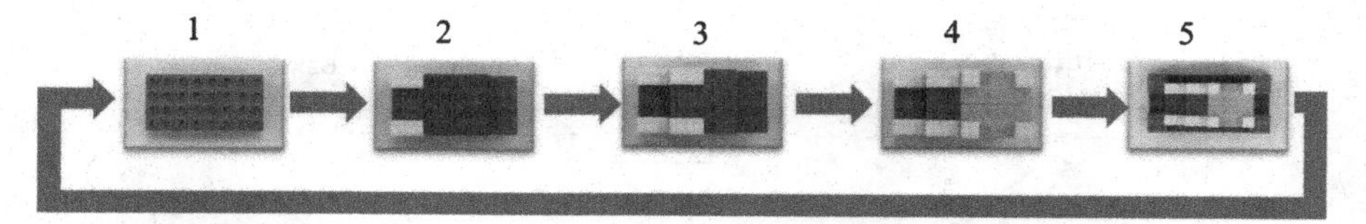

Fig. 7.5 Sequence of assembly parts that Baxter shows to the participant as instruction via its head screen

7.6.2.2 Controlled Behavioral Study

To understand the impact of the robot and human performance on the trust evolution, a 2 (robot performance—low flexibility, high flexibility) × 2 (human performance—non-fatigue, fatigue) mixed experimental design is employed under each control mode. In the high robot performance condition, the robot speed changes in accordance with participant's hand speed without any delay while in the low robot performance condition, the robot speed changes with some random delay plus some sudden stops of the robot. Note that the sudden stops of the robots are the faults of the robot while the random delays are the inflexibility of the robot. Here human fatigue refers to the psychically caused fatigue that commonly occurs in an assembly associate as discussed in Sect. 7.2.3.

7.6.2.3 Imposing Fatigue

Assembly tasks usually require prolonged low-level repetitive work of the associates which causes psychical fatigue. However, in the laboratory setting, it is difficult for a participant to perform long 9-h experiment to study the fatigue condition. It has been shown in Iridiastadi and Nussbaum (2006) that the greatest effort level of shoulder muscle is required when the associate holds a typical hand tool weighting around $15 - -20N$ in abducted shoulder posture (90° to vertical). A similar method as in Iridiastadi and Nussbaum (2006) is used to impose fatigue in the experiments. In the fatigue condition, the participant is asked to warm up and then perform 10 min of exercises. Before doing the exercises, we need to measure the MVC as shown in Eq. (7.3) in Sect. 7.2.3. The MVC level for 90° shoulder posture for the dominant hand shoulder muscle of each participant is measured using a hand dynamometer. In order to measure the MVC level, the participant is asked to sit down on a chair and extend his arm fully and put his hand in the hand dynamometer (fixed under the table in front of the participant) and push it up as much as possible. The hand dynamometer value shows the maximal force which is the MVC value at the start of the experiment. We collect the data three times and use the average value. We then ask the participant to hold a weight around 30 % of their MVC during the exercises. The exercises consist of five 2-min intermittent static arm abduction cycles. For each cycle, the contraction duration is 90 s followed by 30 s rest. We used 166 s cycle time

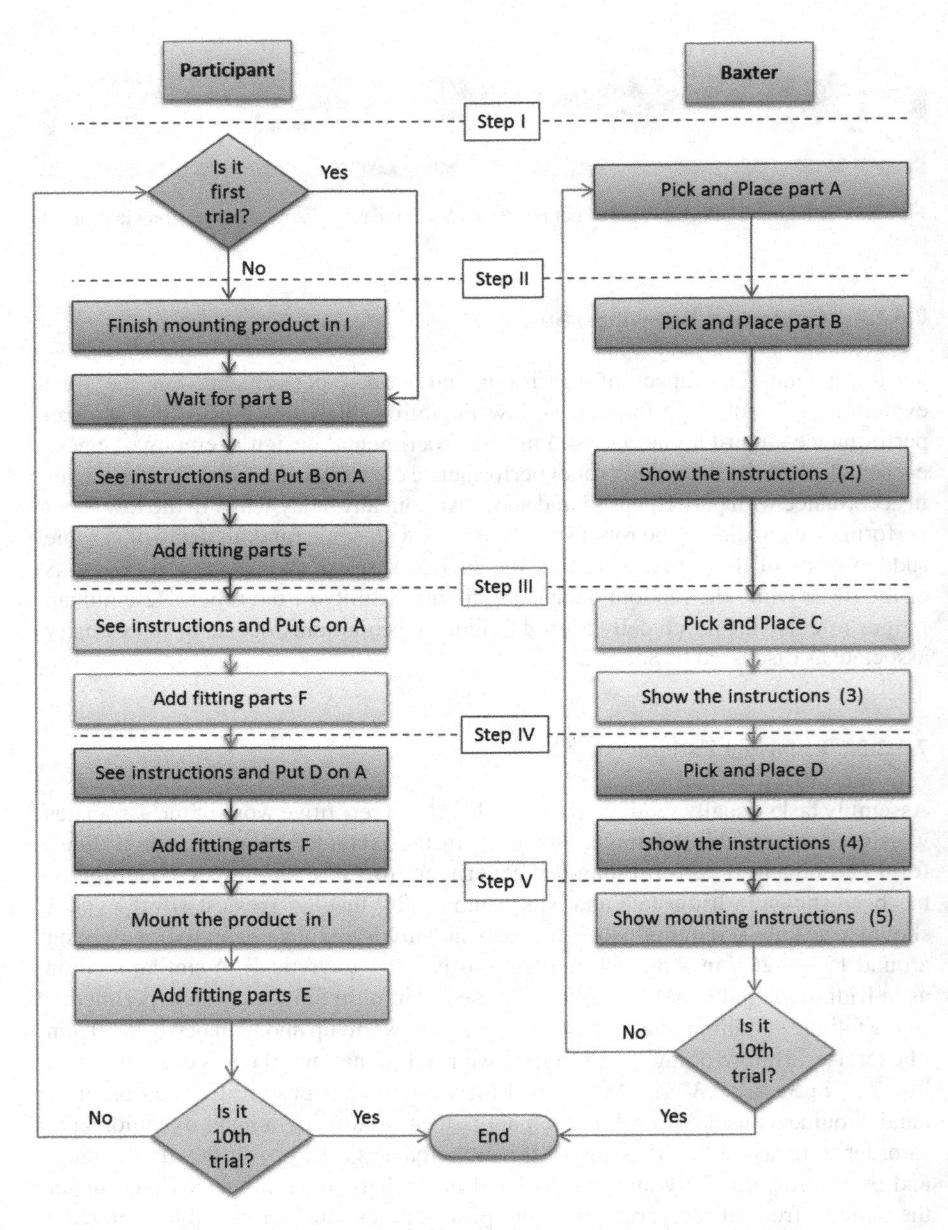

Fig. 7.6 Task flowchart of one cycle of the human-robot collaborative assembly task

similar to the high cycle condition in Iridiastadi and Nussbaum (2006) in our pilot study but the participants complained that it was very hard and we reduced the cycle time to 120 s in the final study. Note that the abduction cycle is different from the experiment cycle discussed in Sect. 7.6.2.1. The maximum isometric force of the participant's shoulder is also measured after every 10 trials.

7.6.2.4 Experiment Procedure

A participant is asked to read a written instruction on how to complete the assembly task. Verbal instructions are also given and the participant is instructed that no data will be collected during the training session. The training session consisted of 10 trials of an assembly task different from the actual experiment task. During the training session, the participant is able to change the speed of the robot using the up or down arrow keys of the keyboard at anytime. In the experiment, the robot speed can be adjusted manually as well as autonomously. The adjustment of the robot speed in the manual mode during the experiment task is similar to the training session. In the autonomous control mode, the robot adjusts its speed and the participant cannot change it. In the collaborative control mode, the robot adjusts its speed autonomously while the participant is also able to change the robot speed whenever he/she wants.

The experiments were conducted in 3 days. In the first day, after the training, the participant performed the experiments in manual mode. The non-fatigue high robot flexible and non-fatigue low flexible conditions are the first and second experiments, respectively. Next, in order to run the experiments in the fatigue condition, the participant was asked to do the fatigue exercise as described in Sect. 7.6.2.3. The participant is then asked to perform the experiments under the fatigue high flexible and fatigue low flexible conditions as the third and fourth experiment, respectively. The data obtained in the manual mode is used to train the neural network based on the explanations in Sect. 7.3. We train the artificial neural network for all of the conditions in manual mode. The trained networks are used for the corresponding condition in the autonomous and collaborative modes. The experiments conducted in the second and third days are for the autonomous and collaborative modes, respectively.

7.6.2.5 Measurements and Scales

At the start of the first day of experiment, the participant was asked to fill out a subjective demographic questionnaire. Moreover, at the beginning of each day, the participant was asked to rate his/her trust to Baxter. A 7-point Likert scale is used for measuring real-time subjective trust of the participant in the robot. The participant is instructed that extreme values of the trust scale—'1' and '7'—mean that they do not trust robot at all or they trust the robot completely. The real-time trust value is measured during the experiment using a separate laptop screen other than Baxter head screen. A message on Baxter head screen pops out and asks the participant to evaluate his/her trust at the end of each trial. Moreover, the participant is informed that he can increase or decrease the trust value anytime during the experiment using the right or left arrow keys of the keyboard on the laptop. Once a participant finishes all 10 trials, we ask him to fill out a survey. The survey measures the overall workload based on the NASA TLX (Hart and Staveland 1988) scale.

7.6.3 *Experimental Results*

7.6.3.1 Trust Model Identification Procedure

We use the Autoregressive Moving Average (ARMA) Model in the MATLAB System Identification Toolbox (Ljung 2007) to identify the parameters of time-series trust model based on the experiment data (i.e. A, B_1, B_2, C_1, C_2, D_1, and D_2 in Eq. (7.1)). The tracking system shown in Fig. 7.3b is used to measure working speed of the human associate, V_H for calculating the robot flexibility in Eq. (7.2). Robot speed is the command that is sent to the robot by the computer. The real-time trust measurements are collected during the experiment.

7.6.3.2 Manual Mode

The results of the experiments are shown in Fig. 7.7. Note that we have normalized the trust level for the sake of comparison but the 7-point Likert scale can be used for analysis without difficulty. As can be seen in this figure, for the first (non-fatigue high robot flexibility) and second (non-fatigue low robot flexibility) sets of experiments, the human is not fatigued so his performance is maximum, i.e. $P_H = 1$. However, after imposing fatigue during the third (fatigue high robot flexibility) and fourth (fatigue low robot flexibility) sets of experiments, his performance decreases. In the first experiment when there is no fault, the participant's trust increases but it drops after occurrence of faults in the second experiment. In the absence of the faults within the third experiment the trust recovers. Note that the level of trust increases with higher rate as compared to the first experiment with the same robot flexibility condition. In the fourth experiment with low-flexible robot performance, the trust decreases but it decreases with lower pace as compared to the case with higher human performance (non-fatigue condition). The quantified trust model for this mode is

$$\begin{aligned} T(k) = \quad & 0.991T(k-1) + 0.014P_R(k) + 0.127P_R(k-1) + 0.046P_H(k) \\ & - 0.143P_H(k-1) - 0.075F(k) + 0.003F(k-1), \end{aligned} \tag{7.11}$$

For this mode, the fit value for the ARMA model is 70.61 % which shows that the model fits the data well. Equation (7.11) indicates that with low values of P_R or high values of P_H trust declines and vice versa. We also observe that since $A = 0.991$, almost seven times the weight of the second largest parameter, the current trust is

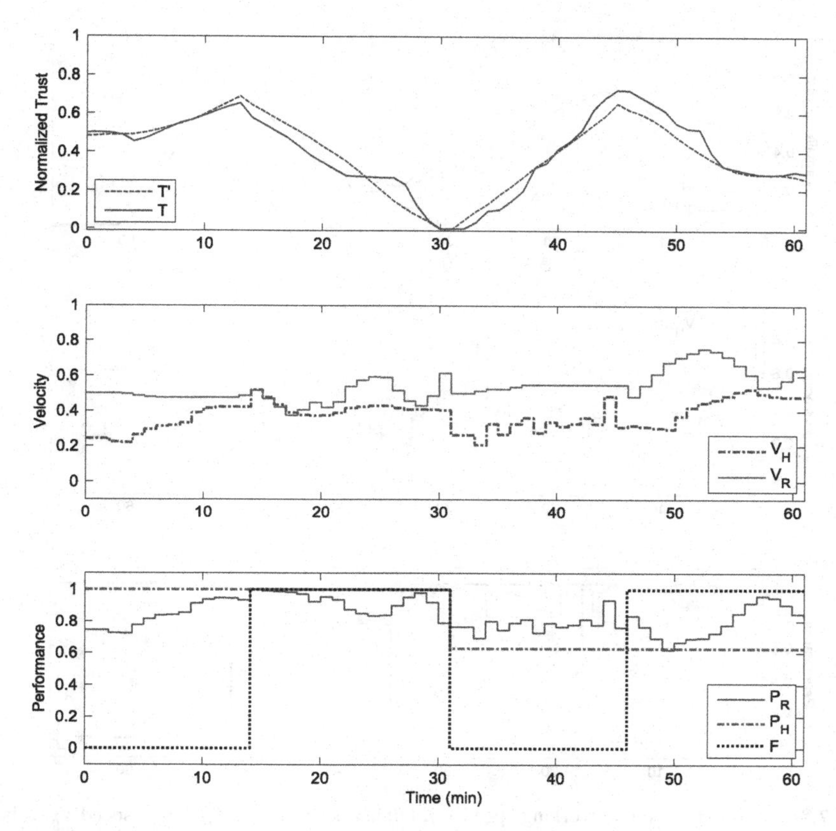

Fig. 7.7 Evolution of human working speed V_H, human performance P_H, robot speed V_R, robot performance P_R, fault, trust T, and trust estimation T' using Eq. (7.11) under the manual mode

mainly dependent on the previous trust if no dramatic performance change occurs. This is consistent with the intuition that trust is highly related with prior trust and only changes when there is a large performance variation.

7.6.3.3 Autonomous Mode

The results of the experiments are shown in Fig. 7.8. As can be seen in this figure, the human and robot performance as well as the changes in the trust value are similar to that of in manual mode. For this mode, the fit value for the ARMA model is 62.34 %. The time-series trust model for this mode is

$$\begin{aligned} T(k) = \quad & 0.959T(k-1) + 0.021P_R(k) + 0.015P_R(k-1) + 0.078P_H(k) \\ & - 0.064P_H(k-1) - 0.045F(k) - 0.013F(k-1), \end{aligned} \tag{7.12}$$

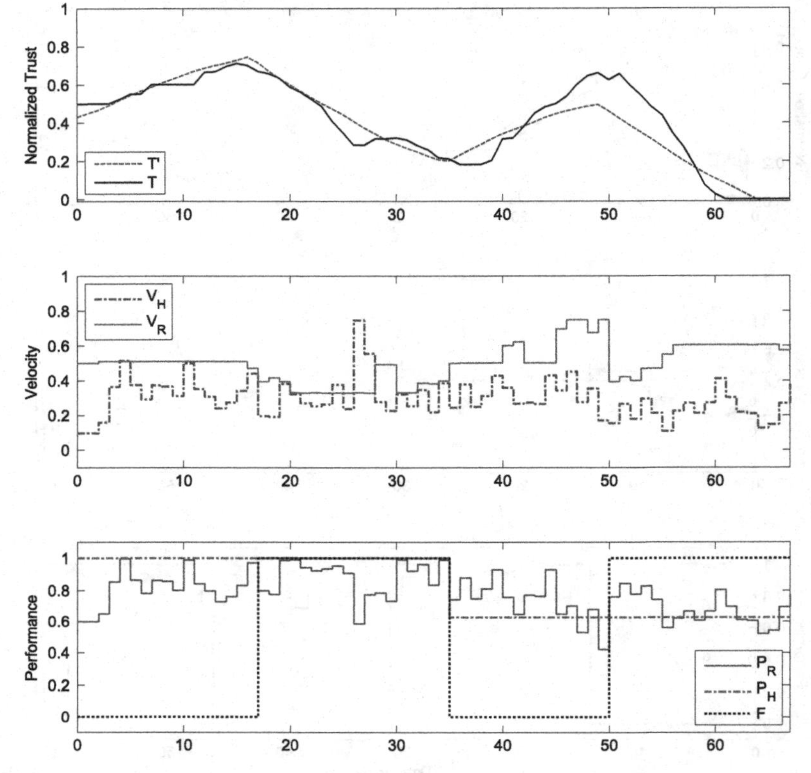

Fig. 7.8 Evolution of human working speed V_H, human performance P_H, robot speed V_R, robot performance P_R, fault, trust T, and trust estimation T' using Eq. (7.12) under autonomous mode

7.6.3.4 Collaborative Mode

The results of the experiments are shown in Fig. 7.9. The fit value for the ARMA model is 45.65 %. As it can be seen in Fig. 7.9, the trust value increases slowly at the start of the experiment from 0.5 to around 0.75. Fault occurrences cause a rapid trust degradation to the level of less than 0.1. Next, the participant's trust to the robot increases sharply after eliminating the faults and it decreases again after the faults occur toward the end of the experiment. Note that for the first and second half phase of the experiment, although the increasing trend of trust without faults and the decreasing trend of trust with faults are consistent, the intensity of these variations within these two phases is very different. In the former phase, trust increases very slowly but drops very fast; While in the latter phase, trust recovers very sharply and declines gradually. This can justify why the fit value is smaller in the collaborative mode compared to the other modes. Future work will seek models with better fitness based on validated human factor research. The time-series trust model for this mode is

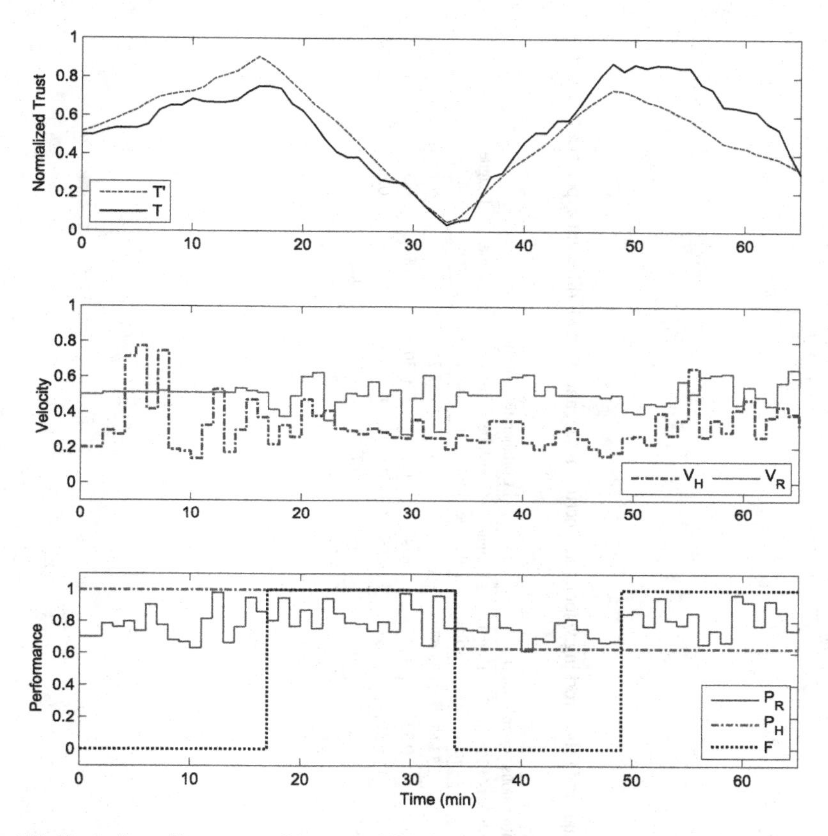

Fig. 7.9 Evolution of human working speed V_H, human performance P_H, robot speed V_R, robot performance P_R, fault, trust T, and trust estimation T' using Eq. (7.13) under collaborative mode

$$\begin{aligned} T(k) = \quad & 0.991T(k-1) + 0.099P_R(k) + 0.033P_R(k-1) - 0.039P_H(k) \\ & - 0.033P_H(k-1) - 0.062F(k) - 0.022F(k-1), \end{aligned} \tag{7.13}$$

7.6.4 Comparison and Conclusion

We measure the participant workload with NASA TLX index after each experiment. Moreover, we calculate the average values of robot speed, human speed, robot performance, human performance and trust in all of these conditions. Table 7.1 shows the comparison of these values for different experiment conditions. As can be seen in this table, for the fresh (non-fatigue) flexible condition, the overall workload of the participant is similar in all of the three control modes and it is lower as

Table 7.1 Comparison between workload, average human and robot working velocity and performance, and trust for different experiment conditions and modes

Mode	Manual (mean values)				Autonomous (mean values)				Collaborative (mean values)			
Condition	Fresh flexible	Fresh inflexible	Fatigue flexible	Fatigue inflexible	Fresh flexible	Fresh inflexible	Fatigue flexible	Fatigue inflexible	Fresh flexible	Fresh inflexible	Fatigue flexible	Fatigue inflexible
Workload	47	67.7	46	63.3	47	53.67	38.67	59	48.3	57.7	52.7	59.7
V_H	0.32	0.42	0.32	0.43	0.32	0.32	0.32	0.24	0.37	0.30	0.26	0.36
V_R	0.48	0.49	0.54	0.64	0.51	0.37	0.59	0.55	0.51	0.47	0.53	0.52
P_R	0.84	0.92	0.78	0.80	0.81	0.88	0.73	0.69	0.77	0.82	0.73	0.82
Trust	4.24	2.88	3.01	3.64	4.59	3.41	3.48	2.32	4.75	3.19	3.80	5.08
Δ Trust	0.9	−3.5	4.4	−2.6	1.2	−2.7	3.1	−4.0	1.5	−4.3	4.9	−3.4

compared to the fresh inflexible condition for every control mode. Moreover, this value is lower for the fatigue flexible condition as compared to fatigue inflexible in all modes. In this table, for each experiment with a certain condition under a specific mode, ΔTrust shows the difference between the initial and final trust. The general trend of changes of this value for all the control modes are similar: it goes up in the flexible mode and goes down in the inflexible mode. However, it can be seen that the influences of robot and human performances on trust vary for different control modes. For the fresh flexible and inflexible conditions, although the robot performances in the manual mode are higher than those in the autonomous and collaborative modes, the trust increments are lower as compared to the corespondent values in other modes.

7.7 Conclusion

In this paper, we proposed a time-series trust model for a human associate and his/her robot coworker in a collaborative manufacturing task. We developed a performance model for robot flexibility based on the difference between the human and robot working speed. Since the tasks in manufacturing usually are repetitive kinesthetic tasks, we used the muscle fatigue and recovery model to capture the human performance. We used three methods to control the robot performance. These methods are manually by the human, autonomously by a neural network based robust intelligence controller, or collaboratively using both manual and autonomous inputs. We provided both numerical simulations and experiment validations to demonstrate the effectiveness of the proposed trust model and robust intelligent control scheme. Based on the well-known human factors result we adopted a linear trust model in this paper. Future works will investigate the applicability of the linear model in general HRC manufacturing and modifying accordingly for specific scenarios to increase the model fitness.

Acknowledgements This research is supported in part by the National Science Foundation under grant No. CMMI-1454139. The authors would also like to thank the BMW US Manufacturing Company for loaning Baxter to the Interdisciplinary and Intelligence Research (I^2R) laboratory in the Department of Mechanical Engineering (ME) at Clemson University.

References

Bischoff R, Kurth J, Schreiber G, Koeppe R, Albu-Schäffer A, Beyer A, Eiberger O, Haddadin S, Stemmer A, Grunwald G, et al (2010) The KUKA-DLR lightweight robot arm-a new reference platform for robotics research and manufacturing. In: Robotics (ISR), 2010 41st international symposium on and 2010 6th German conference on robotics (ROBOTIK). VDE, Munich, Germany, pp 1–8

Charalambous G (2013) Human-automation collaboration in manufacturing: identifying key implementation factors. In: Contemporary ergonomics and human factors 2013: proceedings of the international conference on ergonomics and human factors 2013, Cambridge, p 59

Cichocki A, Unbehauen R (1996) Robust neural networks with on-line learning for blind identification and blind separation of sources. IEEE Trans Circuits Syst I Fundam Theory Appl 43(11):894–906

Fayazi S, Wan N, Lucich S, Vahidi A, Mocko G (2013) Optimal pacing in a cycling time-trial considering cyclist's fatigue dynamics. In: American Control Conference (ACC), 2013, pp 6442–6447

Gao J, Lee JD (2006) Extending the decision field theory to model operators' reliance on automation in supervisory control situations. IEEE Trans Syst Man Cybern Syst Hum 36(5): 943–959

Goodrich MA, Schultz AC (2007) Human-robot interaction: a survey. Found Trends Human-Comput Interact 1(3):203–275

Guizzo E, Ackerman E (2012) How rethink robotics built its new Baxter robot worker. IEEE Spectrum. http://spectrum.ieee.org/robotics/industrial-robots/rethink-robotics-baxter-robot-factory-worker. Retrieved 21 Jan 2014

Hagan M, Demuth H (1999) Neural networks for control. In: American Control Conference, 1999. Proceedings of the 1999, vol 3, pp 1642–1656. doi: 10.1109/ACC.1999.786109

Hancock PA, Billings DR, Schaefer KE, Chen JY, De Visser EJ, Parasuraman R (2011) A meta-analysis of factors affecting trust in human-robot interaction. Hum Factors J Hum Factors Ergon Soc 53(5):517–527

Hart SG, Staveland LE (1988) Development of NASA-TLX (task load index): results of empirical and theoretical research. Adv Psychol 52:139–183

Hoff KA, Bashir M (2014) Trust in automation: integrating empirical evidence on factors that influence trust. Hum Factors J Hum Factors Ergon Soc 57(3):407–434. doi:10.1177/0018720814547570

Iridiastadi H, Nussbaum MA (2006) Muscle fatigue and endurance during repetitive intermittent static efforts: development of prediction models. Ergonomics 49(4):344–360

Itoh M, Tanaka K (2000) Mathematical modeling of trust in automation: trust, distrust, and mistrust. In: Proceedings of the Human Factors and Ergonomics Society Annual Meeting, vol 44. Sage, New York, pp 9–12

Krüger J, Lien T, Verl A (2009) Cooperation of human and machines in assembly lines. CIRP Ann Manuf Technol 58(2):628–646

Krüger J, Katschinski V, Surdilovic D, Schreck G (2010) Flexible assembly systems through workplace-sharing and time-sharing human machine cooperative. In: Robotics (ISR), 2010 41st international symposium on and 2010 6th German conference on robotics (ISR/ROBOTIK). VDE, Munich, pp 1–5

Lee J, Moray N (1992) Trust, control strategies and allocation of function in human-machine systems. Ergonomics 35(10):1243–1270

Lee JD, See KA (2004) Trust in automation: designing for appropriate reliance. Hum Factors J Hum Factors Ergon Soc 46(1):50–80

Lewandowsky S, Mundy M, Tan G (2000) The dynamics of trust: comparing humans to automation. J Exp Psychol Appl 6(2):104

Liu JZ, Brown RW, Yue GH (2002) A dynamical model of muscle activation, fatigue, and recovery. Biophys J 82(5):2344–2359. ISSN 0006-3495

Ljung L (2007) System identification toolbox for use with {MATLAB}. The MathWorks, Inc., Natick

Ma L, Chablat D, Bennis F, Zhang W (2009) A new simple dynamic muscle fatigue model and its validation. Int J Ind Ergon 39(1):211–220

Ma L, Chablat D, Bennis F, Zhang W, Guillaume F (2010) A new muscle fatigue and recovery model and its ergonomics application in human simulation. Virtual Phys Prototyping 5(3): 123–137

Mehrotra K, Mohan CK, Ranka S (1997) Elements of artificial neural networks. MIT, Cambridge

Merritt SM, Ilgen DR (2008) Not all trust is created equal: dispositional and history-based trust in human-automation interactions. Hum Factors J Hum Factors Ergon Soc 50(2):194–210

Moray N, Inagaki T, Itoh M (2000) Adaptive automation, trust, and self-confidence in fault management of time-critical tasks. J Exp Psychol Appl 6:44

Ostergaard EH (2012) Lightweight robot for everybody [industrial activities]. IEEE Robot Autom Mag 19(4):17–18

Robotics R (2013) http://www.rethinkrobotics.com/

Sadrfaridpour B, Burke J, Wang Y (2014a) Human and robot collaborative assembly manufacturing: trust dynamics and control. In: RSS 2014 Workshop on Human-Robot Collaboration for Industrial Manufacturing

Sadrfaridpour B, Saeidi H, Wang Y, Burke J (2014b) Modeling and control of trust in human and robot collaborative manufacturing. In: 2014 AAAI Spring Symposium Series

Shi J, Jimmerson G, Pearson T, Menassa R (2012) Levels of human and robot collaboration for automotive manufacturing. In: Proceedings of the Workshop on Performance Metrics for Intelligent Systems, PerMIS '12. ACM, New York, pp 95–100. ISBN 978-1-4503-1126-7

Stopp A, Horstmann S, Kristensen S, Lohnert F (2001) Towards interactive learning for manufacturing assistants. In: Proceedings of 10th IEEE International Workshop on Robot and Human Interactive Communication, 2001. IEEE, New York, pp 338–342

Tan JTC, Duan F, Zhang Y, Watanabe K, Kato R, Arai T (2009) Human-robot collaboration in cellular manufacturing: design and development. In: IEEE/RSJ International Conference on Intelligent Robots and Systems, 2009 (IROS 2009). IEEE, New York, pp 29–34

White H (1992) Artificial neural networks: approximation and learning theory. Blackwell, Cambridge

Widrow B, Rumelhart DE, Lehr MA (1994) Neural networks: applications in industry, business and science. Commun ACM 37(3):93–105. ISSN 0001-0782

Chapter 8
Investigating Human-Robot Trust in Emergency Scenarios: Methodological Lessons Learned

Paul Robinette, Alan R. Wagner, and Ayanna M. Howard

8.1 Introduction

Today, robots are being actively deployed in scenarios that help humans achieve tasks ranging from cleaning floors to bomb disposal; however such tasks either present low risk to humans (e.g., cleaning a floor) or are tightly controlled by human experts (e.g., bomb disposal). To increase the potential for autonomous robots to aid humans in additional high-risk tasks, robots must recognize the factors that affect human trust in their abilities. Understanding trust decisions can be difficult for humans, so care must be taken to properly imbue this ability to robots and other intelligent systems.

In our prior work, we have explored using robots to aid humans in emergency evacuations (Robinette and Howard 2011; Robinette et al. 2012) as well as the trust decisions a human would have to make in regard to these robots (Robinette and Howard 2012; Robinette et al. 2013, 2014a, b, c). This application provides a high-risk, time critical situation with real-world implications for human-robot trust.

Unfortunately, few research protocols exist for investigating human-robot trust. The methods that do exist have largely focused on very narrow aspects of the trust phenomenon and/or situations (Sabater and Sierra 2005). Further, by definition, the presence of trust implies risk on the part of the person or the robot. Placing study participants at risk is challenging from an ethical point of view and presents logistic problems. For example, an experiment may ask participants to move around a building while a fire is simulated using artificial smoke, visible flame, and fire

P. Robinette (✉) • A.M. Howard
Georgia Institute of Technology, North Ave NW, GA 30332, Atlanta
e-mail: probinette3@gatech.edu; ayanna.howard@ece.gatech.edu

A.R. Wagner
Georgia Tech Research Institute, 430 10th St NW, GA 30332, Atlanta
e-mail: Alan.Wagner@gtri.gatech.edu

© Springer Science+Business Media (outside the USA) 2016

R. Mittu et al. (eds.), *Robust Intelligence and Trust in Autonomous Systems*,
DOI 10.1007/978-1-4899-7668-0_8

alarms. One participant may view this experience as completely artificial and thus feel no risk, while another participant may panic and injure himself or herself in this exact same situation. Moreover, measuring trust is inherently subjective and strongly influenced by factors outside of the experimenters' control. These factors make the investigation of human-robot trust extremely difficult.

Our approach for handling these challenges has been refined over numerous different experiments involving 770 participants collectively. This article presents the lessons that we have learned over the course of conducting these studies with the aim of informing future human-robot trust research. A brief listing of our major experimental milestones can be found in Table 8.1. We began with experiments that used written narratives to explore trust situations (described in Sect. 8.4) and then expanded into experiments that asked participants to make trust decisions in one and two round simulations with guidance robots (described in Sect. 8.5). Throughout these experiments, we tested a variety of metrics, motivations, and behaviors.

This chapter is organized as follows: the next section discusses our conceptualization of trust, followed by related work in the fields of human-robot trust, human-robot interaction and human-subject experimentation. Next, we present lessons that we have learned from studies that require people to read narratives involving trust. The section that follows discusses simulation experiments involving a guidance robot. We conclude with specific recommendations for human-robot interaction researchers conducting experiments related to trust.

8.2 Conceptualizing Trust

Numerous researchers have proposed conceptions of trust that range from computational implementations of cognitive processes (Castelfranch and Falcone 2010), to neurological changes in reciprocity games (King-Casas et al. 2005), to a probability distribution over an agent's actions (Gambetta 1990). Other researchers consider trust to have multiple forms, depending on the actors and environment (Hoffman et al. 2013). After a review of the available literature, Lee and See conclude that trust is *the attitude that an agent will help achieve an individual's goals in a situation characterized by uncertainty and vulnerability* (Lee and See 2004). Building from Lee and See's definition of trust Wagner states that trust is "a belief, held by the trustor, that the trustee will act in a manner that mitigates the trustor's risk in a situation in which the trustor has put its outcomes at risk" (Wagner 2009a). This definition is meant to serve as an operationalized version of Lee and See's definition for trust.

Outcome matrices are a useful tool for formally conceptualizing social interaction. These matrices (or normal-form games in the game theory community) explicitly represent the individuals interacting as well as the actions they are deliberating over. The impact of each pair of actions chosen by the individuals is represented as a scalar number or outcome. For interactions involving trust, it is common to label one individual as a trustor and the other as a trustee. This

Table 8.1 A list of our major experimental milestones related to our study of human-robot trust

Type of experiment	Name	Number of participants	Measurement	Motivation for participants
Narrative	Pilot experiment 1	20	Agreement with definition	N/A
Narrative	Pilot experiment 2	32	Agreement with definition	N/A
Narrative	Full experiment	128	Agreement with definition	N/A
Single round robot guidance	Trust matrix experiment 1 (small mazes)	30	Decision to use robot and self-reported trust	Monetary bonus
Single round robot guidance	Trust matrix experiment 2 (large mazes)	57	Decision to use robot and self-reported trust	Monetary bonus
Single round robot guidance	Series of pilot experiments (varied questions and outcome matrices)	64	Decision to use robot and self-reported trust	Monetary bonus
Single round robot guidance	Equal outcome matrix experiment	59	Decision to use robot and self-reported trust	Monetary bonus
Single round robot guidance	Trust and equal outcome matrix experiment	120	Decision to use robot and self-reported trust	Time constraint in emergency scenario
Double round robot guidance	Unsuccessful robot behavior pilot experiment	25	Decision to use robot and self-reported trust	Monetary bonus
Double round robot guidance	Experiment 1	106	Decision to use robot and self-reported trust	Monetary bonus
Double round robot guidance	Experiment 2	129	Decision to use robot and self-reported trust	Time constraint in emergency scenario

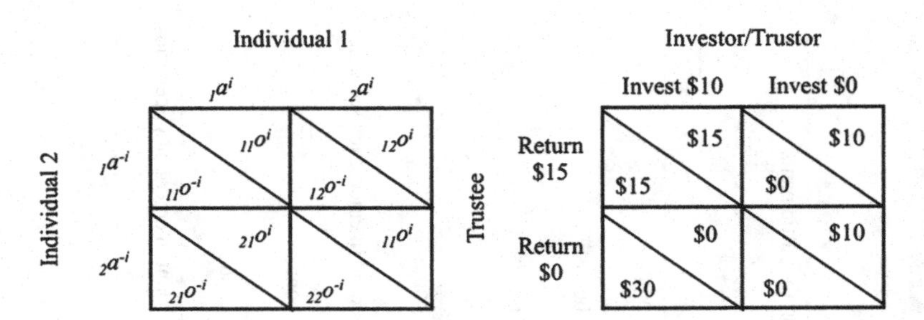

Fig. 8.1 An example outcome matrix is depicted formally and as an investment game. The risk associated with the trustee's action can be approximated by subtracting the values on the right in the invest $10 columns. See Fig. 8.2 for a more detailed explanation of the outcome matrices as used in this work

representation affords a natural means for quantifying the risk associated with an action choice as the difference in outcome across rows or columns. An example is presented in Fig. 8.1.

Given a definition for trust, the outcome matrix representation can be used to conceptualize this definition in terms of social interaction. For example, one can create scenarios that are generally agreed to contain trust and attempt to formally represent these scenarios as matrices. One can then search for similarities across several trust scenarios in an attempt to deduce a definition for trust. Alternatively, one can, as we have, formulate a particular definition of trust as an outcome matrix. Much of our subsequent research has focused on testing whether or not our definition accurately matches people's intuitions of trust.

As we have demonstrated in related work, outcome matrices can be created from the models a robot has learned about the people it interacts with (Wagner 2009b). These models inform the outcome matrix representation by suggesting which actions are available to a person in a particular environmental context as well as predicting which action an individual will select. With respect to trust, these partner models act as reputation models of the trustee allowing the trustor to gauge the risk associated with relying on this particular individual. Stereotypes can be used to bootstrap the process of assessing the trustworthiness of newly encountered individual by assigning them and comparing them to a category of partner model types (Wagner 2012).

8.2.1 Conditions for Situational Trust

The outcome matrix representation can be used to formally represent the types of situations that have long been used in trust research (Kelley and Thibaut 1978; Axelrod 1984). The investment game, for example, presents an investor with some amount of money. The investor must decide whether to invest the money with a

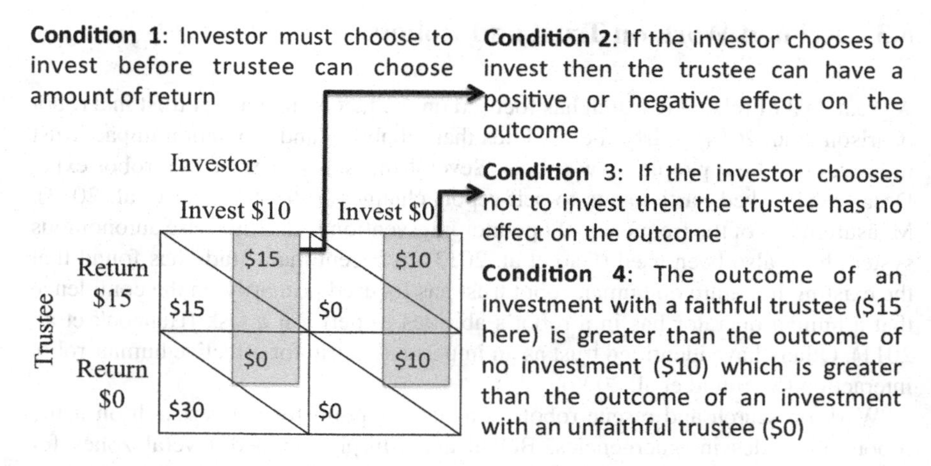
Condition 1: Investor must choose to invest before trustee can choose amount of return
Condition 2: If the investor chooses to invest then the trustee can have a positive or negative effect on the outcome
Condition 3: If the investor chooses not to invest then the trustee has no effect on the outcome
Condition 4: The outcome of an investment with a faithful trustee ($15 here) is greater than the outcome of no investment ($10) which is greater than the outcome of an investment with an unfaithful trustee ($0)
Investor
Invest $10
Invest $0
Trustee
Return $15
Return $0
$15
$15
$0
$10
$0
$10
$30
$0
$0

Fig. 8.2 The conditions for trust derived from our operational definition for trust are shown above with examples from the Investor-Trustee game

trustee or not. If the investor chooses to invest, the money invested appreciates to some larger amount. The trustee must then decide what amount to return to the investor. Figure 8.1 depicts an example game. Investment games such as this have become the de facto method for investigating trust by the trust research community (King-Casas et al. 2005).

In the matrix presented in Fig. 8.1, for example, the investor has a choice: she can choose to invest or not to invest. Likewise, the trustee can choose to return some amount of money or not to return any money. Although the matrix is a specific example of a situation involving trust, we can easily abstract away the actions that the actors are deliberating over to develop a series of conditions for trust. In prior work we show that these conditions (see Fig. 8.2), which are derived from our working definition for trust, can be used to segregate outcome matrices into those that require trust on the part of a trustor and those that do not.

The use of these conditions results in a binary indication of whether or not a situation demands trust. This is not to say that trust is a binary phenomenon. Rather, the conditions simply indicate that the selection of a particular action by the trustor demands trust. Further, many real and important situations present people with a decision that has little middle ground. For instance, being a passenger in an autonomous car is a binary decision in the sense that one chooses either to be a passenger or not be a passenger. The situation affords little opportunity for a third, middle risk option.

8.3 Related Work on Trust and Robots

A great deal of related research has focused on the factors that affect trust in a robot (Carlson et al. 2014). Carlson et al. finds that reliability and reputation impact trust in surveys of how people view robots. Several measures of trust in a robot exist. Desai et al. asked participants to self-report changes in trust (Desai et al. 2013). Measurements of the frequency of operator intervention in an otherwise autonomous system have also been used (Gao et al. 2013). A recent meta-study has found that the existing literature on human-robot trust has focused primarily on the confidence that a human operator has in a robot's abilities to perform a task (Hancock et al. 2011). Others have identified trust as an important metric for effective human-robot interaction (Steinfeld et al. 2006).

Work on search and rescue robots is a starting point for our research on using robots as guides in emergencies. Bethel and Murphy defined several zones for robot interaction: the intimate zone (0–0.46 m), the personal zone (0.46–1.22 m, maximum distance for communication), the social zone (1.22–3.66 m) and the public zone (further than 3.66 m) (Bethel and Murphy 2008; Murphy 2004). More recent work has extended this to UAVs (Duncan and Murphy 2013).

8.4 Crowdsourced Narratives in Trust Research

Crowdsourcing has become a popular method to increase the number and diversity of participants in human-computer interaction and even human-robot interaction experiments (Kittur, Chi, and Suh 2008). We chose to crowdsource our experiment in order to broaden the pool of people from which our data was generated. The greater the variety in our participant pool, the greater the generality of our results. Crowdsourcing uses the combined resources of a large group of people connected over the internet, to accomplish a goal or perform a task. Studies have examined the use of crowdsourcing as a means for garnering experimental subjects and found that the validity of experiments utilizing crowdsourcing is not otherwise compromised (Gosling et al. 2004). With respect to robotics, simulated robots and environments designed within game engines allow subjects to interact with robots in a multitude of ways. These interactions are mediated by the participant's web browser.

Services such as Amazon's Mechanical Turk can be used to recruit and financially compensate large numbers of participants. Studies have found that Mechanical Turk provides a more diverse participant base than traditional human studies performed with university students (Paolacci et al. 2010; Buhrmester et al. 2011; Berinsky et al. 2012; Horton and Chilton 2010). These studies found that the Mechanical Turk user base is generally younger in age but otherwise demographically similar to the general population of the United States (at the time of those studies, Mechanical Turk was only available in the USA).

In order to guarantee quality work, only workers with overall acceptance rates 95 % and above were used. To ensure diversity no participant was allowed to enroll in the study more than once. Workers that attempted to enroll in the study more that once were warned that their data would be rejected and pay refused. We also rejected responses that included incomplete answers and comments. This research was approved by the Georgia Institute of Technology Internal Review Board.

Our initial research goal was to evaluate our definition for trust and the conditions derived from this definition. To accomplish this goal we needed a clear and understandable way to present different matrices to participants. We decided to use textual narratives (i.e., stories) as a way to present the matrices in a manner that most people could understand. We felt that narratives allowed a great deal of flexibility for creating situations that closely matched the original matrix. Moreover, the use of narratives only required basic reading skills in order to participate in the study. Finally, because outcome matrices are often described as short stories (e.g., prisoner's dilemma, stag hunt game) the use of narratives was a natural fit (Axelrod 1984).

In order to empirically evaluate our conditions for trust, we needed to create narratives that matched outcome matrices that met and did not meet the conditions. We were able to further divide the matrices that violated the definition of trust into sub-categories based on the way the definition was violated. For instance, a matrix that contains equal outcome values did not put the trustor at risk and hence violates our definition for situational trust. Table 8.2 depicts the different matrix types. The first matrix in Table 8.2 represents a situation that requires trust and meets our conditions for trust. The other four matrices violate at least one condition of trust. The **Equal Outcomes** matrix violated all conditions by providing a situation where the trustor risked nothing in the interaction. The **Trustor-Dependent Trustee-Independent** matrix presented a situation where only the trustor's actions affected the outcome, thus the trustor was not placing any risk in the hands of the trustee. This violates the second and fourth conditions. Likewise, the **Trustor-Independent, Trustee-Dependent** matrix represents a situation where the trustor has no control whatsoever. If the trustor is not able to make a decision then the situation does not meet our definition of trust. This matrix violates conditions three and four. Finally, the **Inverted Trust** matrix presents a scenario where the trustor receives the worst reward when the trustee intends to fulfill trust and the best reward when the trustee intends to break trust. Thus the trustor would wish that the trustee would act in a manner that breaks trust, rather than maintains it. This matrix violates the fourth condition.

Each participant was asked to read and evaluate twelve scenarios. Care was taken to present each participant with as many different versions of the experimental variables being tested as possible. Participants were paid $1.67 for the completion of their survey.

Table 8.2 The categories and descriptions of trust and no trust situations tested along with an example outcome matrix for each

<table>
<tr><th>Category</th><th>Example</th><th>Description</th></tr>
<tr><td>Trust Matrix</td><td><table><tr><td></td><td></td><td colspan="2">Trustor</td></tr><tr><td></td><td></td><td>a^i</td><td>a^i</td></tr><tr><td rowspan="2">Trustee</td><td>a^{-i}</td><td>$2000</td><td>$400</td></tr><tr><td>a^{-i}</td><td>$0</td><td>$400</td></tr></table></td><td>Fulfills trust according to the definition and its conditions.</td></tr>
<tr><td>Equal Outcomes</td><td><table><tr><td></td><td></td><td colspan="2">Trustor</td></tr><tr><td></td><td></td><td>a^i</td><td>a^i</td></tr><tr><td rowspan="2">Trustee</td><td>a^{-i}</td><td>$2000</td><td>$2000</td></tr><tr><td>a^{-i}</td><td>$2000</td><td>$2000</td></tr></table></td><td>Violates all conditions of trust by removing all risk to the trustor.</td></tr>
<tr><td>Trustor-Dependent, Trustee-Independent</td><td><table><tr><td></td><td></td><td colspan="2">Trustor</td></tr><tr><td></td><td></td><td>a^i</td><td>a^i</td></tr><tr><td rowspan="2">Trustee</td><td>a^{-i}</td><td>$2000</td><td>$0</td></tr><tr><td>a^{-i}</td><td>$2000</td><td>$0</td></tr></table></td><td>Only allows the trustor to affect the situation. The trustor does not risk any outcomes on the actions of the trustee.</td></tr>
<tr><td>Trustor-Independent, Trustee-Dependent</td><td><table><tr><td></td><td></td><td colspan="2">Trustor</td></tr><tr><td></td><td></td><td>a^i</td><td>a^i</td></tr><tr><td rowspan="2">Trustee</td><td>a^{-i}</td><td>$2000</td><td>$2000</td></tr><tr><td>a^{-i}</td><td>$0</td><td>$0</td></tr></table></td><td>Only allows the trustee to affect the outcomes of the trustor. The trustor has no choice in thescenario.</td></tr>
<tr><td>Inverted Trust Matrix</td><td><table><tr><td></td><td></td><td colspan="2">Trustor</td></tr><tr><td></td><td></td><td>a^i</td><td>a^i</td></tr><tr><td rowspan="2">Trustee</td><td>a^{-i}</td><td>$0</td><td>$400</td></tr><tr><td>a^{-i}</td><td>$2000</td><td>$400</td></tr></table></td><td>Presents a situation where the trustor wishes for the trustee to break trust in order to get the best outcome.</td></tr>
</table>

8.4.1 *Iterative Development of Narrative Phrasing*

The narratives that we created were based on several different scenarios that we felt offered some flexibility in terms of storytelling. One was an investment scenario meant to verbalize the investment game depicted in Fig. 8.1. A second scenario described a navigation task based on our interest in emergency evacuation. The final scenario was a hiring decision. The narratives were written to be as simple as possible while still allowing the flexibility to test each of our outcome matrices. The names Alice and Bob were consistently used to represent the characters in the scenario. The narratives began with a sentence or two introducing the scenario. Next, each of the four potential actions and outcomes are described. The narrative ends with a statement describing the decision and resulting action that was taken by Alice or Bob and a question asking the subject whether or not they believed that the chosen action indicated trust. In order to rule out potential confounding factors, half of the narratives displayed a positively stated action and the other half displayed a negative action ("Bob chooses to hire Alice" versus "Bob chooses NOT to hire Alice"), the ordering of the narratives, and the outcome amounts were all randomized. Participants were asked to explain each individual answer.

Best practices were used when developing the narrative surveys (Gehlbach and Brinkworth 2011) including the creation of several pilot studies, examination of within-subject reliability, use of randomization to eliminate biases, and measurement and evaluation of potential confounding variables. Figures 8.3 and 8.4 depict the evolution of these narratives.

Not surprisingly, early pilot studies indicated that the wording of the narratives could influence participant decisions. This can be seen in Fig. 8.5, where 86 % of responses agreed with our definition when presented with a Trust Matrix, but only 49 % agreed when an Equal Outcomes matrix was presented. For example, initially subjects were asked if the selection of an action indicated that one individual did not trust the other individual (Fig. 8.3). Examining participants' explanations for their answers indicated that they generally understood the narrative and the actions taken by the trustor in the narratives, but some did not notice that a negative trust question had been asked. For some participants the negative phrasing led to confusion. We found that questions such as, "Does this decision indicate that Bob does NOT trust Alice?" could be interpreted in several ways. One interpretation is that trust is not involved or present during the situation. Another is that Bob distrusts Alice. Participants offered explanations such as "There was nothing for Alice to gain. So there was no need for her to trust. No distrust is indicated" and "It indicates neither trust nor distrust." After careful consideration, we eliminated the negatively stated trust questions believing that our working definition for trust and associated conditions could be adequately investigated with positive statements. This pilot study demonstrated that most individuals do not have clear delineations between notions such as "not trust", "distrust", "mistrust", and "trust is not required". Although our research is only interested in how people define "trust" rather than the various terms that indicate no trust, this may be a fruitful area of future research.

In an additional pilot study, some participants seemed to latch on to key words, such as "invest," "follow," and "hire." This can be seen in Fig. 8.5 where 93 % of participants agree with our definition when a Trust Matrix is presented, but only 79 % agree when an Equal Outcomes matrix is presented. Explanations by participants, such as, "In this case, even though the outcomes are the same regardless of Alice's decision, I would say that her choice to hire Bob is a sign of trust," "There is no situation where she 'loses' any money from either investing/not investing, she must believe that he can do good with the money," and "Since Bob decides to follow Alice's directions, this indicates that he trusts Alice. Though he will arrive at the destination regardless if he trusts her or not. If he knows this, it potentially makes it easier to trust Alice," clearly indicate anchoring bias (Tversky and Kahneman 1974).

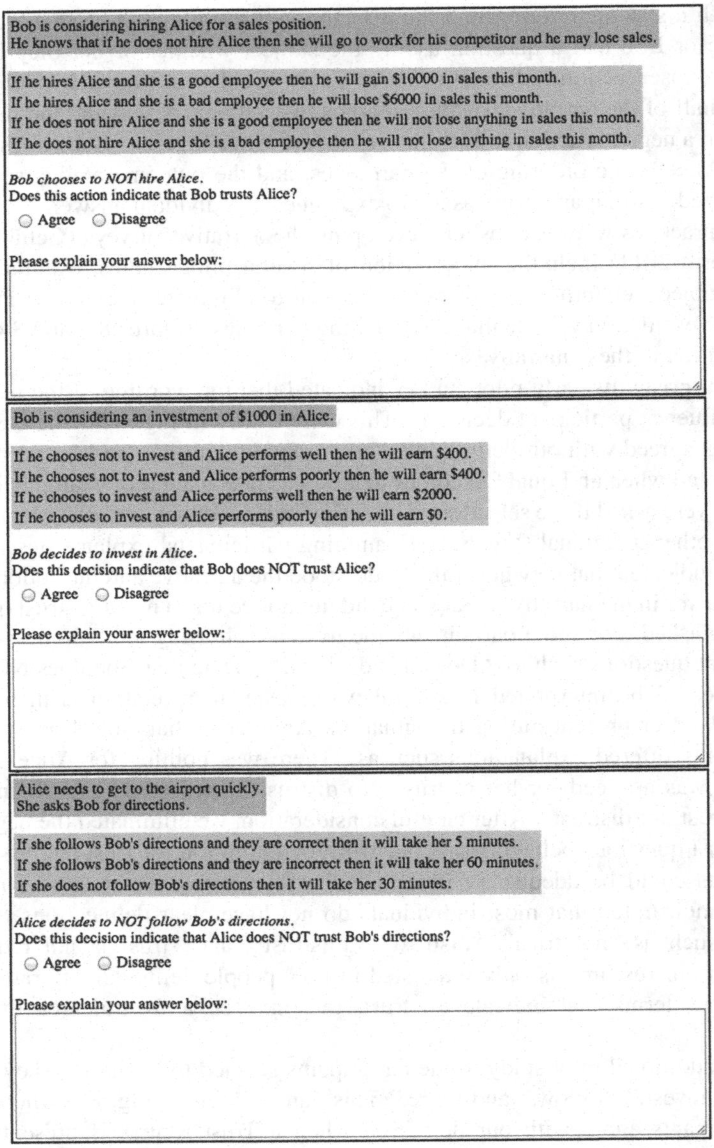

Bob is considering hiring Alice for a sales position.
He knows that if he does not hire Alice then she will go to work for his competitor and he may lose sales.

If he hires Alice and she is a good employee then he will gain $10000 in sales this month.
If he hires Alice and she is a bad employee then he will lose $6000 in sales this month.
If he does not hire Alice and she is a good employee then he will not lose anything in sales this month.
If he does not hire Alice and she is a bad employee then he will not lose anything in sales this month.

Bob chooses to NOT hire Alice.
Does this action indicate that Bob trusts Alice?

○ Agree ○ Disagree

Please explain your answer below:

Bob is considering an investment of $1000 in Alice.

If he chooses not to invest and Alice performs well then he will earn $400.
If he chooses not to invest and Alice performs poorly then he will earn $400.
If he chooses to invest and Alice performs well then he will earn $2000.
If he chooses to invest and Alice performs poorly then he will earn $0.

Bob decides to invest in Alice.
Does this decision indicate that Bob does NOT trust Alice?

○ Agree ○ Disagree

Please explain your answer below:

Alice needs to get to the airport quickly.
She asks Bob for directions.

If she follows Bob's directions and they are correct then it will take her 5 minutes.
If she follows Bob's directions and they are incorrect then it will take her 60 minutes.
If she does not follow Bob's directions then it will take her 30 minutes.

Alice decides to NOT follow Bob's directions.
Does this decision indicate that Alice does NOT trust Bob's directions?

○ Agree ○ Disagree

Please explain your answer below:

Fig. 8.3 The first iteration of the narratives

Anchor bias describes the human tendency to focus heavily on early and/or specific pieces of information and disregard later information. Because of this bias, we chose to replace specific actions that people were focusing on with less specific terms. For example, the statement "Bob is considering an investment of $1000 in Alice."

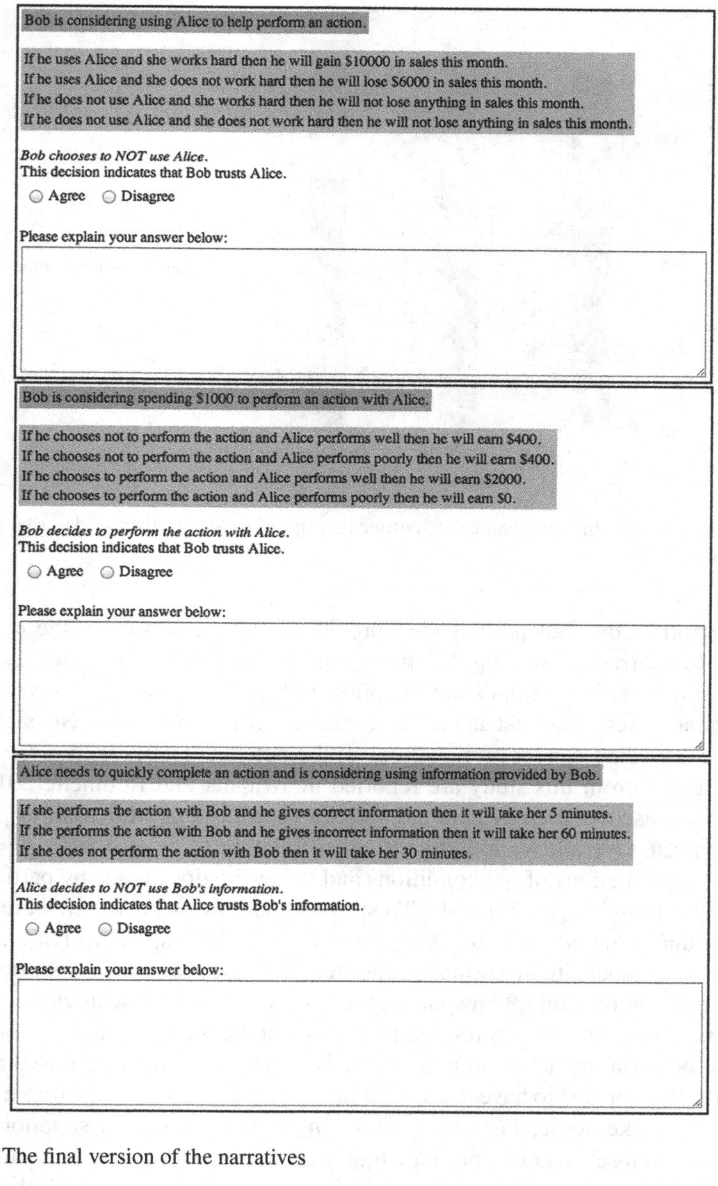

Bob is considering using Alice to help perform an action.

If he uses Alice and she works hard then he will gain $10000 in sales this month.
If he uses Alice and she does not work hard then he will lose $6000 in sales this month.
If he does not use Alice and she works hard then he will not lose anything in sales this month.
If he does not use Alice and she does not work hard then he will not lose anything in sales this month.

Bob chooses to NOT use Alice.
This decision indicates that Bob trusts Alice.

○ Agree ○ Disagree

Please explain your answer below:

Bob is considering spending $1000 to perform an action with Alice.

If he chooses not to perform the action and Alice performs well then he will earn $400.
If he chooses not to perform the action and Alice performs poorly then he will earn $400.
If he chooses to perform the action and Alice performs well then he will earn $2000.
If he chooses to perform the action and Alice performs poorly then he will earn $0.

Bob decides to perform the action with Alice.
This decision indicates that Bob trusts Alice.

○ Agree ○ Disagree

Please explain your answer below:

Alice needs to quickly complete an action and is considering using information provided by Bob.

If she performs the action with Bob and he gives correct information then it will take her 5 minutes.
If she performs the action with Bob and he gives incorrect information then it will take her 60 minutes.
If she does not perform the action with Bob then it will take her 30 minutes.

Alice decides to NOT use Bob's information.
This decision indicates that Alice trusts Bob's information.

○ Agree ○ Disagree

Please explain your answer below:

Fig. 8.4 The final version of the narratives

became "Bob is considering spending $1000 to perform an action with Alice." The final iteration of the narratives used in this experiment removed all keywords, such as "invest" or "hire," and replaced them with less specific phrases, such as "perform the action." This allowed us to reduce anchor bias. The exact wording for the three narratives can be found in Fig. 8.4.

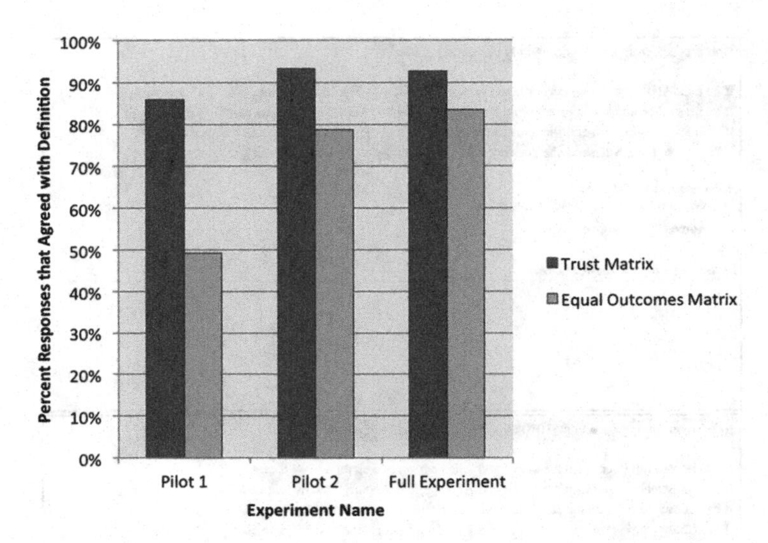

Fig. 8.5 Results from the pilot and full experiments using textual narratives to describe potential trust scenarios

In the full study, 128 participants' provided 1920 responses to the questions asked by the narrative. See Fig. 8.5 for a comparison between this study's results and the corresponding results from the pilot studies. The scenarios showed minor, insignificant differences that appear attributable to random error. No significant difference regarding gender or magnitude of the outcome matrix values was found. The full results from this study are reported in (Wagner and Robinette 2015), but some of our discoveries may also aid future experimenters who wish to perform similar studies. Overall, we found a strong correlation ($\phi = +0.592$, $p < 0.01$) between the predictions of our conditions and the evaluations made by participants. Participants strongly agreed that the Trust Matrix narratives presented were indeed situations that required trust (93 % agreement over 640 responses) but had some disagreements about situations that did not require trust according to our definition (66 % agreement over all 896 responses for designated no trust scenarios).

In some cases, participants invented reasons that the trustor would choose or not choose to perform the action in order to make sense of a situation. Based on their comments, this appears to have occurred when they were confronted with a narrative that did not make sense. For example, when confronted with a situation where Bob decides to lose \$2000 by participating in an action with Alice, one participant explains, "Bob trusts Alice because his decision has nothing to do with the money just his friendship with Alice." There is no mention in any of the narratives about a friendship or past relationship between the agents, yet the participant believes that there must be some reason Bob has chosen to lose this money and thus provides additional details so that the situation makes sense. With respect to the data, these peculiar narratives appear to have influenced participant trust evaluations more when the matrix did not meet our conditions for trust and may hint to a limitation of the use of narratives.

Overall, the use of crowdsourced narratives to examine trust offered several advantages and disadvantages. Advantages include the ability to reach a large and diverse population of subjects, flexibility in terms of describing trust scenarios, and an ability to develop narratives that closely matched the matrices from which they were derived. For example, it is difficult to examine the trust involved in a hiring decision without using some type of narrative. Because we believe that our framework can be used to represent most situations involving trust, it was important to capture results from several different scenarios. This approach is not without its limitations: it was difficult to manage or eliminate all psychological biases, the narrative approach was disconnected from our larger goal of exploring human-robot trust, and translating these matrices into narratives resulted in some peculiar descriptions of situations.

8.5 Crowdsourced Robot Evacuation

As mentioned above, a key disadvantage of the narrative approach to investigating trust is its disconnection to robotics. In this section we describe experiments designed to test trust using an environment designed for human-robot interaction.

Because a diverse set of participants was desired, crowdsourcing was once again utilized as a means for recruiting and paying study subjects. We developed a simulator that allowed participants to interact with a virtual robot using a web browser. A task that we believed would be significantly influenced by trust—robot guided emergency evacuation—was chosen. For this task participants were asked to choose whether or not they would like to use a robot for guidance when evacuating from a building. The building environment was modeled after a maze with corridors, dead ends, and no visual landmarks. Each simulation used the Unity 3D game engine to simulate the virtual maze and the virtual robot. Three-dimensional models for the game engine were created in Blender and Unity 3D. Participants were paid between $1 and $4, depending on the exact study.

Motivating participants can be a difficult challenge. We discuss our approaches to this challenge in greater detail below, but throughout all experiments we either used performance-based monetary bonuses to encourage participants to exit quickly or we informed them that they would not "survive" the simulated emergency if they did not evacuate within a short time frame.

8.5.1 Single Round Experimental Setup

Initially, we conducted additional experiments related to our working definition of trust and conditions for trust. In this case, however, we wanted participants to make a decision for themselves rather than reason about the actions of others. Further, by using either monetary or time incentives tied to the speed at which the maze environment was navigated we believed that the subjects would feel at risk.

Our initial experiments required a single trip through the maze. Each experiment began by thanking the person for participating in the experiment. Next the subject was provided information about the evacuation task. In some experiments this included presenting the environment and robot to the subjects in videos, images, or text and providing information that allowed the participant to evaluate the risk associated with choosing to follow the robot. Participants were shown examples (again in the form of videos, pictures, and/or text) of good and bad robot performance (e.g., robots that are fast and efficient and robots that are not) and participants were given an idea of the complexity of the maze. Also as part of this introduction, participants were given the chance to experiment with the controls in a practice environment. The practice environment was a simple room with three obstacles and no exit.

After this introduction, participants were given the choice to use the robot or not. With the exception of two pilot studies, participants were told that their choice to use the robot would not affect their compensation for this experiment. Participants were then placed at the start of the virtual maze. If they chose to use the robot it would start out directly in front of their field of view and immediately begin moving towards its first waypoint. The robot would move to a new waypoint whenever the participant approached. If the participant elected to not use the robot then no robot would be present and the participant would have to find the exit on his or her own.

After the maze-solving round was complete, participants answered a short survey about the round and about themselves. The exact questions asked in the survey varied considerably over the course of developing the pilot studies and the final experiment.

The results from this experiment are discussed in (Wagner and Robinette 2015). After several experiments with varying motivations and simulation environments, we found a strong correlation between participant responses as to whether a situation required trust and our definition of trust ($\phi = +0.406$, $p < 0.001$). When conditions for trust were met, 74.0 % of participants indicated that they trusted the robot, compared with only 32.9 % when conditions were not met.

8.5.2 *Multi-Round Experimental Setup*

Our conceptualization of trust relates risk to a model of the trustee's actions. Thus, in order to explore aspects of trust repair we needed to create a multi-round robot evacuation experiment. This was accomplished by modifying our simulation to perform two rounds of evacuations. This experiment required the participant to navigate two different mazes. They were offered the opportunity to use the robot twice and asked to complete a survey after each maze evacuation. The experimental setup for the single round experiment had been developed and tested to the point that we were confident in the procedure and survey questions. Both rounds were identical in all other ways.

Experimentally, a multi-round paradigm allowed us to violate the participant's trust in the first round and to then evaluate the impact on the person's decision to use the robot and self-reported trust during the second round. Using this procedure we were able to measure the change in trust across rounds as well as the correlation between self-reported trust and the decision to follow.

The experimental setup was similar in most respects to the single round procedure. Participants began each experiment clicking a link to a Unity 3D Web Player executable. Next they viewed an introductory message that described the navigation task they were to perform. This page included photos of an exit and the guidance robot. The guidance robot varied depending on the experimental conditions. They were then offered the opportunity to practice navigating in a maze. They had a first-person view of the practice environment and used their keyboard arrow keys to move. After the practice session, they were presented with illustrative examples of prior human-robot performances in the maze. The nature of these examples varied depending on the particular experiment. The participant was then asked to decide whether or not they would like a robot to provide guidance during the first round of the experiment. After making their choice, the person then navigated the maze and completed a short survey. They were then offered another opportunity to decide if they wanted to use the guidance robot in a second, different, maze. They then navigated the maze in the second round and completed a short survey about their second round decision. The robot's guidance performance in the second round always matched its performance in the first round. The experiment concluded with a final survey that collected demographic information.

The results from this experiment are presented in (Robinette et al. 2014c). Overall, we found that participants reported a significant decrease in self-reported trust between a robot performing well and a robot performing poorly (53 % decrease).

8.5.3 *Asking About Trust*

Our initial experiments found that participants would occasionally act as if they trusted the robot while reporting that they did not. This led to us to closely examine our method of asking about trust. Initially participants were asked: "When you made your decision to follow or not follow the robot, did you trust the robot as a guide in this scenario?" This produced good results when a Trust Matrix was used to design the experiment, but mixed results in other cases. Pilot studies were performed immediately afterwards, focusing on the Equal Outcomes matrix (see Table 8.2) and the trust question. We analyzed what it means for the participant to answer these questions. It was not initially clear if participants were stating whether they trusted the robot, the robot's ability to lead them to an exit, or something else. Additional pilot studies were performed with different wordings of the trust question. For example, we asked, "Did you trust the robot?", "Did your decision to follow or not follow the robot indicate that you trusted the robot?" and also varied responses available to the participants to include the option "Trust was not involved in the decision", in addition to "Yes," and "No." Overall, we found very little difference in the data resulting from these changes in wording.

In later single-round experiments, the issue of trust question wording was revisited. This time, participants who chose to use the robot were asked to agree or disagree with the statement "My decision to use the robot shows that I trusted the robot." Participants who chose to not use the robot were asked, "My decision to not use the robot shows that I trusted the robot." Each group was also asked if they trusted the robot itself. We again found very little difference in responses. Ultimately, we concluded that the wording of the question itself did not matter when compared with changes we made to the scenario.

8.5.4 Measuring Trust

Many methods for measuring trust have been proposed. In the field of human-robot interaction, for example, trust has been measured by asking participants to report their real-time change in trust in an autonomous vehicle (Desai et al. 2013) and by measuring the number of times an operator corrects the movements of an autonomous robot (Gao et al. 2013).

Our method for measuring trust varied with the type of experiment. In the narrative experiments, trust was indirectly measured by asking participants to evaluate the interactions of fictional people. For the simulation experiments, two measures of trust were considered. The primary method of measuring trust was participant self-reports in the form of survey questions. In various ways, these questions asked participants whether or not they trusted the robot. In some experiments we asked two trust-related questions. In our double round experiments, the participant's decision whether or not to follow the robot during the second round was a second method for evaluating trust. Results from our emergency evacuation experiment indicate a large positive correlation ($\phi(129) = +0.661,\ p < 0.001$ for round 1 and $\phi(90) = +0.745, p < 0.001$ for round 2) between the participants' decision to follow the robot and their self-report of trust. In this experiment, interaction with a robot that failed to provide guidance to an exit led to a 50 % drop in robot usage during the second round as well as a drop in self-reported trust of 53 %.

8.5.5 Incentives to Participants

As described above, investigations of trust require that the subject perceive or believe that they are at risk. Of course, there are ethical boundaries to the types and ways that an experimenter can make participants feel at risk. The use of financial incentives is a common way to put subjects at risk without the possibility of physical or emotional harm (King-Casas et al. 2005).

Because participants were recruited via Amazon's Mechanical Turk service, we assumed that they were strongly motivated by money. Participants were typically able to complete our experiments quickly, so our payments often gave a higher than average hourly wage, even though the payments were between $1 and $4. For comparison, subjects were paid a flat rate of $1.67 for completing the narrative study. For the maze evacuation studies, subjects were paid a base payment of approximately $2 for completing the study. They were then offered a $1 bonus for completing the maze quickly. We assumed that a 50 % bonus would serve as considerable motivation for completing the maze in a timely fashion. The amount of time required to complete the maze was impacted by the quality of guidance provided by the robot, if they elected to use it. Using written text and videos, they were informed that they could expect to receive $1 if the robot guides them efficiently, $0 if the robot is a bad guide, and some number in between otherwise. This outcome matrix was modeled after the Trust Matrix defined in the Narrative Experiment section (see Table 8.2).

Participants had very little control over their bonus; so all bonuses were paid out in full after the experiment, regardless of performance. They had no knowledge of this before the experiment. See Fig. 8.6 for a screenshot of the interactive portion of the experiment showing the method for displaying the bonus remaining.

Fig. 8.6 A screenshot from the robot evacuation experiment is presented above. The amount of time that has elapsed is pictured to the left and the amount of the participant's bonus is pictured to the right. The amount of the bonus decreased as the time taken to navigate the maze increased

As mentioned above, we used both self-reports and the participant's decision to follow the robot as two different measures of trust. We hypothesized that the use of financial incentives would result in a decrease in both measures when the robot provided poor guidance and no decrease in trust when the robot provided good guidance. This prediction was wrong. We found that although people tended to self-report a loss of trust, they nevertheless continued to follow the robot in the second round. As an example, 74 % of participants in the single round experiment with a Trust Matrix outcome who chose to use the robot reported that they trusted the robot. This result was promising by itself, however 50 % of participants who chose to use the robot in an Equal Outcomes experiment also reported trust in the robot. This result caused us to question our use of monetary bonuses as a motivational technique.

We conducted additional studies that included an expanded set of survey questions exploring each individual's motivation for participating in the experiment. We found that 53 % of participants reported that the bonus was the most important motivation, 24 % noted that completing the study quickly was most important, and 23 % claimed enjoyment was their primary motivation. Based on the comments from participants, we determined that those who did not trust the robot continued to use it in the second round because they considered it better than no source of guidance at all. As described below, various types of behavior were tested in order to communicate that the robot had failed in its navigation task. Some of these behaviors, however, still lead the participants to an exit, even if none of the bonus was preserved. Even when the robot's navigation behavior did not find an exit, participants often still continued to use it in the hope that it would eventually find an exit. In fact, a few participants erroneously believed that having a poor performing robot in the first round increased the likelihood of having a fast robot in the second round.

Based on the results from these motivation surveys, we began to explore other ways to motivate participants. Because of our interest in search and rescue, we modified our scenario to be an emergency evacuation. In this modified scenario, participants were told that our goal was to discover how people leave a building in an emergency. Instead of receiving a bonus for a fast completion, they were told that they would only survive if they found the exit in time. As before, a countdown timer appeared in the middle of their view to tell them the remaining time. Participants were compensated $1.00 for their participation in single round experiments and $2.00 for their participation in double round experiments (the same as the base pay in the monetary bonus experiments). Figure 8.7 depicts the interactive portion of the experiment with the emergency evacuation motivation.

Presenting the maze navigation task as an emergency resulted in self-reports of trust that closely matched the participant's decision to follow the robot during the second round. Again using the single round experiment as an example, 74 % of participants reported trust in the robot when the conditions for trust were met, but only 33 % reported trust when the conditions for trust were not met. Participant's comments also indicated that the emergency motivation strongly influenced both the self-report and the decision to follow.

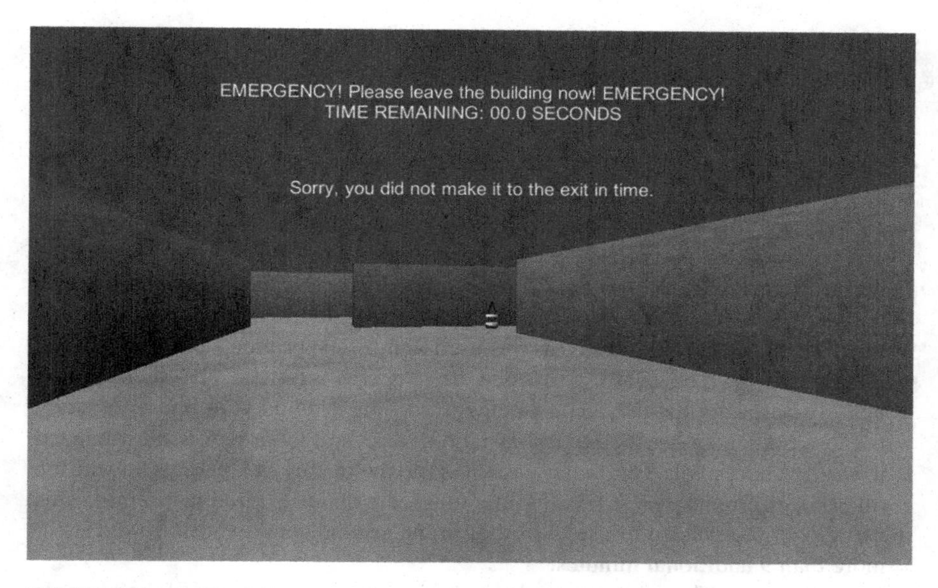

Fig. 8.7 A screenshot from the robot evacuation experiment using an emergency as the participant's motivation is presented. The participants were told that their task was to act as if they were in an emergency evacuation and had to find an exit within 30 s in order to survive

8.5.6 Communicating Failed Robot Behavior

While understanding a participant's motivations is an important factor related to the evaluation of human-robot trust, the robot's ability to communicate must not be neglected. The actions of the robot inform the human of the robot's ability to be trusted in future interactions. In pilot studies we evaluated several different types of robot guidance failures. The failures were intended to mimic the types of failures likely to occur during real evacuation procedures. All but two of these failure behaviors were eliminated because participants were unable to determine that the robot had failed and hence resulted in extremely long experiment completion times.

In each of the behaviors below, the robot followed a predefined set of waypoints throughout the environment. Waypoints were set near corners or occlusion points so that travel between points was linear. The robot moved faster than the participant in order to lead the person. It then waited at each waypoint for the participant to catch up before moving to the next waypoint. The robot was always in view of the participant if the participant moved along the lines between waypoints. Hence, the amount of time it took participants to reach the exit depended on the environment and on the participant. We created the following evacuation behaviors for the robot:

- **Fast navigation**—The robot proceeded directly to the exit at a high rate of speed. Robots that acted in this manner were capable of finding the exit in any maze within 30 s. This was the only behavior that resulted in rapid evacuation.

- **Slow navigation**—The robot explored many different routes before eventually finding the exit. Robots that acted in this manner were capable of finding the exit of any maze in 90 s. This type of robot behavior was used in most experiments presented here.
- **Failed navigation**—The robot proceeded directly to a corner of the environment that is not near the exit and stopped. This behavior was meant to emulate the behavior of a robot that has incorrect information about the exit location. It was believed that participants would view a robot governed by this behavior as less trustworthy than a robot that navigated slowly. No significant difference, however, was found in either the decision to use the robot or in self-reported trust. This type of robot behavior was used in most experiments presented here.
- **Small Loops**—The robot proceeded to a nearby obstacle and then circled that obstacle continuously. This behavior was only included in one pilot study because participants tended to follow it around several loops before they realized it was unsuccessful. The worst-case participant in this behavior followed the robot for approximately 3.6 min even though the bonus expired in 1.5 min. This participant then chose to use the robot in the second round and followed it for more than 9 additional minutes.
- **Large Loops**—The robot proceeded to a large obstacle and circled the obstacle continuously. As with the small loops behavior, several circumnavigations were required before participants realized that the robot was not leading them to an exit. At a minimum, participants followed the robot around one loop, which required a considerable amount of time. The worst-case participant followed the robot for more than 5 min. This participant also chose to use the robot again in the second round, but quickly understood its behavior was the same and abandoned it after just 36 s.
- **Continuous Searching**—The robot exhaustively searched every corner of the map except the hallway leading to the location of the exit. Once this search was completed it started again, following the exact same waypoints. Again, participants in the pilot study failed to realize that the robot was performing poorly with one participant following the robot for almost 12 min. This corresponded to over three complete searches of the entire environment. This behavior was only tested in a single pilot study because it took too long for participants to realize that the robot had failed.
- **Wall Collision**—The robot approached the correct hallway to the exit but then collided with a wall. It continued to collide with the wall instead of providing further guidance. This robot behavior was meant to emulate a working navigation system and a broken obstacle avoidance system. This behavior was only tested in one pilot study because participants did not recognize that the robot had failed, instead interpreting its bizarre collision behavior as a signal that it had found the exit. In reality, the exit was just out of sight.

In all, we tested six robot behaviors that were meant to convey failure and one robot behavior that was meant to convey success. Of the behaviors tested in full studies, we generally found significantly different results between successful (i.e., fast navigation) and the slow or failed navigation behaviors. Surprisingly, our data shows that the type of unsuccessful navigation behavior used by the robot did not have a significant effect on the decision to follow the robot in the second round or on the participant's self-reports of trust. The slow and failed behaviors produced nearly identical results during experimentation. Based on these experiments, we conclude that communicating unsuccessful robot behavior is a significant challenge.

Our pilot studies with various unsuccessful robot behaviors raise concerns about typical humans over-trusting robots to perform their stated task. While many people seem predisposed to not trust any new form of technology, others seem to instantly trust a new technology to perform its task, regardless of evidence to the contrary. Participants were willing to follow a robot in what we consider to be an obviously unsuccessful search for an exit for almost 12 min for $2 in compensation (the bonus had expired by this time). Even in our most successful test, 50 % of participants who had experienced an unsuccessful robot chose to use the robot again in a second round (for comparison to situations where human leaders are asked to be trusted, see the meta-study in (Dirks and Ferrin 2002) where the authors found a small but significant effect on trust by job performance metrics). This issue shows the need for a robot to realize when it has failed and inform nearby human stakeholders to find another means of accomplishing the given task.

8.6 Conclusion

We have performed multiple studies involving 770 participants in order to validate our working definition and conditions for trust in human-human and human-robot interactions. In these studies, participants have examined outcome matrices and decided the extent to which the interactive situation described by these matrices demands trust. In the first study, written narratives were used to sculpt these situations. In the second, simulated evacuations through a maze were used to convey risk and force the participant to make a decision.

Overall, 11 experiments involving a total of 770 participants have taught us the following lessons related to the empirical evaluation of trust:

- Pick the method that aligns best with the scenario being tested. Some scenarios may be better tested in a simulation environment, rather than as a narrative.
- In the same vein, pick a method of compensation or motivation that aligns well with the scenario. Sometimes money is a sufficient incentive, but oftentimes incentives besides money will motivate participants.
- Avoid the anchoring bias. If narratives are to be used, watch for words that will overwhelm all other considerations a participant might have about the scenario.
- It is difficult for a non-expert human to understand when a robot has failed at a task. For example, if the robot is built to be a navigation guidance robot, participants will expect it to be a good guide.

- Use complementary methods to achieve both generality and grounded, empirical evaluations.

Our experiences demonstrate that it can be difficult to impress an exact outcome matrix on a participant, even if the numbers are clearly stated in the form of a narrative or in the form of examples. For example, words like "invest" and "follow" can bias a participant to ignore the given numerical outcomes and assume that the situation requires trust. Moreover, situations that are difficult to comprehend may cause participants to augment scenarios with invented information. Finally, communicating failure is not as straightforward as might be expected. In spite of these challenges, the methods outlined in this chapter have successfully been used to explore the topic of trust in a manner that is not tied to a single specific scenario and has been verified with results based on data from a large and diverse population. As methods for investigating trust take shape, it is important that results of studies apply broadly. Crowdsourcing allows testing with such broad and diverse populations.

Using crowdsourcing as a means for testing hypotheses related to trust demands methods for conveying interactive scenarios to the participants. Narratives offer a general, yet less grounded way of doing so. Simulated scenarios immerse subjects in a test environment, but make drawing general conclusions more daunting. Thus, our most valuable lesson learned is that utilizing different yet complementary methods can provide both generality and grounded empirical evaluations of human-robot trust.

Acknowledgement This work was funded by award #FA95501310169 from the Air Force Office of Sponsored Research.

References

Axelrod R (1984) The evolution of cooperation. Basic, New York

Berinsky AJ, Huber GA, Lenz GS (2012) Evaluating online labor markets for experimental research: Amazon.com's mechanical turk. Polit Anal 20(3):351–368

Bethel CL, Murphy RR (2008) Survey of non-facial/non-verbal affective expressions for appearance-constrained robots. IEEE Trans Syst Man Cybernet Part C 38(1):83–92

Buhrmester M, Kwang T, Gosling SD (2011) Amazon's mechanical turk: a new source of inexpensive, yet high-quality, data? Perspect Psychol Sci 6(1):3–5

Carlson MS, Desai M, Drury JL, Kwak H, Yanco HA (2014) Identifying factors that influence trust in automated cars and medical diagnosis systems. In: Proceedings of the AAAI spring symposium on the intersection of robust intelligence and trust in autonomous systems, Palo Alto

Castelfranch C, Falcone R (2010) Trust theory: a socio-cognitive and computational model. Wiley, New York

Desai M, Kaniarasu P, Medvedev M, Steinfeld A, Yanco H (2013) Impact of robot failures and feedback on real-time trust. In: Proceedings of the 8th ACM/IEEE international conference on human-robot interaction, Tokyo, pp 251–258

Dirks KT, Ferrin DL (2002) Trust in leadership: meta-analytic findings and implications for research and practice. J Appl Psychol 87(4):611–628

Duncan BA, Murphy RR (2013) Comfortable approach distance with small unmanned aerial vehicles. RO-MAN, 2013 IEEE, pp 786–792

Gambetta D (1990) Can we trust trust? In: Gambetta D (ed) Trust, making and breaking cooperative relationships. Basil Blackwell, New York

Gao F, Clare AS, Macbeth JC, Cummings ML (2013) Modeling the impact of operator trust on performance in multiple robot control. AAAI spring symposium: trust and autonomous systems, Palo Alto

Gehlbach H, Brinkworth ME (2011) Measure twice, cut down error: a process for enhancing the validity of survey scales. Rev Gen Psychol 15(4):380–387

Gosling SD, Vazire S, Srivastava S, John OP (2004) Should we trust web-based studies? A comparative analysis of six preconceptions about internet questionnaires. Am Psychol 59(2):93–104

Hancock PA, Billings DR, Schaefer KE, Chen JYC, Visser EJD, Parasuraman R (2011) A meta-analysis of factors affecting trust in human-robot interaction. Hum Factors 53(5):517–527

Hoffman RR, Johnson M, Bradshaw JM, Underbrink A (2013) Trust in automation. Intell Syst 28(1):84–88

Horton JJ, Chilton LB (2010) The labor economics of paid crowdsourcing. In: Proceedings of the 11th ACM conference on electronic commerce, pp 209–218

Kelley HH, Thibaut JW (1978) Interpersonal relations: a theory of interdependence. Wiley, New York

King-Casas B, Tomlin D, Anen C, Camerer CF, Quartz SR, Montague PR (2005) Getting to know you: reputation and trust in two-person economic exchange. Science 308:78–83

Kittur A, Chi EH, Suh B (2008) Crowdsourcing user studies with mechanical turk. In: Proceedings of the SIGCHI conference on human factors in computing systems

Lee JD, See KA (2004) Trust in automation: designing for appropriate reliance. Hum Factors 46:50–80

Murphy RR (2004) Human-robot interaction in rescue robotics. IEEE Trans Syst Man Cybernet Part C 34(2):138–153

Paolacci G, Chandler J, Ipeirotis PG (2010) Running experiments on Amazon mechanical turk. Judgment Decision Making 5(5):411–419

Robinette P, Howard AM (2011) Incorporating a model of human panic behavior for robotic-based emergency evacuation. RO-MAN, 2011 IEEE, pp 47–52

Robinette P, Howard AM (2012) Trust in emergency evacuation robots. In: 2012 IEEE international symposium on safety, security, and rescue robotics (SSRR), pp 1–6

Robinette P, Vela PA, Howard AM (2012) Information propagation applied to robot-assisted evacuation. In: 2012 IEEE international conference on robotics and automation (ICRA), pp 856–861

Robinette P, Wagner AR, Howard A (2013) Building and maintaining trust between humans and guidance robots in an emergency. In: AAAI spring symposium, Stanford University, Palo Alto, pp 78–83

Robinette P, Wagner AR, Howard A (2014a) Assessment of robot guidance modalities conveying instructions to humans in emergency situations. In: Proceedings of the IEEE international symposium on robot and human interactive communication, RO-MAN 14, Edinburgh

Robinette P, Wagner AR, Howard A (2014b) Modeling human-robot trust in emergencies. In: AAAI spring symposium, Stanford University

Robinette P, Wagner AR, Howard AM (2015) The effect of robot performance on human-robot trust in time-critical situations. Tech Rep. GT-IRIM-HumAns-2015- 001, Georgia Institute of Technology, Institute for Robotics and Intelligent Machines

Sabater J, Sierra C (2005) Review of computational trust and reputation models. Artif Intell Rev 24:33–60

Steinfeld A, Fong T, Kaber D, Lewis M, Scholtz J, Schultz A, Goodrich M (2006) Common metrics for human-robot interaction. In: Proceedings of the first ACM SIGCHI/SIGART conference on human-robot interaction, pp 33–40

Tversky A, Kahneman D (1974) Judgment under uncertainty: heuristics and biases. Science 185(4157):1124–1131

Wagner AR (2009a) The role of trust and relationships in human-robot social interaction. Ph.D. Dissertation. School of Interactive Computing, Georgia Institute of Technology, Atlanta

Wagner AR (2009b) Creating and using matrix representations of social interaction. In: Human-robot interaction (HRI), San Diego, pp 125–132

Wagner AR (2012) Using cluster-based stereotyping to foster human-robot cooperation. In: Proceedings of IEEE international conference on intelligent robots and systems, IROS 2012, Villamura, pp 1615–1622

Wagner AR, Robinette P (2015) Towards robots that trust: human subject validation of the situational conditions for trust. Interact Stud 16:89–117

Chapter 9
Designing for Robust and Effective Teamwork in Human-Agent Teams

Fei Gao, M.L. Cummings, and Erin Solovey

9.1 Introduction

With the development of automation technology, operators' tasks often shift from manual control of a single task to supervising multiple tasks and agents, which can require monitoring, coordination, and complex decision-making. However, the required cognitive load for working with multiple agents could easily exceed the capacity of a single operator, even with high levels of automation. There is an increasing demand for teams of humans to perform tasks that are less efficiently done or impossible to do by individual humans.

Teams have the potential of offering greater adaptability, productivity, and creativity than any one individual can offer and provide more complex, innovative, and comprehensive solutions (Gladstein 1984). However, working as a team imposes extra workload related to coordination and communication, and teams can fail for many reasons (Salas and Fiore 2004). Factors such as a poor combination of individual efforts, a breakdown in internal team processes (e.g., communication),

F. Gao (✉)
Massachusetts Institute of Technology, 77 Massachusetts Avenue, E40-206, Cambridge, MA 02139, USA
e-mail: feigao@mit.edu

M.L. Cummings
Department of Mechanical Engineering and Materials Science, Department of Electrical and Computer Engineering, Duke University, Box 90300, 144 Hudson Hall, Durham, NC 27708, USA
e-mail: m.cummings@duke.edu

E. Solovey
College of Computing and Informatics Drexel University, 3141 Chestnut Street, UC108, Philadelphia, PA 19104, USA
e-mail: erin.solovey@drexel.edu

© Springer Science+Business Media (outside the USA) 2016

R. Mittu et al. (eds.), *Robust Intelligence and Trust in Autonomous Systems*,
DOI 10.1007/978-1-4899-7668-0_9

and an improper use of available information have been identified as potential sources of team failure (Salas et al. 2005).

Effective teamwork in highly dynamic environments requires a delicate balance between giving agents the autonomy to act and react on their own and restricting that autonomy so that the agents do not work at cross purposes (Work et al. 2008). To achieve robust and effective teamwork, we must understand the nature of such teamwork, including team structure, team processes and dynamics, and their impact on team performance. In this study, we investigated the teamwork across multiple operators working together with multiple heterogonous autonomous vehicles using two experiments. In Experiment 1, the impact of team structure on team performance under different levels of uncertainty was investigated. Reasons for inefficient coordination were identified. In order to improve the coordination, in Experiment 2, four interface design conditions were compared using the same testbed to see whether facilitating information-sharing within the team could improve team coordination and team performance.

9.2 Related Work

Autonomous systems affect teamwork in two primary ways. First, autonomous systems affect the way a task can be completed through task interdependence and work assignment. Second, automation affects how information is presented and shared among team members, which further influence the way team members coordinate and communicate. Previous work has identified many important factors that include team structure, shared mental model and team situation awareness, as well as communication.

9.2.1 Team Structure

Team structure is an important factor hypothesized to affect team effectiveness (Lewis et al. 2011). Team structure affects the manner in which the task components are distributed among team members (Naylor and Dickinson 1969), as well as team communication and coordination. The team structure that is suitable for a specific scenario largely depends on the task characteristics and resources available (Macmillan et al. 2004). For a team of operators working together with multiple heterogeneous autonomous vehicles, there are several ways to organize the vehicles. One common method is functional organization, in which individuals specialize and perform certain roles. For example, one person is responsible for searching and another person is responsible for responding to targets. By specialization on the part of each member, groups are able to tackle problems more efficiently. The clear task responsibility also reduces the need for coordination. One major downside of functional organization is the difficulty in shifting workload flexibly to break up unexpected bottlenecks.

Another way to organize the team is divisional organization, in which each working unit can be responsible for all type of tasks. In divisional organization, each member is allocated with some resources of each type. By creating self-contained tasks, it reduces the amount of information processed within an organization when the level of uncertainty is high (Galbraith 1974). For example, a company can have several divisions each responsible for one product. Each division has its own set of functional units like research, design, marketing etc. Divisional structure was designed in order to have a fast response to the market (Macmillan et al. 2004). In one command and control scenario, it was found that the effectiveness of teams using the divisional and functional structures depends on the nature of the tasks to be accomplished and the uncertainty in the situation. Specifically, functional teams perform better when the environment and tasks are predictable. Divisional teams have a higher level of robustness and perform better when the environment and tasks have more uncertainty (Macmillan et al. 2004).

9.2.2 *Shared Mental Model and Team Situation Awareness*

Whether working as an individual or a team, developing and maintaining a high level of situation awareness (SA) is critical in autonomous vehicle control. SA includes perception of the elements in the environment, comprehension of the current situation, and projection of future status (Endsley 1995). Team coordination poses extra SA requirements. Team members need to be aware of their teammates' situation in addition to their own. If two or more team members need to know about a piece of information, it is not sufficient if one knows the information perfectly while others know nothing at all. The degree to which each team member possesses the SA required for his or her responsibilities was defined as team SA (Endsley 1995). To develop team SA, each team member needs to understand the impact of other team members' task status on one's own functions and the overall mission, as well as how their own task status and actions impact on other team members. Based on such comprehensions, team members should also be able to project what fellow team members will do to plan their actions effectively (Endsley and Jones 1997).

The quality of team SA affects team communication, coordination and performance directly or indirectly. Blickensderfer et al. (1997) found that teams that shared expectations regarding member roles and task strategies before a radar tracking task communicated more efficiently during the task and achieved higher overall performance outcomes. Previous research identified several ineffective team SA processes that should be avoided, including one member leading others off, insufficient sharing of pertinent information, failure to prioritize the tasks and adhere to the main goal, and relying on unreliable expectations (Bolstad and Cuevas 2010).

There are several ways to improve team SA. From system design perspective, team SA can be improved by tools to facilitate team communication, shared displays or shared environments, etc.

Situation awareness has become a critical top of concern when designing a human-machine interface. A system that improved situation awareness should provide a proper amount of information accurately based on the user's situation awareness needs. For teams, one important aspect of the interface design is to facilitate information-sharing among team members. Efforts had been made to improve team situation awareness using team displays for command and control teams, forest fire fighting teams, teams in operating rooms as well as in workspace (Biehl et al. 2007; Bolstad and Endsley 1999, 2000; Parush et al. 2011; Parush and Ma 2012). A team display used in forest fire fighting scenario improved situation awareness and performance, particularly when there was a communication breakdown (Parush and Ma 2012). However, it was also found that the use of an abstracted shared display enhanced team performance, while the use of shared displays that completely duplicated the other team members displays decreased performance and increased workload (Bolstad and Endsley 2000). Despite the potential benefits, a team display aiming to enhance team situation awareness should be carefully designed to avoid an overly complex interface.

9.2.3 Communication

Communication, an important coordination mechanism, influences the share of information among team members. Communication relates to building an accurate understanding of team members' needs, responsibilities, and expected actions (Macmillan et al. 2004), which allows them to anticipate one another's needs so that team members can coordinate effectively (Stout et al. 1999). If the team members don't communicate sufficiently, they may not develop a clear understanding of the situation, which may result in delayed actions, errors, and a suboptimal distribution of team resources.

On the other hand, communication takes time and carries a coordination cost. It can represent a type of process loss, which means team performance could be lower than the combination of individual performance due to the extra work on team coordination (Steiner 1972). Research has investigated the negative effects of communication in terms of increased workload and decreased performance. In a team of six persons performing a joint task force mission of air-based and sea-based operations, it was found that a lower need for coordination and a lower communication rate were associated with better performance (Macmillan et al. 2004). In another study, excessive word usage was found to have a negative association with team performance (McKendrick et al. 2013).

The appropriate amount of communication is impacted by factors such as task characteristics, team structure, level of workload, etc. (Bowers et al. 1996; Oser et al. 1991). In general, an ideal balance is to communicate enough to exchange the required information without too much increase on coordination overhead. In order to reduce coordination overhead, a strategy teams often use under high workload is to switch from explicit communication to implicit coordination (Orasanu 1990; Stout et al. 1996). Instead of communicating explicitly to control teammates, such as proposing actions, prompting or requesting information (Entin and Serfaty 1999), implicit coordination is adopted. Some effective implicit coordination strategies include periodic situation assessment, offering information without explicit request, and providing information to indirectly guide teammates' actions are some effective implicit communicate strategies (Entin and Serfaty 1999; Orasanu 1990; Shah and Breazeal 2010; Stout et al. 1996).

9.3 Experiment 1: Team Structure and Robustness

As discussed previously, different team structures have advantages depending on the nature of the task and environment. Human-agent teams often work under uncertainty. One major source of uncertainty is task load. The arrival time and types of tasks are often unpredictable that balancing the tasks and workload among team members can significantly affect outcomes. In Experiment 1, we investigated the communication and coordination process and performance of human-agent teams with different team structures and under different levels of task load uncertainly.

9.3.1 Testbed

The software testbed for our study is Team Research Environment for Supervisory Control of Heterogeneous Unmanned Vehicles (TRESCHU), a video game-like simulation of unmanned vehicle control by a team of three operators. The simulation included three ground control stations, with one operator assigned to each station controlling three vehicles. The three operators were referred to as Alpha (A), Bravo (B), and Charlie (C). The scenario was search and rescue operation in which operators must identify contacts as either friendly or threats, and respond to them appropriately—friendlies must be dropped aid packages, and threats must be neutralized.

Each mission scenario required a team of operators and autonomous vehicles to handle contacts that appeared intermittently over the map. There were three ground control stations, with one operator assigned to each station controlling three vehicles. New contacts appeared on the map as Unknowns. Operators were required to send a scouting vehicle to identify the unknown as either Friendly or Threat, after which they could assign a rescue vehicle or a tactical vehicle to respond. Once assigned, the vehicle would autonomously travel to that particular contact location on the map in a straight line and would continue until either the vehicle reached its assigned destination or the operator re-assigned the vehicle elsewhere.

Once a vehicle arrived at a contact, the operator performed one of three tasks that depended on the vehicle and contact type: scout, rescue, or tactical. All three tasks involved a birds-eye view of the terrain. In the scout task, there were two items of interest presented in the upper left corner of the screen. The operator's task was to select the one item that appears somewhere in the overview map. The rescue task involved controlling the position and movement of crosshairs and dropping aid packages to friendly contacts on the ground. The crosshairs were relatively steady but the projectiles were falling slowly and susceptible to the wind. The tactical task required the operator to center the crosshairs over a stationary threat on the ground and to neutralize it.

The scout task required visual search ability. The rescue and tactical tasks required hand-eye coordination. These three tasks had different levels of difficulty. Rescue tasks were the hardest and took the longest time. Scouting tasks were the easiest and took the shortest time.

Because of the need to first identify the contact before completing one of the other two tasks, two vehicles were required to complete each scenario, one for scouting and one for the rescue or tactical task. The rescue or tactical vehicle could be assigned before or after the scouting task. The timeline for processing a task is shown in Fig. 9.1. The time between the appearance of an unknown contact and the time it was neutralized or aided was called objective completion time. Team performance was measured by averaging the objective completion time (referred to as AOCT later) of all contacts during the mission.

The interface contained four parts: a Map, a Chat Panel, a System Panel and a Monitor Other Vehicles Button (Fig. 9.2). The Map represents the geographical area that the operators were responsible for, the three vehicles under their control,

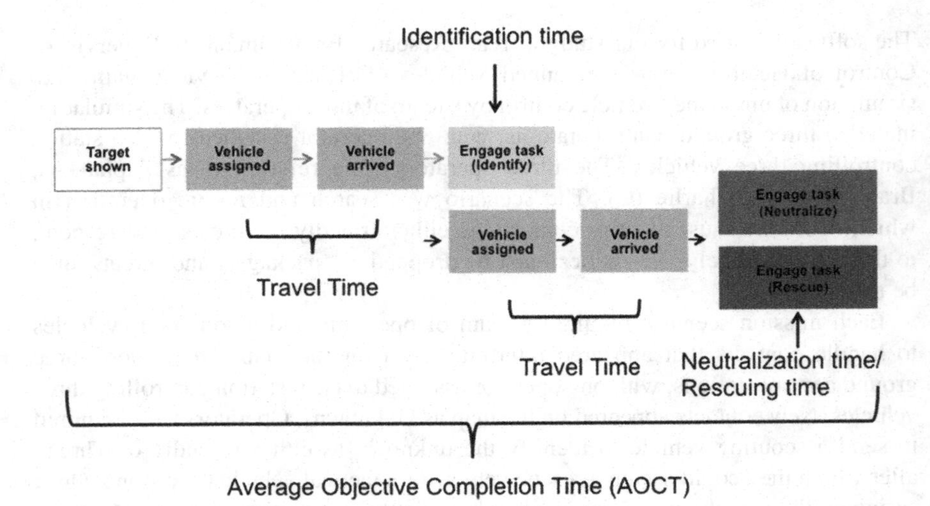

Fig. 9.1 Task workflow

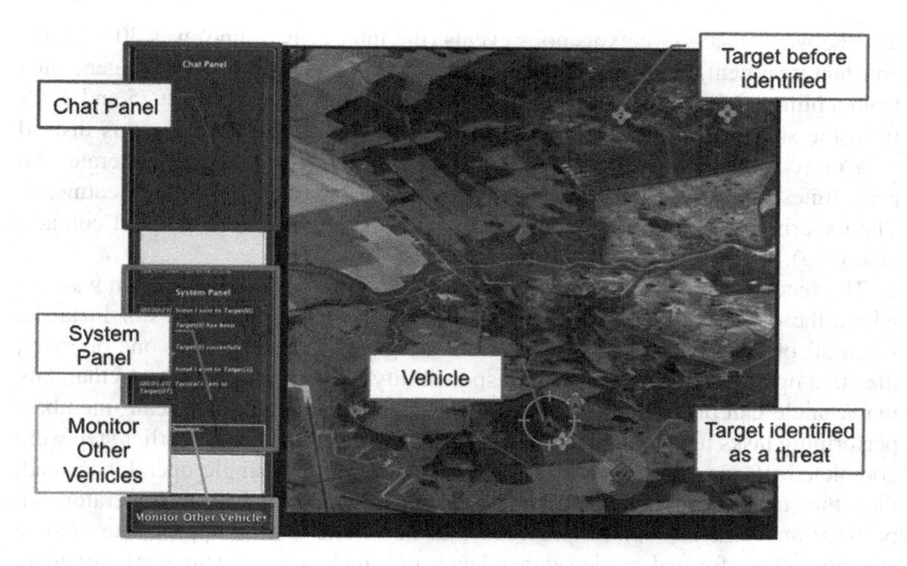

Fig. 9.2 Team research environment for supervisory control of heterogeneous unmanned vehicles (TRESCHU) interface

and all the contacts that need to be handled by the team. TRESCHU has three kinds of contacts (Unknown, Friendly, Threat) and three corresponding vehicles (Scout, Tactical, Rescue). Types of vehicles were differentiated by color, and types of contacts were differentiated by both color and shape. The operators were able to communicate with each other via instant messaging in a chat interface window.

Operators typed messages into the chat, which would then appear on all the other operators' chat panels instantly. Chat messages were labeled with the operators unique IDs, which corresponded to the labels for each operator's vehicle icons. The System Panel would occasionally send messages to a particular operator, such as a confirmation message that the operator had assigned a particular vehicle to travel to a particular location. It also sent the operator an error message when he or she attempted to claim or engage a vehicle already claimed or engaged by other operators. Operators were unable to see the location of other operators' vehicles unless they explicitly commanded the interface to do so with the Monitor Other Vehicles button.

9.3.2 Experiment Design

A 2×2 mixed design experiment was conducted where the independent variables were team structure (divisional, functional) and the inter-arrival time of unidentified contacts (constant, erratic). The two conditions of inter-arrival time were designed to simulate different uncertainty levels in task load for human-agent teams. The time between successive exogenous events (the inter-arrival time) was 30 s for the constant treatment. For the erratic factor level, the inter-arrival times were generated from a bimodal distribution where the means of the modes were set at 75 and 225 s from the start of the trial, with a standard deviation of 15 s. We use this instead of a more random arrival process (e.g. passion arrival process) to generate two peak times. Ten teams of three participants each completed all four treatments. The experimental trials had a total of 16 exogenous events (unidentified contacts emerging).

The second independent variable was team structure. A functional team was one where the operators have rigidly defined roles and responsibilities. For instance, when all of the vehicles of one type were assigned to one and only one operator, then that operator was given the full responsibility for performing the tasks that only that vehicle can do. This formed sequential dependency in which team members performing tasks in the later steps had to wait until the tasks in the earlier step were completed. If one of each vehicle type was allocated to a single operator instead, then that team structure would be considered divisional since any operator can perform any task that arises, provided that he or she had an appropriate vehicle available. This formed pooled dependency in which independent works of team members were combined to represent team output (Thompson 1967).

9.3.3 *Results*

Thirty participants participated in the experiment and were tested in groups of three. They went through the four combinations of independent variables in randomized sequence with each session lasted about 15 min. The initial experiment results showed that functional teams performed significantly better than divisional teams ($F\ (1,\ 24) = 1.484,\ p < 0.01$), as shown in Fig. 9.3. The interaction effect was also significant with functional teams performed better with constant arrival, while divisional teams performing better with erratic arrival ($F\ (1,\ 24) = 10.47,\ p = 0.04$) (Mekdeci and Cummings 2009). Although divisional teams showed their robustness against the uncertainty of task arrival, their performance was not as good as desired, especially under the constant arrival process. In this effort, we further investigated the teamwork process to identify several reasons for the poor performance: duplicated work, underutilization of vehicles, and infrequent communication.

9.3.3.1 Duplicated Work

Further analysis on teamwork process shows that teams that had worse performance also tended to have poor team coordination. One example of such poor coordination occurred in divisional teams, where we observed duplicated vehicle assignments. We analyzed who assigned vehicles to each contact after it appeared and later after it was identified. In Fig. 9.4, each column represents each contact. Each square

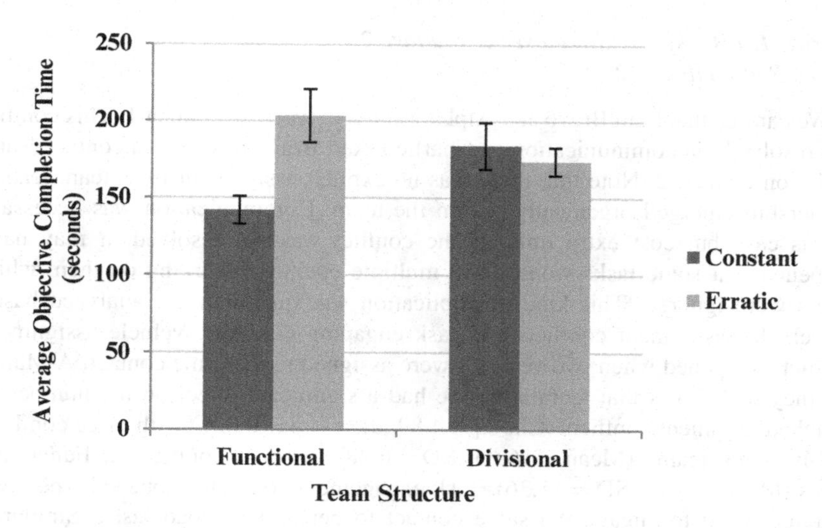

Fig. 9.3 Team performance under different conditions

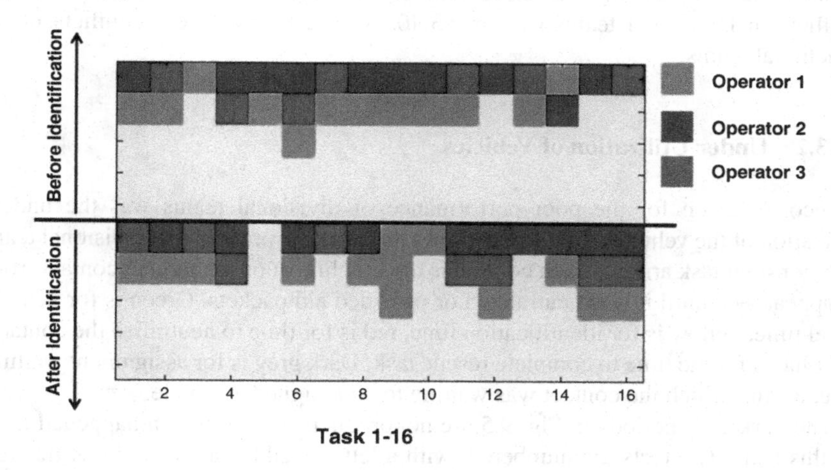

Fig. 9.4 Vehicles assigned to each task by operators

in a column represents a vehicle assigned to this contact. Green, blue and red corresponding to operator 1, 2 and 3. We can see that often multiple operators assigned identification vehicles to the same contact before the contact was identified. Similarly, there were several times that multiple operators assigned vehicles to the same contact after it was identified. This resulted in a waste of resources.

This required conflict resolution within the team by explicit communication, which cost time. In the communication transcript, we observed messages such as these:

Bravo: *B 1 (Meaning Bravo is taking contact 1).*
Alpha: *A 1 (Meaning Alpha is taking contact 1).*

Charlie: *Let B take it—he's closer, A you take 2.*
Alpha: *Redirecting to 2.*

We can see that both Bravo and Alpha wanted to work on contact 1. This conflict was resolved via communication as Charlie asked Bravo to work on contact 1 and Alpha on contact 2. Note that there was no explicit assignment for a team leader. Leadership emerged organically within the team. Communication was necessary in this case but cost extra time. If the conflict was not resolved, it may have happened that some tasks would have multiple operators working on them while others were ignored. This kind of duplication was quantitatively analyzed based on vehicle assignment conflicts and task engaging conflicts. Vehicle assignment conflicts happened when two vehicles were assigned to the same contact. A Mann-Whitney test shows that team structure had a significant effect on the number of vehicle assignment conflicts (Chi-sq = 4.89, df = 1, p = 0.027) with more conflicts in Divisional teams (Mean = 16.65, SD = 6.38) and less conflicts in Functional teams (Mean = 9.75, SD = 11.36). Task engagement conflicts happened when two operators tried to engage the same contact to perform payload tasks. Similarly, team structure had a significant effect (Chi-sq = 23.92, df = 1, $p < 0.001$) with more conflicts in Divisional teams (Mean = 3.40, SD = 2.80) and zero conflicts in all Functional teams.

9.3.3.2 Under Utilization of Vehicles

A second reason for the poor performance of divisional teams was the under-utilization of the vehicles. Figure 9.5 shows the working process of a divisional team with constant task arrival. Each column is the timeline of one emergent contact from its appearance until it was neutralized or provided aid packets. Green is for vehicle travel time, yellow is for identification time, red is for time to neutralize the contact, and blue is for the time to complete rescue task. Dark grey is for assignment waiting time, during which the contact was waiting to be assigned a vehicle.

The dark grey periods in Fig. 9.5 are nonproductive time, which happened a lot for this team. Contacts are numbered, with a letter T added after the identification for threats or F for friendly. For example, 1 is an unidentified contact. It is updated as 1 T if identified as a threat or 1 F if friendly. The longest idle time was highlighted by the black box in Fig. 9.5. We looked at the log of communication and the actions of operators during this time, which are summarized in Table 9.1. Operator Alpha assigned a vehicle to an unknown contact 4, and later to a threat contact 0T and a friendly contact 3F. All three vehicles operated by Alpha were busy. He also reported his actions to his teammates via chat messages. Operator Bravo was working on threat contact 2T. After that, he worked on another threat contact 4T, and later assigned a vehicle to an unknown contact 7. Only one vehicle controlled by Bravo was busy at one time. Operator Charlie assigned vehicles to several contacts (4, 3F, 4T, and 0T) after he finished identifying unknown contact 3. However, all of

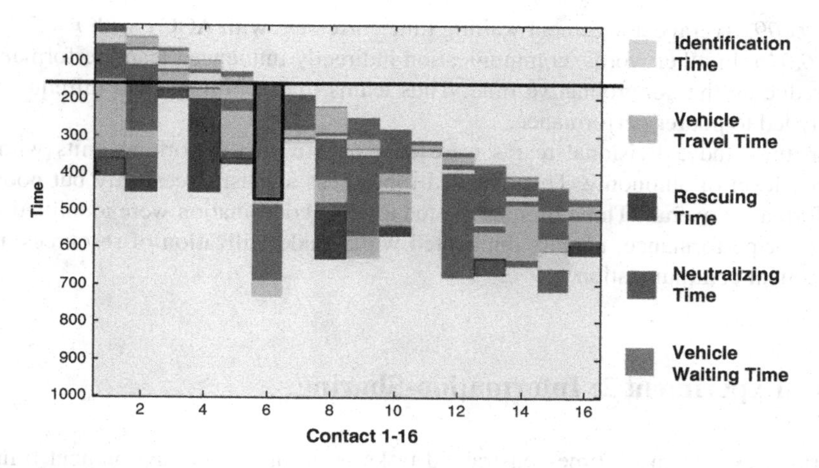

Fig. 9.5 Timeline of task completion in a divisional team

Table 9.1 Coordination and vehicle utilization

Time flow	Alpha	Bravo	Charlie
Start	Assigned to 4	Arriving at 2T	Identifying 3
Process	"ok, I will get 4 too." "and 3F"	"alpha, you take 0T"	Assigned to 4, 3F, 4T, 0T
	Assigned to 0T	Destroy 2T	"who got 0F"
	Assigned to 3F	"i got 4T"	Assigned to 1F
		Assigned to 7	
Result: #Idle	0	2	3

these contacts had already been claimed by the other two team members. Operator Charlie then asked about task allocation and assigned a vehicle to contact 1F. During this time, none of the vehicles operated by him were busy and none of them were assigned to contact 5. From these we can see the vehicles were not used to their full capacity. While there were enough idle vehicles, some tasks had no vehicle assigned to them.

9.3.3.3 Infrequent Communication

We found that chat density had an influence on the task assignment waiting time, which is the nonproductive time between the appearance of a contact and the time it was assigned a vehicle. We conducted a partial correlation analysis for Average Objective Completion Time (AOCT), average assignment waiting time, and the number of chat messages. Team structure, arrival process, and trial sequence were controlled in order to separate the influence of communication. Although chat density did not have a significant correlation with the overall objective completion time, it negatively correlated with average assignment waiting time with $r = -0.427$,

$p = 0.009$. Average assignment waiting time correlated with AOCT with $r = 0.392$, $p = 0.016$. In other words, communication indirectly influenced team performance by reducing the nonproductive time. Thus teams that communicated infrequently likely led to poorer performance.

In this study, divisional teams were designed to create working units with a higher level of autonomy. They showed robustness against uncertainly but poorer performance overall. Three reasons related to team coordination were identified for the poor performance, namely duplicated work, underutilization of resources and infrequent communication.

9.4 Experiment 2: Information-Sharing

Working as a team on time-constrained tasks in an uncertain environment brings many challenges. To achieve high performance, team situation awareness, communication and coordination are critical. It is important that team members understand what their teammates are doing and get the required information in a timely manner. While explicit communication can be time consuming, supporting implicit information-sharing via the user interface could be more effective and efficient.

These considerations motivated a second experiment. Based on the three reasons identified for the poor performance in divisional teams in Experiment 1, we conducted Experiment 2 to study how teams can be structured and supported by technology to be both flexible and efficient. Specifically, we investigated the effect of enhanced information-sharing tools under different uncertainty levels in divisional teams.

9.4.1 Independent Variables

A 2×4 repeated measures experiment was conducted. The first independent variable was the uncertainty level, which was defined by inter-arrival time of unidentified contacts (constant, erratic), as in Experiment 1.

The second independent variable was information-sharing condition. It was designed to see whether team performance could be improved by enhancing team situation awareness, implicit coordination and communication through information-sharing. In Experiment 1, it was found that *divisional* teams were more robust to uncertainty but had overall performance degradation when compared to *functional teams,* due to duplication of task assignment, under-utilization of resources, and infrequent communication. Information sharing (or lack thereof) was the source of this discrepancy.

There are different ways to share information in teams. The most common way is explicit communication, which is supported by chat panel in the testbed. However, explicit communication is time consuming and poses extra workload on human

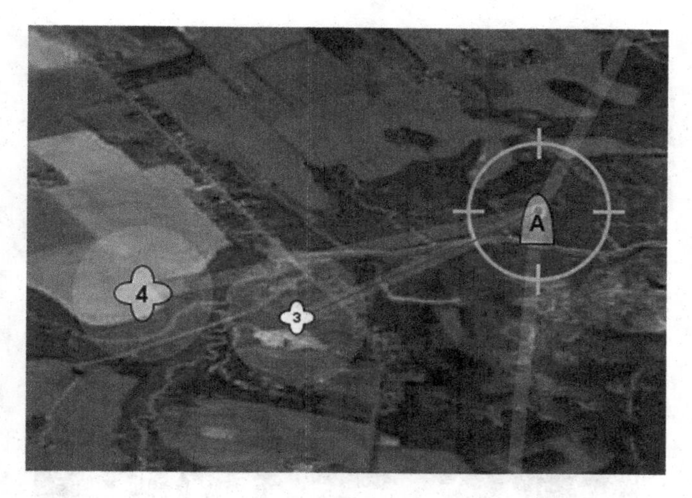

Fig. 9.6 Icon differentiation

memory. In Experiment 1, teams with worse performance also communicated less. Without the aid of an information-sharing feature, team members had to rely on explicit communication, which was inhibited when the task load was high, ultimately resulting in degraded performance. To this end, four conditions were compared in the experiment: baseline, icon differentiation, status list, and both:

- In the baseline condition, no additional information-sharing mechanism was provided.
- In the icon differentiation condition, contacts that had been assigned a vehicle would change color to white and reduce in size, as shown in Fig. 9.6. We wanted to separate the contacts that had already been claimed from others. We used the color white because it is neutral and has enough contrast with the darker background. We also wanted to minimize them so that team members could devote their resources to contacts that had not yet been claimed.
- In status list condition, the IDs of contacts that had not been assigned any vehicle were listed in a table by three categories, as shown in Fig. 9.7. The list could be hidden by clicking the checkbox on top of the list. The status list conveyed the same information as in icon differentiation. However, there were two major differences in terms of visualization. Unlike using white to differentiate the contact, people could still tell the type of contact from the icon color. However, looking at the status list required longer eye movement and extra time in visual search to match an ID in the list to an icon on the map.
- In the last condition, both the icon differentiation and the status list were presented.

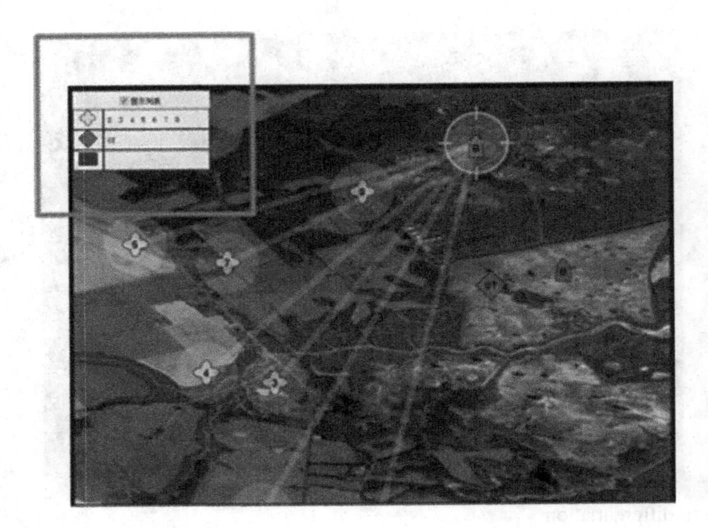

Fig. 9.7 Status list

9.4.2 Dependent Variables

Dependent variables include team performance, measures of team coordination processes, subjective workload and user preference. Team performance was measured using the Average Objective Completion Time (AOCT), as in Experiment 1. Segments of the objective completion time, including identification time, neutralizing task time, rescue task time, vehicle travel time, and assignment waiting time, were also calculated. Measures of team coordination processes included time spent in monitor mode, communication time, vehicle assignment conflicts, and task engagement conflicts. Subjective workload was measured using the NASA-TLX rating (Hart 2006). User preference was user's ranking of the four information-sharing conditions based on their preference.

9.4.3 Participants

A total of 81 participants, participated in the experiment. Participants were tested in groups of three. Data from three groups were removed because the test was not completed due to system errors. The remaining 72 participants were aged 18–28 years old, with an average of 22.3 years old and a standard deviation (SD) of 1.64. Among them, 27 were female and 45 were male. All of the participants were undergraduate or graduate students. Participants were asked for the number of hours they played electronic games per month on average. They also rated their experience on visual searching games, first person shooting games, real time strategy games, and team games respectively on a five-point Likert scale, with high values indicating more game experience. Self-report team game experience was found to have a significant correlation with AOCT ($r = -0.417$, $p = 0.042$), and average workload in the team ($r = -0.494$, $p = 0.014$). In other words, teams that had more team game experience tended to finish the tasks faster and with lower workload. Game experience on other categories was not significantly correlated with either performance or the workload.

9.4.4 Procedure

Participants were tested in groups of three under a single uncertainty level: either constant or erratic arrival process. The participants were in the same room, but could not see other team members' displays. The experiment began with a training session introducing the testbed interface, tasks, and the mission goal. Participants then practiced for a complete session under the baseline condition. After that, they were instructed to discuss their team strategy for 5 min before beginning the four test sessions with different information-sharing conditions. Their sequences were randomized and counter-balanced. Before each test session, the information-sharing feature used in this session was explained. Each trial was completed when all 16 contacts were processed. Subjective workload was measured using NASA-TLX rating at the end of each session. Participants ranked the four conditions based on their preference and provided comments after all the sessions were completed.

9.4.5 Results

Data logged during the experiment were post processed to obtain performance and process data. The results were analyzed based on the four experiment sessions from four aspects: task performance, team coordination measures, subjective workload and user preference. Data in the training session was not included in the analysis.

9.4.5.1 Team Performance

MANOVA was used for the analysis of team performance. No multivariable outlier was found. The assumption of homogeneity covariance matrices was satisfied across the four information-sharing conditions (Box's $M = 72.82$, $df = 63$, $p = 0.43$), but not across the two uncertainty levels (Box's $M = 61.8910$, $df = 21$, $p < 0.001$). However, since all the cells had equal sample size, MANOVA was still used because of the correlation among dependent variables. Significant differences were found among the two uncertainty levels on the dependent variables (Pillai's criterion $= 0.703$, $F(6, 22) = 6.72$, $p < 0.001$). The combined dependent variables

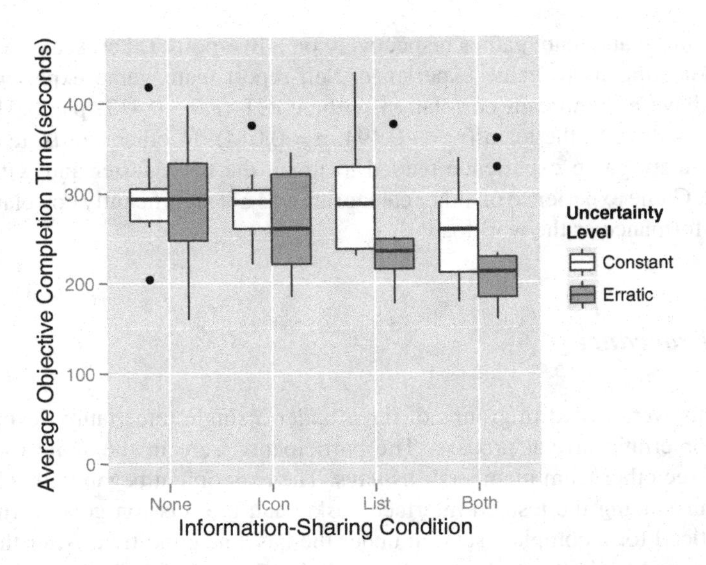

Fig. 9.8 Boxplot of average objective completion time (AOCT)

were also significantly affected by the information-sharing condition (Pillai's criterion = 0.66, F (18, 66) = 2.99, p < 0.001). The interaction effect between information-sharing condition and uncertainty level was not significant.

Univariate analyses of variance (ANOVA) for each dependent variable were conducted as follow-up tests to the MANOVA. Using the Bonferroni method for controlling Type I error rates for multiple comparisons, an alpha level of 0.008 was used. All the time related variables were measured in seconds. For AOCT (Fig. 9.8), information-sharing condition was found to have a significant effect (F (3, 66) = 4.35, p = 0.007). The condition in which both icon differentiation and status list were presented resulted in the fastest objective completion time (Mean = 242.03, SD = 56.97), followed by status list condition (Mean = 262.20, SD = 62.80), icon differentiation (Mean = 277.38, SD = 48.03) and baseline (Mean = 292.01, SD = 59.02). The main effect of uncertainty level and the interaction effect on AOCT were not significant.

Uncertainty level and the information-sharing condition did not have significant impacts on the time to complete payload tasks (identification, rescuing or neutralization), and the assignment waiting time. Total vehicle travel time (Fig. 9.9), which was sum of the time between when a vehicle was assigned to a contact to the time this vehicle arrived, was not significantly affected by uncertainly level. The information-sharing condition had a significant effect on travel time (F (3, 66) = 6.10, p < 0.001). The condition with both the status list and icon differentiation had the shortest travel time (Mean = 39.83, SD = 6.63), followed by status list condition (Mean = 42.29, SD = 9.74), icon differentiation (Mean = 47.30, SD = 9.97) and baseline (Mean = 47.75, SD = 7.22). Since the speed of the vehicles were preset by the system, a decrease on travel time means less

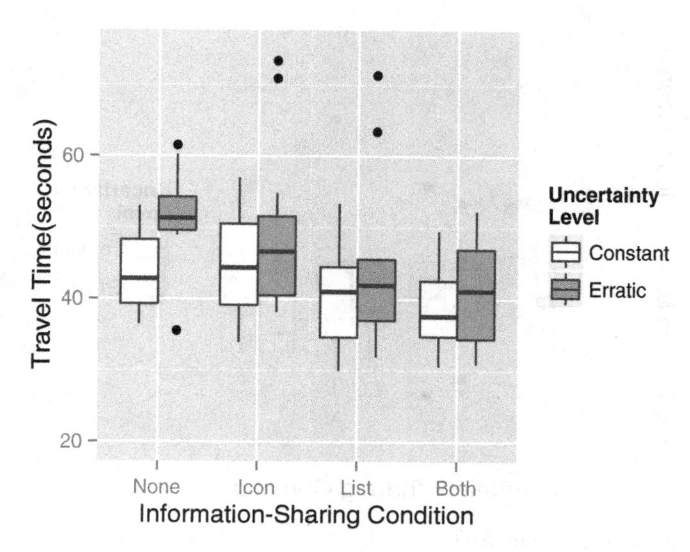

Fig. 9.9 Boxplot of vehicle travel time

distance travelled. The reason was the better coordination enabled by information-sharing tools, which either reduced chances that two vehicles travelled to the same contact or matched the contacts with vehicles better based on the distances between them.

9.4.5.2 Team Coordination

Team coordination was measured from four aspects: time spent in monitor mode, communication time, vehicle assignment conflicts, and task engagement conflicts. A significance level of 0.05 was used.

For total time spent in monitor mode (Fig. 9.10), uncertainty level ($F(1, 22) = 4.57$, $p = 0.044$) and information-sharing condition ($F(3, 66) = 3.25$, $p = 0.027$) both had significant effects. Teams with erratic arrival process spent longer time (Mean = 260.71, SD = 97.71) in monitor mode than those with constant arrival process (Mean = 201.17, SD = 74.01). Information-sharing tools reduced the time participants spent in monitor mode. The condition in which both icon differentiation and status list were presented resulted in the shortest time spent in monitor mode (Mean = 206.04, SD = 89.44), followed by status list condition (Mean = 225.25, SD = 102.11), icon differentiation (Mean = 231.58, SD = 89.42) and baseline (Mean = 260.88, SD = 79.79). This was because the information-sharing tools facilitated task assignment and team coordination. Because of the decision-aiding tools, participants could observe the status of contacts more directly from the interface instead of using the monitor mode to figure it out.

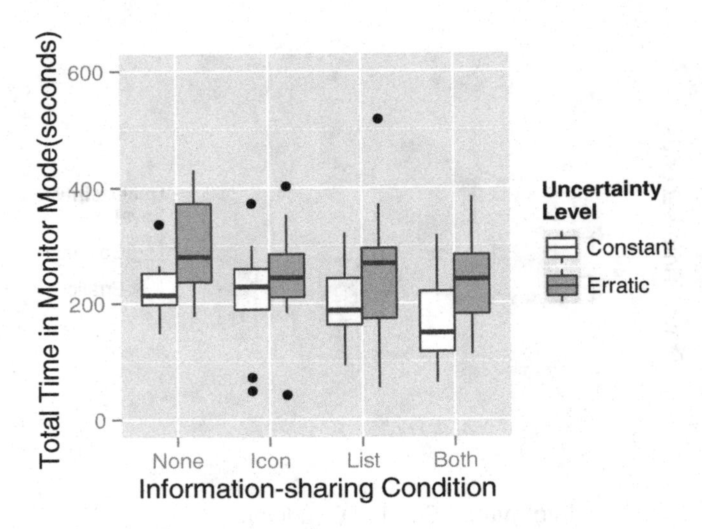

Fig. 9.10 Time spent in monitor mode

Similarly, the information-sharing condition had a significant effect on the amount of communication, as measured by the number of chat messages (Friedman chi-squared = 23.005, df = 3, $p < 0.001$). When there was no information-sharing tool, team members had to communicate more for task assignment and coordination (Mean = 13.75, SD = 12.42). When information-sharing tools were presented, the need for explicit communication was reduced. The average number of chat messages was 5.79 (SD = 6.83) with icon differentiation, 5.21 (SD = 5.44) with status list, and 5.00 (SD = 5.18) when both were presented. Arrival process did not impact the amount of communication significantly.

Content of the communication was coded and categorized into five categories: leadership, information prompt, information request, strategy, and other. Leadership contains requests for another team member to work on a certain contact or area, as well as the confirmation and denial of these requests. Information prompt includes reporting the area or contact one is working on, reporting the places one is going to, and negotiation in case of conflicts. Information request includes asking if there is a team member working on a certain contact and who the team member is. Strategy contains discussion on general strategies, such as which area each team member should be responsible for. All the other communications were included in the last category. These are usually not related to the working process, such as open comments and summary about the mission at the end of trials. For each category, number of chat messages was summed across different teams within each information-sharing condition. As shown in Fig. 9.11, while the baseline condition had the most communication for all categories, its difference with other conditions was the largest for information prompt. In other words, information-sharing tools

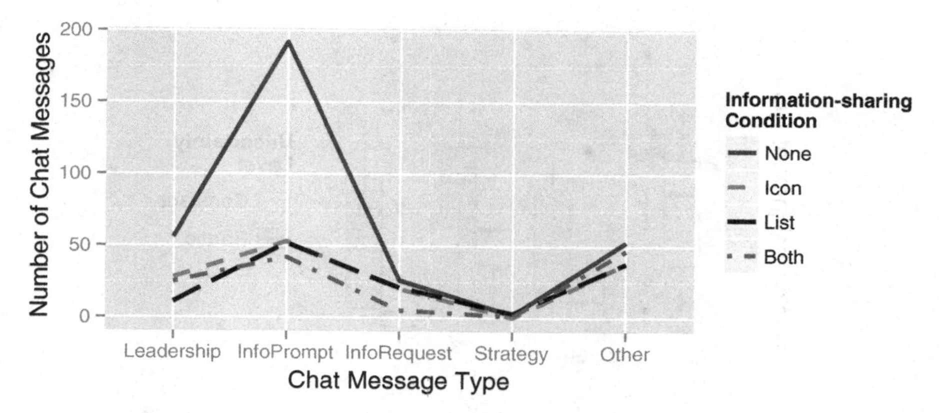

Fig. 9.11 Amount of communication by message type

reduced communication amount mostly by reducing need to report ones' intentions and actions. The presentation of the information-sharing tools was counterbalanced so the impact from any learning effect was limited.

Vehicle assignment conflicts (Fig. 9.12) and task engagement conflicts (Fig. 9.13) reflected the result of team coordination more directly. As shown in Study 1, duplicated work was identified as one reason that contributed to the poor performance of divisional teams. In this study, we found that information-sharing tools significantly affected both vehicle assignment conflicts (Friedman chi-squared = 51.859, df = 3, $p < 0.001$) and task engagement conflicts (Friedman chi-squared = 13.268, df = 3, $p = 0.004$). In other words, information-sharing tools reduced duplicated work in teams, making the teamwork more efficient. Among the four information-sharing conditions, icon differentiation and the one with both icon differentiation and status list presented had the least number of conflicts. Status list did not result in much improvement comparing to baseline condition. Uncertainty level also had a significant impact on vehicle assignment conflicts ($W = 734.5$, $p = 0.002$) and task engagement conflicts ($W = 862$, $p = 0.032$). Erratic arrival process resulted in more conflicts compared to constant arrival process.

Participants could choose to hide the status list by clicking the check box on its top. We calculated the time that the status lists were hidden for the two conditions when status list was presented. On average, teams chose to hide the status list for 20 % of the total mission time (SD = 20.05). The list-hidden time and the total time in monitor mode were positively correlated ($r = 0.38$, $p = 0.007$). In other words, participants that chose to hide the status list spent more time in monitor mode as compensation.

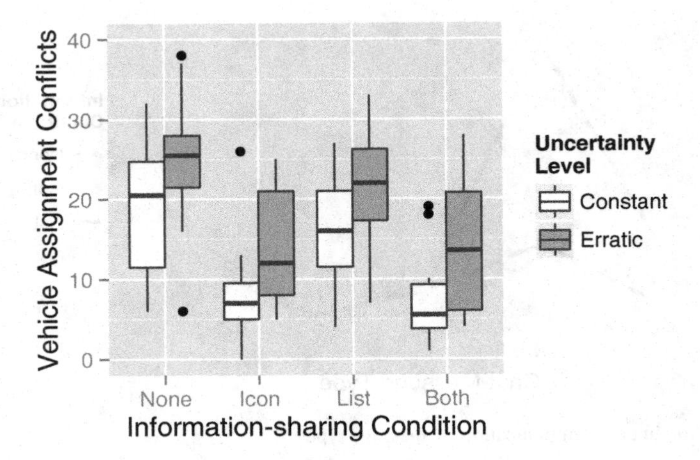

Fig. 9.12 Vehicle assignment conflicts task engagement conflicts

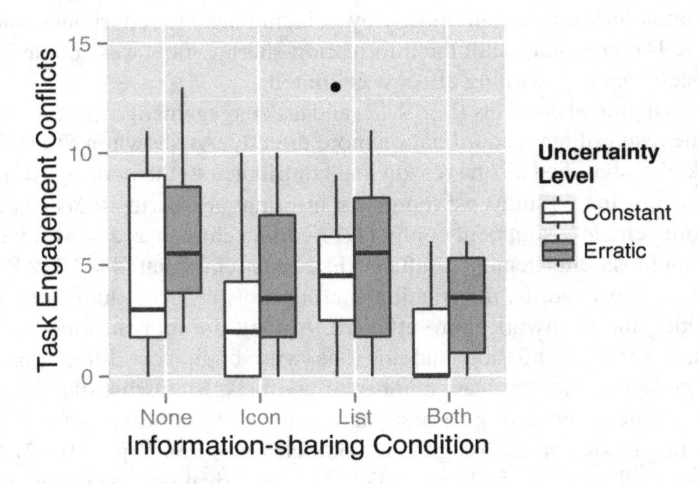

Fig. 9.13 Task engagement conflicts

9.4.5.3 Workload

Subjective workload was rated on a scale from zero to one hundred using the NASA-TLX rating. Information-sharing condition was found to have a significant effect on average subjective workload in teams ($F(3, 210) = 3.57$, $p = 0.015$). The condition in which both icon differentiation and status list were presented resulted in the lowest workload (Mean = 45.87, SD = 12.74), followed by status list condition (Mean = 47.45, SD = 12.08), icon differentiation (Mean = 47.78, SD = 12.42) and baseline (Mean = 49.76, SD = 12.36). The main effects of uncertainty level and the interaction effect were not significant (Fig. 9.14).

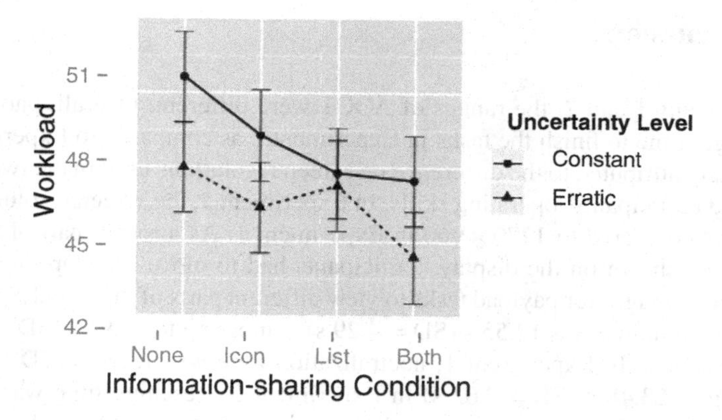

Fig. 9.14 Subjective workload

9.4.5.4 User Preference and Comments

Participants' ranking of the four information-sharing conditions were analyzed using Kruskal-Wallis test. There was a significant difference among the four conditions (Chi-sq = 62.05, df = 3, $p < 0.001$). Icon differentiation and the one with both icon differentiation and status list ranked equally as the top choices, followed by status list, and the baseline condition, which required the participants to use the monitor mode button.

Participants commented that the baseline condition was not convenient, difficult for coordination, easy to have task assignment conflicts, and required extra communication. On the positive aspect, some participants thought this condition was the most interesting because it was challenging. The interface was clearer with no distractions. For icon differentiation, they commented that it was easy to observe which contact has been claimed and assign tasks accordingly. However, it was difficult to determine the type of contacts with the change of color. For the status list, the information was also useful for team coordination, but was less easy to interpret than the icon differentiation. It was most useful for detecting new contacts or checking whether some contacts had been forgotten. The negative side was that it blocked part of the map, although it could be closed. When both icon differentiation and status list were presented, besides the advantages and disadvantages of each tool, some participants felt the information provided in these two could compensate for each other. On the other hand, the interface was more complex and had more distractions. These comments could be used to further improve the design of information-sharing tools.

9.5 Discussion

In Experiments 1 and 2, the ranges of AOCT were different. Overall, participants spent longer time to finish the tasks in Experiment 2 as compared to Experiment 1. This is likely attributed to the difference on screen resolutions used in the two experiments and participants' operating skills. In Experiment 2, the screen resolution was 1024×768 compared to 1270×960 in Experiment 1. As a result, part of the map could not be shown on the display. Participants had to move the map in the main interface and images for payload tasks to view different parts of them. In Experiment 1, identification time was 12.55 s (SD = 4.29 s) comparing to 16.91 s (SD = 4.88 s) in Experiment 2. In Experiment 1, neutralization time was 13.90 s (SD = 6.11 s) comparing to 23.48 s (SD = 7.62 s) in Experiment 2. Rescuing time was 34.74 s (SD = 14.60 s) in Experiment 1 comparing to 52.88 s (SD = 17.43 s) in Experiment 2. Although a direct comparison between the two experiments was not possible in terms of the visual task, the increase of the overall objective completion time should not affect the assessment on the effectiveness of information-sharing tools.

In Experiment 1, we found that teams communicating infrequently had worse performance. In Experiment 2, teams that performed better with information-sharing tools also had less communication. This is because the information-sharing tools served as an implicit communication channel. With these tools, information on contact status and task assignment could be retrieved directly from the interface, reducing the need for time-consuming explicit communication. When such tools were not available, infrequent communication could not provide sufficient information for team coordination, resulting in duplication on task assignment and suboptimal use of team resources.

9.6 Conclusion

In this study, we conducted two experiments to investigate the impact of team structure, uncertainty on task load, and information-sharing tools on team coordination and team performance. In Experiment 1, it was found that divisional teams were more robust against the uncertainty for task arrival processes in terms of team performance. However, this robustness was achieved with an overall worse performance as compared to functional teams. Three reasons for the poor performance were identified, namely duplication on task assignment, under utilization of vehicles, and infrequent communication. In an effort to achieve robust and effective teamwork, the usage of information-sharing tools was explored in Experiment 2. It was found that information-sharing tools reduced the duplication on task assignments, improved overall task performance, and reduced workload in divisional teams.

The conclusions of this study could be useful for the design of human-agent team structure and the development of tools to support teamwork. Consistent with previous research, divisional teams were better able to cope with uncertainty. This reflects on their robustness against different task arrival process. However, divisional teams could have worse performance if the responsibilities of team members are not clear and the communication is not efficient. By providing information-sharing tools for divisional teams, their performance could be improved by reducing the chances of duplicated work and improving coordination, achieving effective and robust teamwork.

The four information-sharing methods resulted in performance improvement at different levels. All reduced the average time required to complete a task and the workload of operators. The best result was achieved when both a status list and icon differentiation were presented. Although we intended to design the two mechanisms to convey the same information, experimental results showed that they actually compensated for each other. People use these two mechanisms in different ways. Icon differentiation was more effective when people wanted to decide whether to work on a specific task. The status list was more effective when people wanted to get an idea of overall progress and strategically allocate tasks among team members. The specific interface design used in this study was not optimized, which could be improved using further usability studies. The key message is that by facilitating information-sharing among team members, the advantage on flexibility and robustness of divisional teams can be maintained while the disadvantages in terms of coordination cost can be limited.

References

Biehl JT, Czerwinski M, Smith G, Robertson GG (2007) FASTDash: a visual dashboard for fostering awareness in software teams. Paper presented at the Proceedings of the SIGCHI conference on human factors in computing systems

Blickensderfer E, Cannon-Bowers JA, Salas E (1997) Training teams to self-correct: an empirical investigation. Paper presented at the 12th annual meeting of the Society for Industrial and Organizational Psychology, St. Louis

Bolstad CA, Cuevas HM (2010) Team coordination and shared situation awareness in combat identification. SA Technologies, Marietta

Bolstad CA, Endsley MR (1999) Shared mental models and shared displays: an empirical evaluation of team performance. Paper presented at the proceedings of the human factors and ergonomics society annual meeting

Bolstad CA, Endsley MR (2000) The effect of task load and shared displays on team situation awareness. Paper presented at the proceedings of the human factors and ergonomics society annual meeting

Bowers CA, Oser RL, Salas E, Cannon-Bowers JA (1996) Team performance in automated systems. In: Parasuraman R, Mouloua M (eds) Automation and human performance: theory and applications. Lawrence Erlbaum, Mahwah, pp 243–266

Endsley M, Jones WM (1997) Situation awareness information dominance & information warfare: DTIC Document

Endsley MR (1995) Toward a theory of situation awareness in dynamic systems. Hum Factors 37(1):32–64

Entin EE, Serfaty D (1999) Adaptive team coordination. Hum Factors 41(2):312–325

Galbraith JR (1974) Organization design: an information processing view. Interfaces 4(3):28–36. doi:10.1287/inte.4.3.28

Gladstein DL (1984) Groups in context: a model of task group effectiveness. Admin Sci Quart 29(4):499–517

Hart SG (2006) Nasa-task load index (NASA-TLX); 20 years later. Proc Hum Factors Ergonomics Soc Annual Meeting 50(9):904–908. doi:10.1177/154193120605000909

Lewis M, Wang H, Chien SY (2011) Process and performance in human-robot teams. J Cogn Eng Decis Making 5(2):186–208

Macmillan J, Entin EE, Serfaty D (2004) Communication overhead: the hidden cost of team cognition. In: Salas E, Fiore SM (eds) Team cognition: understanding the factors that drive process and performance. American Psychological Association, Washington, DC

McKendrick R, Shaw T, de Visser E, Saqer H, Kidwell B, Parasuraman R (2013) Team performance in networked supervisory control of unmanned air vehicles effects of automation, working memory, and communication content. Hum Factors 56:463–475

Mekdeci B, Cummings M (2009) Modeling multiple human operators in the supervisory control of heterogeneous unmanned vehicles. Paper presented at the PerMIS'09, Gaithersburg

Naylor JC, Dickinson TL (1969) Task structure, work structure, and team performance. J Appl Psychol 53:10. doi:10.1037/h0027350

Orasanu J (1990). Shared mental models and crew decision making: DTIC Document

Oser RL, Prince C, Morgan Jr BB, Simpson SS (1991) An analysis of aircrew communication patterns and content: DTIC Document

Parush A, Kramer C, Foster-Hunt T, Momtahan K, Hunter A, Sohmer B (2011) Communication and team situation awareness in the OR: Implications for augmentative information display. J Biomed Inform 44(3):477–485

Parush A, Ma C (2012) Team displays work, particularly with communication breakdown: performance and situation awareness in a simulated forest fire. Paper presented at the proceedings of the human factors and ergonomics society annual meeting.

Salas E, Fiore SM (2004) Team cognition: understanding the factors that drive process and performance. American Psychological Association, Washington, DC

Salas E, Sims DE, Burke CS (2005) Is there a big five in teamwork? Small Group Res 36(5): 555–599. doi:10.1177/1046496405277134

Shah J, Breazeal C (2010) An empirical analysis of team coordination behaviors and action planning with application to human' Äìrobot teaming. Hum Factors 52(2):234–245

Steiner JD (1972) Group process and productivity. Academic, New York

Stout RJ, Cannon-Bowers JA, Salas E (1996) The role of shared mental models in developing team situational awareness: implications for training. Training Res J 2(85–116):1997

Stout RJ, Cannon-Bowers JA, Salas E, Milanovich DM (1999) Planning, shared mental models, and coordinated performance: an empirical link is established. Hum Factors 41(1):61–71

Thompson JD (1967) Organizations in action: social science bases of administrative theory. McGraw-Hill, New York

Work H, Chown E, Hermans T, Butterfield J (2008) Robust team-play in highly uncertain environments. Paper presented at the proceedings of the seventh international joint conference on autonomous agents and multiagent systems, vol. 3, Estoril

Chapter 10
Measuring Trust in Human Robot Interactions: Development of the "*Trust Perception Scale-HRI*"

Kristin E. Schaefer

10.1 Introduction

Robotics technology has vastly advanced in recent years leading to improved functional capability, robust intelligence, and autonomy of the system. However, along with the added benefits of these technical advancements also come changes in the way in which humans will use or interact with the system. The most prevalent change can be seen in the vision for robot design and development for future human-robot interaction (HRI). This vision is now directed toward a greater prevalence of robotic technologies in context-driven tasks that require social-based interaction. More specifically, robots are beginning to shed their more passive tool-based roles and move more towards being an active integrated team member (Chen and Terrence 2009). The intricacies of HRI are bound to change in order to accommodate the integration of a robot as it becomes more of a companion, friend, or teammate, rather than strictly a machine. Thus, this change in direction has a direct translation of the human role as less of an operator and more of a team member or even a bystander. Thereby, the individual's trust in that robot takes a prominent role in the success of any interaction, including the future use of said robot.

This chapter works in conjunction with the rest of the book such that the prevalent focus is on the intersection of robust intelligence (RI) and trust in robotic systems. Therefore, to reduce redundancy, we will limit the background to point to the difficulties relating to trust development and the potential issues that occur when trust is not developed appropriately. First, it is important to note that there are a number of factors that influence trust development. Large scale literature reviews of both HRI and human-automation interaction point to the importance of trust

K.E. Schaefer (✉)
US Army Research Laboratory, Human Research and Engineering Directorate, 459 Mulberry Point Rd., Aberdeen Proving Ground, MD 21005, USA
e-mail: kristin.e.schaefer2.ctr@mail.mil

© Springer Science+Business Media (outside the USA) 2016

R. Mittu et al. (eds.), *Robust Intelligence and Trust in Autonomous Systems*,
DOI 10.1007/978-1-4899-7668-0_10

antecedents relating to the human, the robot, and the environment (Hancock et al. 2011; Schaefer et al. 2014). The key findings from the associated meta-analyses point to the fact there is still much to learn about how trust develops. However, what is prevalent in the literature is the finding that until trust between a human and a robot is solidly established, robotic partners will continue to be underutilized or unused, therefore providing little to no opportunity for trust to develop in the first place (Lussier et al. 2007). This is in part due to the fact that one of the most significant challenges for successful collaboration between humans and robots is the development of appropriate levels of mutual trust in robots (Desai et al. 2009; Groom and Nass 2007). So, regardless of the domain of application, the environment, or the task, a human's trust in their non-human collaborator is an essential element required to ensure that any functional relationship will ultimately be effective.

Research has continued to address the creation and validation of successful evaluation methods for a wide spectrum of HRI issues, including this issue of human-robot trust (Steinfeld et al. 2006). Yet, a limitation in the field has been related to accurate measurement of trust specific to the unique nature of HRI. Human-robot trust is currently measured through subjective assessment. However, these previous studies have been limited by using measurement tools that are a single self-report item (e.g., How much do you trust this robot?) or are an adapted human-interpersonal or human-automation trust scale. The concern with this methodology is that neither of those options truly assesses the full scope of human-robot trust, and brings to question the accuracy of the trust scores. There has been one notable exception: Yagoda and Gillan (2012) developed a subjective human-robot trust scale that is specific to military application. However, the changing vision of HRI continues to press the inclusion of robotic technologies into multiple contextual domains that incorporate varying levels of autonomy, intelligence, and interaction. This calls forth the need for the development of additional trust measurement tools specific to the changing HRI environment.

This chapter summarizes research that was conducted to produce a reliable and validated subjective measure: the *Trust Perception Scale-HRI* (see also Schaefer 2013). The goal of this research was to design a subjective tool specific to the measurement of human-robot trust that could be expressed as an overall percentage of trust. In addition, this scale was designed to effectively measure trust perceptions over time, across robotic domains, by individuals in all the major roles of HRI (operator, supervisor, mechanic, peer, or bystander, as defined by Scholtz 2003), and across various levels of system autonomy and intelligence (see also Beer et al. 2014). To ensure that this new scale was valid, each part of scale development was constructed using the widely-accepted procedures discussed in DeVellis (2003) and Fink (2009). These procedures followed the protocol of large item pool creation, statistical item pool reduction, content validity assessment, and task-based validity testing.

10.2 Creation of an Item Pool

The first step in creating the *Trust Perception Scale-HRI* was to create an Item Pool. An Item Pool is a collection of relevant phrases or items that are associated with trust development. To meet this end, over 700 articles in the areas of human-robot trust, human-automation trust, and human-interpersonal trust were reviewed and analyzed. Theoretical, qualitative, and quantitative relationships were recorded. Potential items were then organized in relation to the *Three Factor Model of Human-Robot Trust* (Hancock et al. 2011). This model was then updated to incorporate potential antecedents of trust (see Fig. 10.1). Specific items to be included in the initial Item Pool were first drawn from these large scale literature reviews.

One major trust-specific finding from these reviews was the importance of design as it related to the robot's physical form and functional capability. While some research had focused on the functional capabilities, limited experimental study had been conducted specifically related to the impact of robot form on trust development. Therefore, two initial experiments were conducted to assess this gap in the literature and further develop the initial Item Pool.

The purpose of the first study was to determine the relationship between physical form and trustworthiness, devoid of any direct information regarding the functional capabilities of the system. One hundred sixty-one participants rated 63 images of real-world industry, military, medical, service, social, entertainment, and therapy robots. These ratings included the degree to which the robot was perceived to be a machine, a robot, and an object, as well as its perceived intelligence (PI), level of

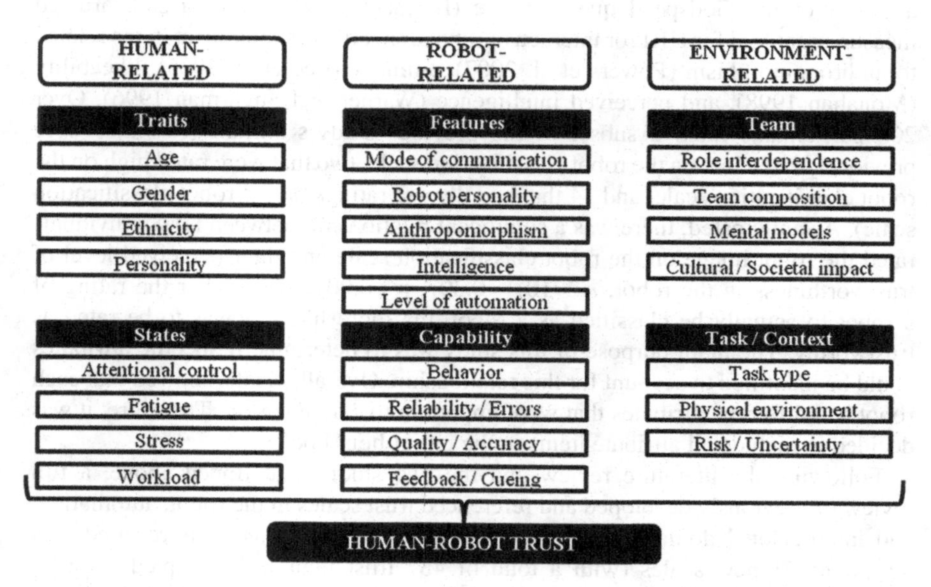

Fig. 10.1 Updated *Three Factor Model of Human-Robot Trust* following an extend literature review of trust in the interpersonal, automation, and robot domains

automation (LOA), trustworthiness, and the degree to which the participant would be likely to use or interact with the robot.

A multiple regression correlation analysis with stepwise entry of variables was conducted to determine the factors that predicted trustworthiness from perceived robot form alone. This was achieved by regressing trustworthiness onto human-related factors (gender, race, age, year in school), personality traits (agreeableness, extroversion, conscientiousness, intellect, neuroticism), negative attitudes toward robots (negative attitudes toward emotions in interactions, negative social influence, and negative situational influence), as well as self-report items of robot form (perceived intelligence, perceived level of automation (LOA), robot classification). The final model included *perceived intelligence (PI), robot classification (RC),* and *negative social influence (SI)* as predictors of *trustworthiness,* $\hat{Y} = 0.825 + 0.651(\text{PI}) + 0.256(\text{RC}) - 0.164(\text{SI})$. It accounted for a significant R^2 of 45.1 % of the variance, $F(3156) = 42.70, p < 0.001$.

These results suggested that preconceived ideas regarding the level of intelligence of a robot are form-dependent and assessed prior to interaction, in much the same way as one individual will assess another individual as a potential teammate. Further, *negative social influence* (e.g., capabilities, functions, etc.) plays a key role in expectation-setting similar to stereotypes of human teammates. Overall, the results of the above-mentioned study provided support that physical form is important to the trust that develops prior to HRI (for additional findings see also Schaefer et al. 2012).

The follow-up study was designed to identify which perceived robot attributes could impact the trustworthiness ratings. Robot attributes were assessed through a subset of the Godspeed questionnaire (Bartneck et al. 2009), a standardized measurement tool for HRI for interactive robots, specifically looking at items related to anthropomorphism (Powers et al. 2007), animacy (Lee et al. 2005), likeability (Monahan 1998), and perceived intelligence (Warner and Sugarman 1996). Over 200 participants rated a subset of the previous study's stimuli (two that were previously rated low on the robot classification scale, two that were rated high on the robot classification scale, and 14 that had diverse ratings on the robot classification scale). As anticipated, there was a significant relationship between how individuals rated the robot image on the robot classification scale and their perceived level of trustworthiness in the robot, $r(2910) = 0.307, p < 0.001$. The higher the rating of a robot to actually be classified as a robot, the more likely it was to be rated as trustworthy. The main purpose of this study was to determine if specific attributes could be identified to account for this relationship. Overall, results showed that each robot had different attributes that were important to classification. Therefore, it was decided to include all attribute items in the initial Item Pool.

Following the literature review and the two studies mentioned above, a full review of previously developed and referenced trust scales in the robot, automation, and interpersonal domains were reviewed to refine the items. This resulted in a review of 51 new scales (with a total of 487 trust items), 22 adapted versions of previously developed scales, and 13 previously developed scales (see also Table 10.1).

Table 10.1 Number of trust scales and trust items reviewed

Number of trust scales assessed	Robot	Automation	Interpersonal
Created new scales	9	30	12
Minimum number of items	1	1	1
Maximum number of items	45	31	29
Adapted previous scales	5	14	3
Previously developed scales	2	9	2
Scale were not discussed	2	11	4

	Strongly Disagree	Disagree	Slightly Disagree	Neutral	Slightly Agree	Agree	Strongly Agree
Most robots make poor teammates.	O	O	O	O	O	O	O
Most robots possess adequate decision-making capability.	O	O	O	O	O	O	O
Most robots are pleasant towards people.	O	O	O	O	O	O	O
Most robots are not precise in their actions.	O	O	O	O	O	O	O

Fig. 10.2 Example items included in the initial Item Pool

The final Item Pool resulted in the creation of 156 initial items. Between two and four items were created for each antecedent, representing equal number of positively and negatively worded (or opposite related) items. Initial scale items were written out as full sentences and referred to a general statement regarding "most robots" on a 7-point Likert-type scale (see Fig. 10.2).

10.3 Initial Item Pool Reduction

The second step in the scale development procedure was to reduce the size of the initial Item Pool using statistical procedures. These procedures began with a Principal Component Analysis (PCA) to identify potential groupings of items, as well as items that were not included in the groupings. Secondary analysis was conducted using paired samples *t*-tests to determine if the positively and negatively worded items were equal and thus could be reduced from the initial Item Pool.

10.3.1 Experimental Method

One hundred fifty-nine undergraduate students (65 males, 94 females) from the University of Central Florida took part in this study via online participation (SurveyMonkey.com). Following informed consent, participants completed the 156 randomized initial trust items. Participants then completed the demographics questionnaire that included gender, age, a mental model question, and prior experience questions. The study took approximately 30 min to complete. Participants' prior experience with robots was assessed to understand previous exposure to robotic technologies. Prior experience has been shown to be related to how an individual forms a mental model of the robot and anticipates future HRI. As expected, the sample population had prior exposure to media representations (N = 156); some minor interaction with real-world robots (N = 36); and some opportunity to control (N = 34) or build (N = 11) a real-world robot during school or club related requirements. Table 10.2 presents results of these questions.

To assess the participants' mental model of a robot, they were asked to describe what a robot looks like with an open-ended question. Mental models refer to structured, organized knowledge that humans possess which describe, explain, and predict a system's purpose, form, function, or state (Rouse and Morris 1986). The responses were coded into categories (see Table 10.3). Seventeen participants directly referenced specific robots from movies or television (e.g., R2D2, C-3P0, iRobot, AI, and Terminator; N = 14); the video game Mass Effect 3 (N = 1); real-world military robots (e.g., Predator, N = 2); and a robotic arm (N = 2).

Table 10.2 Participants prior experience with robots

Prior experience questions	Yes (%)	No (%)
Have you ever watched a movie or television show that includes robots?	98	2
• 1–5 shows (N = 87)		
• 6–10 shows (N = 31)		
• Over 10 shows (N = 18)		
Have you ever interacted with a robot?	23	77
• Museum or theme park animatronics (N = 5)		
• Toys such as Furby (N = 8)		
• Robot vacuum (N = 2)		
• Classroom robots or Battlebots (N = 8)		
• Everyday items such as cell phone, computer, ATM, or Xbox (N = 12)		
• Unclassified (N = 1)		
Have you ever built a robot?	7	93
• Classroom or robotics club robots		
Have you ever controlled a robot?	21	79
• Teleoperation or remote control (N = 21)		
• Speech, Gesture, Commands (N = 3)		
• Computer programmed (N = 6)		

Table 10.3 Coding categories of mental model of a robot

Coding description	N	%
Machine-like (machine, metallic, silver)	121	76.1
Human-like (human-like or specific human features)	49	30.8
Varied (multiple descriptions or ranges of robots)	28	17.6
Tool	4	2.5
Task, Function, or Interaction	30	18.9
Internal Form: Computer, electronics, wires, buttons	25	15.7
External Form: shape, size, rigid, durable	34	21.4
Capabilities: movement	33	20.8
Communication: language	7	4.4
Other (helpful, intelligent, cameras, robot, alien)	7	4.4

10.3.2 Experimental Results

All data were analyzed using IBM SPSS v.19 (SPSS 2010), with an alpha level set to 0.05, unless otherwise indicated. These findings were important as they provided potential cause as to whether to retain or reject specific items from the initial Item Pool.

PCA was performed on the 156 initial trust items. Extraction was used to identify 43 components (using the Kaiser Criterion of Eigenvalue >1 for truncation), accounting for 79.63 % of the variance. Following review of the scree plot, four components were retained. The un-rotated solution was subject to orthogonal varimax rotation suppressed below |0.30|. In the rotated model, the four components accounted for 30.64 % of the variance. In looking at the loadings in the Rotated Component Matrix, 22 items with high loadings (>0.60) were located in Component 1. Based on the loadings of trust items on each of the four components, interpretations can be made about the factors themselves. Component 1 seemed to represent performance-based functional capabilities of the robot. Component 2 seemed to represent robot behaviors and communication. Component 3 may represent task or mission specific items. Finally, Component 4 seemed to represent feature-based descriptors of robots. These components supported the theory addressed by the descriptive *Three Factor Model of Human-Robot Trust* (first described by Hancock et al. 2011). Following PCA, 26 items were considered for immediate removal from the Item Pool.

Means, standard deviations, normality (skewness and kurtosis), correlations, z-scores, and paired samples *t*-tests were conducted to further assess items for retention or removal. To be retained in the scale, items should retain normality. Therefore, 62 items with significant skew and 20 items with significant kurtosis were considered for removal from the Item Pool. In addition, paired samples *t*-tests were conducted on all of the paired items (positive and negatively worded items) to determine if they were interchangeable, thus reducing the item pool. The results of this assessment resulted in 39 paired items that were found to be not significantly different from each other. These results provided a rationale for reducing the scale by an additional 39 items.

Even though some elements might have been considered for removal, the following items were retained for subject matter expert (SME) review due to their importance to trust theory: move quickly, move slowly, require frequent maintenance, autonomous, led astray by unexpected changes in the environment, work in close proximity with people, possess adequate decision-making capability, make sensible decisions, openly communicate, and communicate only partial information. Following the various statistical assessments (PCA, normality assessment, and paired samples *t*-tests), the Item Pool was reduced from 156 items to 73 items.

10.3.3 Key Findings and Changes

Two major changes were made to the scale following this study. First, there were some potential issues that arose with the wording of the items. Two main types of item formation were included in the above version of the scale. Items either began with "Most robots" or "I." This may have impacted the factor creation. Therefore, all items were reduced to a single word or short phrase prior to subject matter expert (SME) review. Secondly, the scale was modified from a 7-point Likert-type scale to a percentage scale with 10 % increments. The decision to make this change in the scale was related to larger purpose to develop a scale that provided a trust rating from no trust (0 %) to complete trust (100 %). This change was supported by research, especially in the interpersonal and e-commerce domains that suggest trust and distrust are viewed as related but separate constructs with differing effects on behavior, consequences, and outcomes (Lewicki et al. 1998; McKnight et al. 2004; Wildman 2011; Wildman et al. 2011).

10.4 Content Validation

The third step in the process to create a reliable and valid subjective scale was content validation. In this step, the goal was to survey SMEs in the area of trust and robotics in order to determine if each item should be retained or removed from the Item Pool. This two-phase semantic analysis included item relevance (content validation) using the protocols described by Lawshe (1975), and the identification of the hypothetical range of differences (e.g., no trust and complete trust differences) for each item.

Table 10.4 Years of experience for the subject matter experts

SME	Robot design	Robot operator	Robot research	HRI	Automation design	Automation research	Trust research
SME 1	0	0	8	8	0	8	0
SME 2	5	4	3	0	0	0	0
SME 3	4	0	4	0	2	2	0
SME 4	7	0	7	0	0	0	0
SME 5	4	8	8	8	0	0	0
SME 6	0	0	0	0	0	7	7
SME 7	11	0	11	11	0	0	3
SME 8	0	0	0	8	0	8	6
SME 9	0	0	4	0	0	4	0
SME 10	7	0	7	0	0	0	0
SME 11	0	0	10	10	20	30	15

All results are reported in years of experience

10.4.1 Experimental Method

Eleven SMEs were included from the United States Army Research Laboratory, United States Air Force Research Laboratory, and faculty members from university research laboratories. All SMEs were considered experts in the fields of trust research, robotics research, or HRI. Table 10.4 provides the SME's years of experience across a variety of robot, automation, and research topics.

SMEs were contacted via email. Upon agreement to participate, they were provided a link to complete an online survey. All data for this experiment were collected through an online tool (SurveyMonkey.com). SMEs were provided background information, purpose, and a brief review of trust theory prior to beginning the multi-part study. In Part 1, SMEs were provided background information, purpose, and a brief review of trust theory prior to beginning the multi-part study. In Part 1, SMEs completed an expertise questionnaire. In Part 2, SMEs were given instructions to complete the 73 item Trust Scale with the instructions "Please rate the following items on how a person with little or no trust in a robot would rate them." In Part 3, SMEs were given instructions to complete the 73 item Trust Scale with the instructions "Please rate the following items on how a person with complete trust in a robot would rate them." All items in Part 2 and Part 3 were randomized. Part 4 was the Content Validation questionnaire based on Lawhe (1975) content analysis protocols. SMEs rated each item on a 3-point Likert-type scale as either "extremely important to include in scale," "important to include in scale," or "should not be included in scale." SMEs could also mark if they felt an item was domain specific (e.g., military robotics, social robotics, etc.). A comment box was available to provide any clarification about why they rated the item a specific way, to provide additional recommendations to the scale design, or to suggest items that may be missing from the scale. The total survey took approximately 30 min to complete.

10.4.2 Experimental Results

Items were analyzed using the Content Validity Ratio developed by Lawshe (1975). The Content Validity Ratio (CVR), depicted in Eq. 10.1, is a commonly used method of analyzing scale items (see also Yagoda and Gillan 2012). The CVR equation was derived from a 3-point Likert scale (1 = Should not be included in scale, 2 = Might be important to include in scale, and 3 = Extremely important to include in scale).

$$CVR = (n_e - N/2) \ / \ (N/2) \tag{10.1}$$

CVR = Content Validity Ratio
n_e = Number of SMEs indicating that an item is Extremely Important
N = Total number of SMEs

Lawshe's protocol suggested that 11 SMEs with a criterion set to 0.59 are needed to ensure that the SME agreement is unlikely to be due to chance. The formula yielded values ranging from +1 to −1. Positive values indicated that at least half of the SMEs rated the item as Extremely Important. Table 10.5 reports the CVR values for the 14 items recommended by the SMEs.

CVR values were also calculated for the items that were rated as "important to include in the scale." This resulted in 37 additional items to consider for inclusion in the finalized scale. The scores from the hypothetical range of differences were used to further evaluate these 37 items. The hypothetical range of differences was assessed from the SMEs completion of the Trust Scale in Part 2 and Part 3 of the study. Paired samples *t*-tests were conducted to identify the hypothetical range of differences (see Table 10.6).

Table 10.5 The 14 items recommended by SMEs as "Extremely Important"

Item	CVR values
1. Function successfully	1.00
2. Act consistently	1.00
3. Reliable	1.00
4. Predictable	1.00
5. Dependable	1.00
6. Follow directions	0.82
7. Meet the needs of the mission	0.82
8. Perform exactly as instructed	0.82
9. Have errors[a]	0.82
10. Provide appropriate information	0.82
11. Malfunction[a]	0.64
12. Communicate with people	0.64
13. Provide feedback	0.64
14. Unresponsive[a]	0.64

$CVR \geq 0.59$
[a]Represents reverse coded items

Table 10.6 The 37 "Important Items" separated by retained and removed items

	CVR	Complete trust		No trust		Range	
		Mean	SD	Mean	SD	*t*	*p*
Items retained							
1. Operate in an integrated team environment	1.00	75.00	21.73	28.00	25.73	3.62	0.006
2. Autonomous	1.00	69.09	20.23	38.89	23.15	3.18	0.013
3. Good teammate	0.82	87.27	11.04	11.82	11.68	16.60	<0.001
4. Performs a task better than a novice human user	0.82	69.09	20.71	24.55	24.23	6.69	<0.001
5. Led astray by unexpected changes in the environment	0.82	71.00	19.12	26.00	15.78	5.78	<0.001
6. Know the difference between friend and foe	0.82	71.00	27.67	14.55	16.95	5.89	<0.001
7. Make sensible decisions	0.82	84.00	11.74	21.00	20.79	8.62	<0.001
8. Clearly communicate	0.82	83.00	11.60	19.00	15.24	12.30	<0.001
9. Warn people of potential risks in the environment	0.82	83.00	11.60	23.00	18.29	7.75	<0.001
10. Incompetent	0.82	85.45	14.40	39.00	33.15	5.24	0.001
11.Possess adequate decision-making capability	0.82	71.11	16.91	20.00	21.60	5.57	0.001
12. Are considered part of the team	0.82	79.00	12.87	35.00	31.36	3.36	0.010
13. Will act as part of the team	0.82	74.00	19.55	30.00	29.06	3.28	0.010
14. Perform many functions at one time	0.82	68.00	23.48	36.00	25.47	4.40	0.002
15. Protect people	0.82	77.00	24.52	23.00	22.14	3.92	0.003
16. Openly communicate	0.82	80.00	15.81	34.00	28.36	3.79	0.005
17. Responsible	0.82	66.36	30.42	27.27	34.67	2.76	0.020
18. Built to last	0.82						
19. Work in close proximity with people	0.82	65.00	19.58	35.56	26.03	2.34	0.047
20. Supportive	0.64	66.00	16.47	18.00	11.35	8.67	<0.001
21. Work best with a team	0.64	71.00	18.53	34.00	28.36	3.41	0.008
22. Tell the truth	0.64	86.36	21.57	46.00	31.69	2.90	0.018
23. Keep classified information secure	0.64	84.00	15.06	55.45	36.43	2.69	0.025
24. Require frequent maintenance	0.64	74.00	20.66	44.00	28.75	2.37	0.042
Items removed from scale							
25. Responsive	1.00	84.00	9.66	32.22	21.67	7.50	<0.001
26. Poor teammate	0.82	84.55	9.66	45.45	36.43	3.64	0.005
27. Are assigned tasks that are critical to mission success	0.82	58.00	32.25	25.00	35.67	1.84	0.098

(continued)

Table 10.6 (continued)

	CVR	Complete trust		No trust		Range	
		Mean	SD	Mean	SD	t	p
28. Communicated only partial information	0.82	53.00	28.30	35.00	28.77	1.03	0.331
29. Instill fear in people	0.73	83.64	18.59	50.91	34.48	3.13	0.011
30. Likeable	0.64	60.00	33.54	14.00	10.75	4.61	0.001
31. Easy to maintain	0.64	60.00	33.54	35.00	31.71	2.89	0.018
32. Responsible for its own actions	0.64	64.00	32.04	32.73	36.63	2.22	0.054
33. Given complete responsibility for the completion of a mission	0.64	59.00	37.55	19.00	37.84	2.16	0.059
34. Monitored during a mission	0.64	49.00	29.61	20.00	28.67	2.08	0.067
35. Are considered separate from the team	0.64	68.00	24.40	36.00	32.39	1.92	0.091
36. Difficult to maintain	0.64	70.00	15.63	50.00	34.64	1.60	0.143
37. Work best alone	0.64	48.00	23.94	41.00	33.48	0.69	0.506

Twenty-four items were retained for further scale validation. The remaining 13 items were removed from the scale for the following reasons: non-significant findings on the paired samples t-tests for eight items, and SMEs comments assisted in the removal of the remaining five items. The first comment was a general comment stating that some items (e.g., easy/difficult to maintain) were repetitive in nature. To address this comment, only one of the repetitive items was included in the revised scale. *Responsive, easy to maintain,* and *poor teammate* were removed. In addition *instill fear in people* was removed due to its nature of distrust more than trust. Finally, the item *likeable* was considered to be too general an item for the scale. An additional comment suggested that two items *are given complete responsibility for the completion of the mission*, and *are assigned tasks that are critical to mission success* represented situational factors that may be a separate issue from trust. Even though CVR analysis revealed that SMEs felt that they might be important items to include in the scale, their responses to the theoretical range of scores did not show a significant change in trust. This added support for the SMEs' comments recommending removal of these two items.

Twenty-two items did not meet the CVR criterion and were reviewed for removal from the scale. Of these 22 items, the four items regarding movement (move quickly, move slowly, mobile, move rigidly) were removed. To further support this decision, there was a comment from a SME suggesting that trust and speed were orthogonal. SMEs further recommended that three items representing robot personality (kind, caring, have a relationship with their human users or operators) were recommended for removal. An additional nine of these items (human-like, fake, alive, dead, offensive, organic, ignorant, apathetic, and make decisions that affect me personally) were removed from the scale due to the lack of change in scores on the theoretical range of responses. However, out of those 22 items that did not meet the CVR criterion, a number of SMEs felt that some items may have particular importance to trust development as robots advance further into socially relevant relationships. Therefore, the following four items were retained in the scale: friendly, pleasant, conscious, and lifelike.

Semantic analysis of the trust scale items reduced the scale from 73 items to 42 items. It further identified 14 items that were extremely important to trust measurement, with an additional 24 items that could be important to trust measurement. Four domain specific items (friendly, pleasant, conscious, and lifelike) were also retained on the scale based on SME recommendations. No additional items or item revisions were recommended by the SMEs. In addition, no changes to the scale design were recommended.

10.5 Task-Based Validity Testing: Does the Score Change Over Time with an Intervention?

Trust has typically been measured following an interaction. However, trust is dynamic in nature, as ongoing interactions and relational history continuously influence trust levels at any given point in time. Consequently, trust before, during, and after an interaction may not be identical. Further, trust in the same partner will likely change over time as the relationship progresses (Bloomqvist 1997). Research in the areas of automation (see Merritt and Ilgen 2008), as well as interpersonal trust (see McAllister 1995), recommended that trust should be measured at multiple times, specifically before and after the task or interaction. Within the *Three Factor Model of Human-Robot Trust*, pre-interaction measurement has been used to identify initial trust perceptions that are influenced by human traits, robot features, and the individual's perception of the environment and perceived robot capabilities. Post-interaction measurement was used to identify changes in trust related to human states and trust perceptions following interactions. Obtaining the most reliable and accurate reflection of the dynamic nature of trust in an interaction may necessitate measuring trust multiple times.

In order to examine this concept of dynamic trust, an experiment was designed to assess the 42 Item Trust Scale's capability to measure changes in perceived trust. Computer-based simulation was used to develop a monitoring task specific to a "screen the back" scenario for a Soldier-robot team (Army Research Laboratory 2012). Trust, as measured using the 42 Item Trust Scale, was assessed both pre-interaction, and post-interaction in the high trust condition (the robot provided 100 % reliable feedback on target detection), as well as post-interaction in the low trust condition (the robot provided 25 % reliable feedback). It was hypothesized that there are mean differences in trust that occur over time, with respect to changes in robot reliability. More specifically, trust will increase from pre-interaction to after experiencing a 100 % reliable interaction; trust will decrease after first experiencing a 100 % reliable interaction and then experiencing a 25 % reliable interaction.

10.5.1 Experimental Method

Participants included 81 undergraduate students (25 males, 56 females; $M = 22.57$ years, $SD = 3.95$) from an undergraduate psychology course in Science and Pseudoscience at the University of Central Florida. Participants had varied backgrounds with respect to robot familiarity. All participants previously watched movies or television shows that incorporated robots: 1–5 shows (N = 30), 6–10 shows (N = 9), and over 10 shows (N = 42). Forty-four participants previously interacted with robots, ranging from the Roomba™ vacuum cleaner to a bomb disposal robot. Participants also reported previously controlling robots through a variety of modalities: voice (N = 11), game controller (N = 24), gestures or pictures (N = 3), or a radio control (RC) system (N = 35). In addition, five participants previously built robots for class-based projects.

All materials were administered through paper and pencil versions. The Negative Attitudes toward Robots Scale (NARS; Nomura et al. 2004), and demographics questionnaire were included to identify potential individual difference ratings. The NARS has three subscales: NARS_S1 represents negative attitudes toward situations of interaction with robots; NARS_S2 represents negative attitudes toward social influence of robots; and NARS_S3 represents negative emotions in interactions with robots. Trait-based trust was measured through the Interpersonal Trust Scale (ITS; Rotter 1967). State-based trust was measured through the 42 item Trust Scale, administered pre- and post-interaction. Participants' states were not assessed in this experiment due to the nature of the task (monitoring only).

All HRI scenarios were developed using the Robotic Interactive Visualization & Experimental Technology (RIVET) computer-based simulation system developed by General Dynamics Robotic Systems (GDRS; Gonzalez et al. 2009), in collaboration with the US Army Research Laboratory. RIVET uses an adapted Torque Software Development Kit (SDK) development and runtime environment through a Client/Server networking model. Development of the virtual environment (VE) was accomplished through a TorqueScript language, similar to C^{++}. The VEs used a base environment developed previously for Army research activities. This included the layout of the physical environment (e.g., ground, roadways, buildings, and lighting), as well as inclusion of the Soldier and terrorist non-player characters. Task-specific customization of the environment was accomplished through scripting syntax. Examples of specific customization included entering objects, obstacles, and creation of paths.

An experimental scenario was created using the RIVET computer-based simulation system of a Soldier-Talon™ robot team completing a "screen the back" mission (Army Research Laboratory 2012). Participants monitored the Talon robot as it navigated to the back of the building, repositioned behind a barrel, monitored the back of the building for human targets, and provided a speech-based response when a target was detected. A single video was created from the camera-view on the Talon™ robot. The video of the simulation was created using FRAPS® real-time video capture and benchmarking program with a 30 frame rate/s .avi file.

The .avi file was converted into the .mp4 file format to add auditory feedback. The robot stated "target detected" in a male computer synthetic voice. There were eight possible targets included in each scenario. No false alarms were included.

The study was conducted in a single session. Participants were provided a copy of the informed consent while the experimenter read it aloud. Participants then viewed an image of the Talon robot and completed the 42 Item Trust Scale (Time 1; pre-interaction). Next, participants were instructed about the human-robot task they were about to monitor. Following Video 1, participants completed the 42 Item Trust Scale (Time 2; post-interaction, 100 % reliable feedback). Participants then received the second task instructions, monitored Video 2, and completed the 42 Item Trust Scale (Time 3; post-interaction, 25 % reliable feedback). Following these monitoring tasks, participants completed the ITS, NARS, and demographics questionnaire. The study took approximately 90 min in its entirety.

10.5.2 Experimental Results

All data were analyzed using IBM SPSS Statistics v.19 (SPSS 2010), with an alpha level set to 0.05, unless otherwise indicated.

10.5.2.1 Individual Item Analysis

The first step of analysis was to determine if each item changed over time. A one-way within-subjects repeated measures analysis of variance was conducted for each of the 42 items of the Trust Scale. The factor was "time" and the dependent variable was the individual score. Thirty-four items showed a significant mean difference for the condition of "time." Using Post-hoc analysis, six items were found to only have significant differences between Time 1 (Pre-Interaction) and Time 2 (post-interaction). These items included: *lifelike, perform many functions at one time, friendly, know the difference between friend and foe, keep classified information secure,* and *work in close proximity to people.* These results may have occurred due to a significant change in the mental model from pre- to post-interaction; therefore, the items were retained in the scale. Confidence interval analysis was conducted on the remaining two items (*operate in an integrated team environment, built to last*) and showed no significant change between Time 1, Time 2, or Time 3. These items were removed from the scale. Additional analyses were conducted using the 40 retained items.

10.5.2.2 Trust Score Validation

First, a general score of trust was created for each of the three time periods. Following reverse coding of specific items, the 40 items were summed and divided by 40 to formulate a score between 0 and 100. To assess the impact of time on trust development, a one-way within-subject repeated measures analysis of variance was conducted. The results indicated a significant effect of time, $F(2, 79) = 119.10$, $p < 0.001$, $\eta\rho^2 = 0.75$. Post-hoc analysis using the Fisher LSD revealed significant mean difference ratings for all three times the trust scale was administered. It revealed that trust was significantly greater in Time 2 (post-interaction, 100 % reliable condition) than Time 1 (pre-interaction) and Time 3 (post-interaction, 25 % reliable condition), thus supporting the hypothesis. In addition, mean trust scores at Time 1 were significantly greater than at Time 3 (see Fig. 10.3).

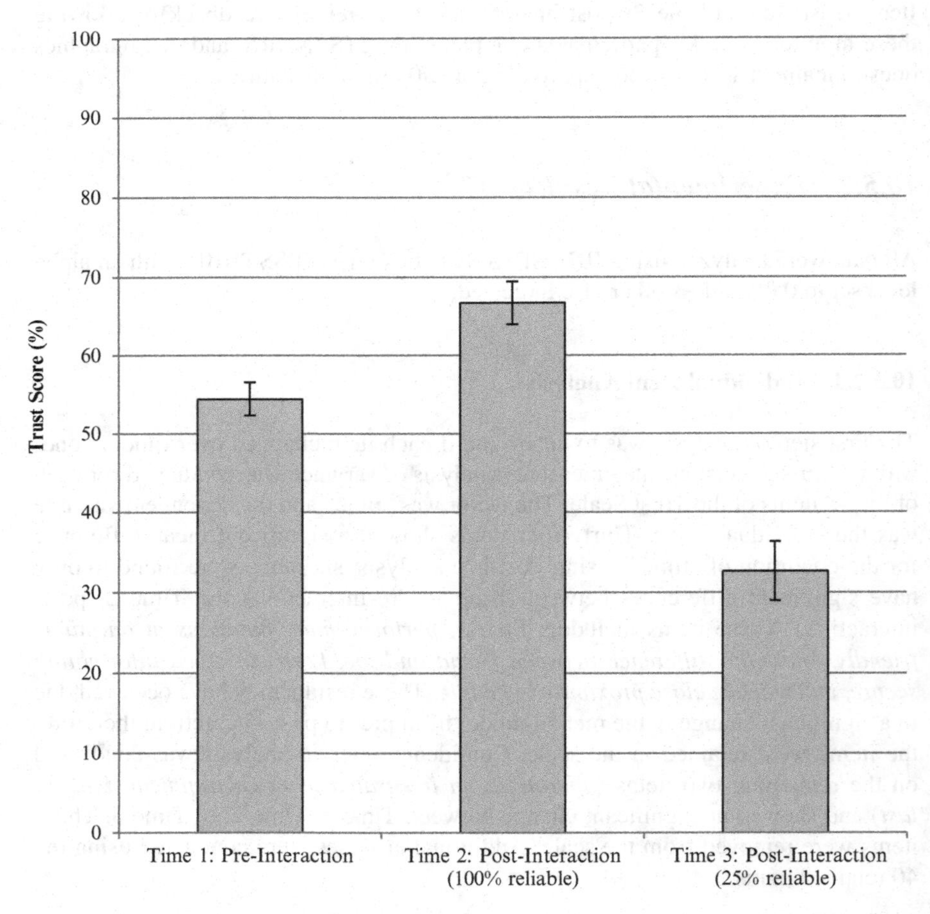

Fig. 10.3 Bar graph representing significant mean differences of trust over time, with 95 % confidence interval error bars

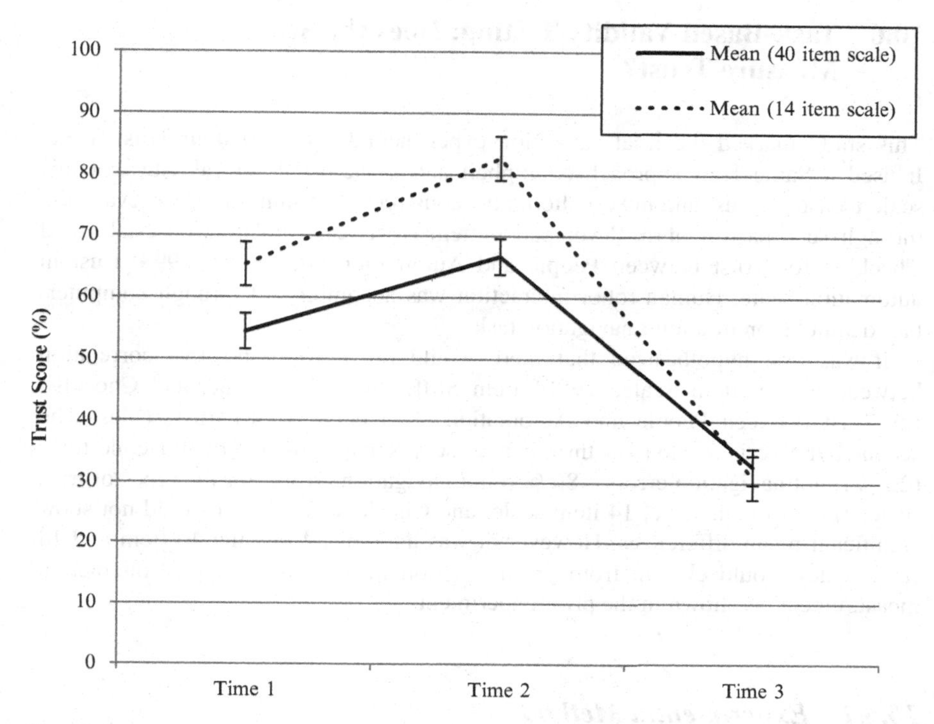

Fig. 10.4 Trust scores for the 40 item and the 14 item scale across time

10.5.2.3 40 Items Versus 14 Items

Additional analyses were also conducted to identify the differences between the 40 Item Trust Scale and the 14 Item SME recommended scale. A 2 Scale (40 items, 14 items) × 3 Time (Time 1, Time 2, Time 3) repeated measures analysis of variance was conducted (see Fig. 10.4).

Results showed a significant effect of Time, $F(2240) = 186.59$, $p < 0.001$, $\eta p^2 = 0.609$; Scale $F(1240) = 273.61$, $p < 0.001$, $\eta p^2 = 0.533$; and an interaction between Time and Scale, $F(2240) = 108.84$, $p < 0.001$, $\eta p^2 = 0.476$. Review of the confidence intervals showed a significant difference between the scales at Time 1 and Time 2, but not Time 3. While findings revealed significant differences between the two scales, graphical representations showed similar patterns in the results. Taking into account both the individual analyses of each item measured over Time, as well as the comparative results of the two scales, it appeared that the total trust score of the 40 Item Trust Scale provided a finer level of granularity and thus a more accurate trust rating.

10.6 Task-Based Validity Testing: Does the Scale Measure Trust?

This study marked the final validation experiment for the 40 Item Trust Scale. It used a Same-Trait approach (Campbell and Fiske 1959) to validate that this scale measured trust and not an alternative construct. The same-trait was evaluated through a comparison of the developed 40 Item Trust Scale, and the well-established Checklist for Trust between People and Automation (Jian et al. 1998) trust in automation scale. Human-robot interaction was accomplished through computer-based simulation of a joint navigation task.

It was first hypothesized that there would be a strong positive correlation between the 40 Item Scale, the 14 Item SME selected subscale, and Checklist for Trust between People and Automation (Jian et al. 1998) trust scales. The second hypothesis was that the three change scores in the post-interaction conditions (20 % robot navigation errors—80 % robot navigation errors) for the types of trust scales (i.e., 40 item scale, 14 item scale, and Checklist for Trust) would not show significant mean differences. However, it was anticipated that the 40 Item and 14 Item scales would change from pre-interaction measurement to post-interaction measurement, as shown in the prior experiment.

10.6.1 Experimental Method

Twenty-one undergraduate students from the University of Central Florida (12 males, 9 females) participated in two Soldier-robot team-based computer simulations to provide the next level of task-based validity testing. Multiple scales were included to measure subjective trust, personality traits, demographics, and human states. Subjective trust was measured through the developed 40 Item Trust Scale. A partial measure represented by the 14 Items recommended by SME's was also assessed. The well-established Checklist for Trust between People and Automation (Jian et al. 1998) was included for Same-Trait analysis. Items including the word 'automation' were adapted to 'robot.' The Interpersonal Trust Scale (Rotter 1967) was also included. The 7-point Mini-IPIP personality assessment (Donnellan et al. 2006) was used to measure the Big 5 personality traits: agreeableness, extraversion, intellect, conscientiousness, and neuroticism. The Dundee Stress State Questionnaire (DSSQ; Matthews et al. 1999) was included to measure human states (i.e., mood state, motivation, workload, and thinking style) before and after a task.

The virtual environment (VE) for the present procedure was developed in RIVET and used a base environment of a Middle Eastern town developed by GDRS in collaboration with the US Army Research Laboratory. Independent task-specific customization of the physical environment was accomplished through scripting syntax. Specific customization included entering objects, obstacles, and creation of paths, etc. Scripting files were created for both the training session and the task conditions. The task was to assist an autonomous robot from a set location to a rendezvous point. Participants controlled a Soldier avatar throughout the Middle Eastern town using a keyboard and mouse interface. It was possible that the robot could become stuck on an obstacle and required the participant's assistance. Participants could move certain obstacles out of the way of the robot by simply walking into the obstacle. The mission ended when both the Soldier and robot reached the rendezvous location. In Simulation A, the robot autonomously navigated around four out of the five obstacles. In Simulation B, the robot only navigated around one of the obstacles. Each simulation was approximately 1 min in length. The order of simulation presentation was counterbalanced and determined prior to participation.

Following completion of informed consent, participants completed three questionnaires: the demographics questionnaire, the Mini-IPIP personality inventory, and the ITS. Participants were then shown a picture of the Talon robot and completed the 40 Item Trust Scale, the Checklist for Trust between People and Automation, and the DSSQ to acquire baseline information. Participants then completed the two simulated tasks, followed by completion of the two trust scales and the post-task DSSQ following each task. The simulated tasks were recorded using FRAPS® real-time video capture and benchmarking program with a 30 frame rate/s .avi file. Video was recorded from the Soldier character's perspective. It was saved with a unique identifier to maintain participant confidentiality. The entire study took approximately 1 h to complete.

10.6.2 Experimental Results

All data were analyzed using IBM SPSS Statistics v.19 (SPSS 2010), with an alpha level set to 0.05, unless otherwise indicated. Initial analyses were conducted to assess changes in human states over time. Results demonstrated no significant difference in mood state or motivation subscales. The thinking style subscales of self-focused attention and concentration showed a significant difference between pre-interaction and post-interaction, but no difference between the two post-interaction conditions. A similar result was found for the thinking content subscale, task interference. Due to these findings, no additional analyses were conducted assessing human states. The Same-Trait methodology compared the developed 40 Item Trust Scale, the SME's recommended 14 Item Trust Scale, and the Checklist for Trust between People and Automation.

10.6.2.1 Correlation Analysis of the Three Scales

The first step in this validation was to identify the relationships between the three trust scales. In support of Hypothesis 1, significant positive Pearson correlations were found between all three scales (see Table 10.7).

Table 10.7 Same-trait trust scale correlations over time

	M	SD	1	2	3
Pre-interaction trust					
1. 40 Item Scale	60.60	13.58	1		
2. 14 Item Scale	72.45	12.70	0.829**	1	
3. Checklist for Trust between People and Automation	71.71	16.27	0.620**	0.745**	1
Post-interaction (20 % errors)					
1. 40 Item Scale	49.92	19.59	1		
2. 14 Item Scale	60.61	20.19	0.918**	1	
3. Checklist for Trust between People and Automation	71.54	16.22	0.857**	0.854**	1
Post-interaction (80 % errors)					
1. 40 Item Scale	46.70	22.43	1		
2. 14 Item Scale	57.42	24.93	0.934**	1	
3. Checklist for Trust between People and Automation	72.27	17.16	0.855**	0.852**	1

** represent significance at the .01 level

The second step in this validation process was to determine if there was a significant mean difference between the post-interaction change scores (20 % robot navigation errors—80 % robot navigation errors) for the three scales. A within-subjects repeated measures analysis of variance was conducted. In support of Hypothesis 2, there was not a significant mean difference between the post-interaction change scores for the three trust scales, $F(2,19) = 2.64$, $p = 0.097$. This result provided additional support that the developed scale measures the construct of trust.

10.6.2.2 Pre-post Interaction Analysis

Additional analyses were conducted to determine the differences between the pre-post interaction scores between the three scales. Paired samples t-tests were conducted to assess the change in trust pre-post interaction. A significant change in trust was found between the pre-post interaction trust measurement for the 40 Item Trust Scale, $t(40) = 3.87$, $p < 0.001$, and the 14 Item Trust Scale, $t(40) = 3.86$, $p < 0.001$. However, the Checklist for Trust between People and Automation trust scale did not change ($p = 0.932$). This finding suggested that the developed trust scale does indeed measure something additional to the previously developed trust scale.

10.6.2.3 Differences Across Scales and Conditions

To further explore these scale differences, a 3 Trust Scales (40 item, 14 item, and Checklist for Trust) × 3 Conditions (pre-interaction, 20 % robot error, and 80 % robot error) repeated measures analysis of variance was conducted. There was a main effect of scale, $F(2,59) = 105.16$, $p < 0.001$ $\eta p^2 = 0.781$, but not condition ($p = 0.191$). Results are depicted in Fig. 10.5. Confidence interval analysis of the mean trust scores demonstrated that there was no significant difference between the three scales that were recorded pre-interaction. This finding suggested that all three scales provide similar trust scores prior to HRI. Further there were no significant differences between the 14 Item Trust Scale and the well-established Checklist for Trust between People and Automation (Jian et al. 1998). This finding is not surprising as the items from both the 14 Item Trust Scale and the Jian et al. scale referenced the capability of the system. The important finding was the significant differences between the 40 item Trust Scale and the Jian et al. scale during the two post-interaction conditions.

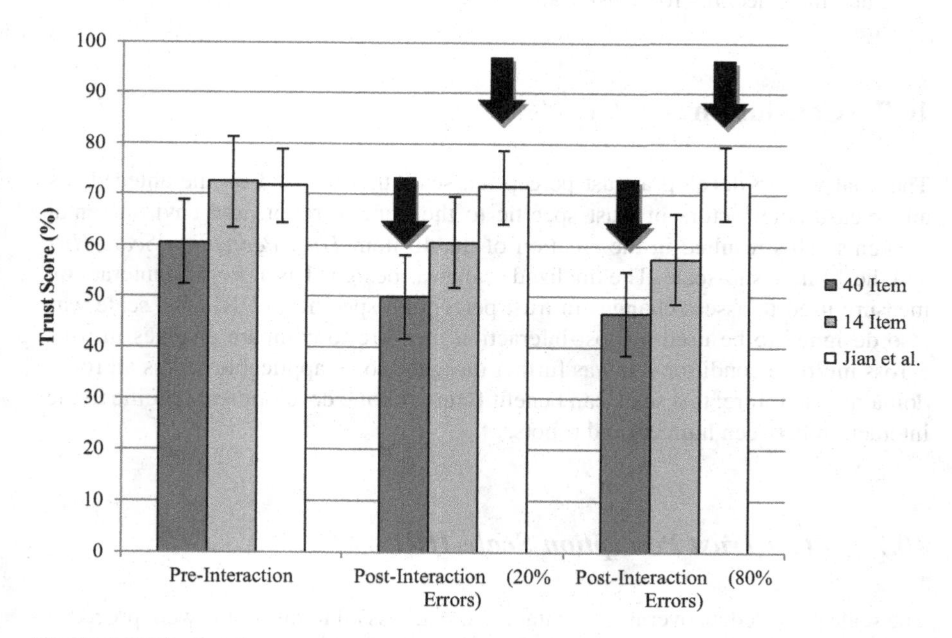

Fig. 10.5 Differences between the 40 Item, 14 Item, and Checklist for Trust between People and Automation (Jian et al. 1998) trust scales

10.6.3 Experimental Discussion

This study demonstrated that the developed trust scale assessed the construct of trust. In addition, it provided support for additional benefits of the developed 40 Item Trust Scale above and beyond previously used scales (i.e., Checklist for Trust between People and Automation; Jian et al. 1998). First, there were strong positive correlations between the three scales. Second, mean analysis showed significant differences between the 40 item and Jian et al. scale in the post-interaction conditions. Both the 40 Item and the 14 Item scales showed a significant change in trust from pre-interaction to post-interaction; however the Checklist for Trust did not change. This change in trust was mirrored in the previous study, and is supported by the trust theory. Therefore, it can be postulated that the developed Trust Scale accounted for the relationship between the change in mental models and trust development that occurs after HRI. In addition, findings from this study, together with the findings from the previous study provide support that the 40 Item Trust Scale had more accurate trust scores than both the 14 Item SME recommended scale and the Checklist for Trust scale.

10.7 Conclusion

The goal was to develop a trust perception scale that focused on the antecedents and measurable factors of trust specific to the human, robot, and environmental elements. This resulted in the creation of the 40 item *Trust Perception Scale-HRI* and the 14 item sub-scale. The finalized scale was designed as a pre-post interaction measure used to assess changes in trust perception specific to HRI. The scale was also designed to be used as post-interaction measure to compare changes in trust across multiple conditions. It was further designed to be applicable across all robot domains. Therefore, this scale can benefit future robotic development specific to the interaction between humans and robots.

10.7.1 The Trust Perception Scale-HRI

The scale provided an overall percentage score across all items. Items were preceded by the question "What percentage of the time will this robot . . . " followed by a list of the items. The finalized 40 item scale is provided in Table 10.8, and took between 5 and 10 min to complete.

Table 10.8 Finalized Trust Perception Scale-HRI

	0 %	10 %	20 %	30 %	40 %	50 %	60 %	70 %	80 %	90 %	100 %
What % of the time will this robot be . . .											
1. Considered part of the team	O	O	O	O	O	O	O	O	O	O	O
2. Responsible	O	O	O	O	O	O	O	O	O	O	O
3. Supportive	O	O	O	O	O	O	O	O	O	O	O
4. Incompetent[a]	O	O	O	O	O	O	O	O	O	O	O
5. Dependable[b]	O	O	O	O	O	O	O	O	O	O	O
6. Friendly	O	O	O	O	O	O	O	O	O	O	O
7. Reliable[b]	O	O	O	O	O	O	O	O	O	O	O
8. Pleasant	O	O	O	O	O	O	O	O	O	O	O
9. Unresponsive[a,b]	O	O	O	O	O	O	O	O	O	O	O
10. Autonomous	O	O	O	O	O	O	O	O	O	O	O
11. Predictable[b]	O	O	O	O	O	O	O	O	O	O	O
12. Conscious	O	O	O	O	O	O	O	O	O	O	O
13. Lifelike	O	O	O	O	O	O	O	O	O	O	O
14. A good teammate	O	O	O	O	O	O	O	O	O	O	O
15. Led astray by unexpected changes in the environment	O	O	O	O	O	O	O	O	O	O	O
What % of the time will this robot . . .											
16. Act consistently[b]	O	O	O	O	O	O	O	O	O	O	O
17. Protect people	O	O	O	O	O	O	O	O	O	O	O
18. Act as part of the team	O	O	O	O	O	O	O	O	O	O	O
19. Function successfully	O	O	O	O	O	O	O	O	O	O	O
20. Malfunction[a]	O	O	O	O	O	O	O	O	O	O	O
21. Clearly communicate	O	O	O	O	O	O	O	O	O	O	O
22. Require frequent maintenance[a]	O	O	O	O	O	O	O	O	O	O	O
23. Openly communicate	O	O	O	O	O	O	O	O	O	O	O
24. Have errors[a]	O	O	O	O	O	O	O	O	O	O	O
25. Performa a task better than a novice human user	O	O	O	O	O	O	O	O	O	O	O
26. Know the difference between friend and foe	O	O	O	O	O	O	O	O	O	O	O
27. Provide feedback[b]	O	O	O	O	O	O	O	O	O	O	O
28. Possess adequate decision-making capability	O	O	O	O	O	O	O	O	O	O	O
29. Warn people of potential risks in the environment	O	O	O	O	O	O	O	O	O	O	O
30. Meet the needs of the mission/task[b]	O	O	O	O	O	O	O	O	O	O	O

(continued)

Table 10.8 (continued)

	0 %	10 %	20 %	30 %	40 %	50 %	60 %	70 %	80 %	90 %	100 %
31. Provide appropriate information[b]	O	O	O	O	O	O	O	O	O	O	O
32. Communicate with people[b]	O	O	O	O	O	O	O	O	O	O	O
33. Work best with a team	O	O	O	O	O	O	O	O	O	O	O
34. Keep classified information secure	O	O	O	O	O	O	O	O	O	O	O
35. Perform exactly as instructed[b]	O	O	O	O	O	O	O	O	O	O	O
36. Make sensible decisions	O	O	O	O	O	O	O	O	O	O	O
37. Work in close proximity with people	O	O	O	O	O	O	O	O	O	O	O
38. Tell the truth	O	O	O	O	O	O	O	O	O	O	O
39. Perform many functions at one time	O	O	O	O	O	O	O	O	O	O	O
40. Follow directions[b]	O	O	O	O	O	O	O	O	O	O	O

[a]Represents the reverse coded items for scoring
[b]Represents the 14 item sub-scale items

10.7.2 *Instruction for Use*

When the scale is used as a pre-post interaction measure, the participants should first be shown a picture of the robot they will be interacting with or provided a description of the task prior to completing the pre-interaction scale. This accounts for any mental model effects of robots and allows for comparison specific to the robot at hand. For post-interaction measurement, the scale should be administered directly following the interaction. To create the overall trust score, 5 items must first be reverse coded. The reverse coded items are denoted in the above table. All items are then summed and divided by the total number of items (40). This provides an overall percentage of trust score.

While use of the 40 Item scale is recommended, a 14 Item subscale can be used to provide rapid trust measurement specific to measuring changes in trust over time, or during assessment with multiple trials or time restrictions. This subscale is specific to functional capabilities of the robot, and therefore may not account for changes in trust due to the feature-based antecedents of the robot. Trust score is calculated by first reverse coding the 'have errors,' 'unresponsive,' and 'malfunction' items, and then summing the 14 item scores and dividing by 14. The 14 items are marked in Table 10.8.

10.7.3 Current and Future Applications

This scale was developed to provide a means to subjectively measure trust perceptions over time and across robotic domains. In addition, it can be used by individuals in all the major roles of HRI: operator, supervisor, mechanic, peer, or bystander. Therefore, there are many potential avenues for future research using *Trust Perception Scale-HRI*. Current and near-term research studies highlight the expansion of human-robot trust antecedents, which were first described in the *Three Factor Model of Human-Robot Trust* (Hancock et al. 2011; Schaefer et al. 2014). These include the exploration of human-robot trust as it relates to: Soldier, bystander, and robot proximity in high-risk military tasking; transparent system communication and feedback for Soldier-robot teaming in high-risk environments; the development of natural language processing in human-robot teams; and dual-task engagement with an autonomous vehicle designed for on-base personnel transport. While a number of these studies are currently under review or in press, a few preliminary results are discussed below.

First, Sanders et al. (2014) used the *Trust Perception Scale-HRI* to assess the impact of the amount of communication feedback from the robot (constant, contextual only, minimal) and modality of information (visual, text, audio) on trust development. Their initial results found a greater increase in the change in trust (post-interaction minus pre-interaction) for a constant stream of information compared to contextual information only and minimal information across three different communication modalities (text, auditory, and visual) during a Soldier-robot team surveillance task clearing an area of weapons and locating civilians to be safely evacuated from a hostile zone. These researchers are continuing to use the *Trust Perception Scale-HRI* for future studies in transparent communication, human roles (team member, bystanders), and social dynamics including proxemics, as part of the US Army Research Laboratory's Robotics Collaborative Technology Alliance tasking related to determinants of shared cognition and social dynamics in future Soldier-robot teams.

Second, a recent study using the *Trust Perception Scale-HRI* was conducted where participants monitored a simulated and a live robot surveying an environment, locating an object, and touching the object (Schafer et al. 2015). Results supported previous findings that individuals trust a reliable robot significantly more than the unreliable robot. Also, the results showed that the scale is effective for measuring trust in both simulated and live HRI experimentation.

A third area of on-going and near-term work using the *Trust Perception Scale-HRI* is exploring trust development with respect to the development of robotic passenger vehicles and transparent passenger user interfaces (Schaefer 2015). The design of this set of computer-based simulation studies is in line with the goals of the US Army Tank Automotive Research, Development and Engineering Center's ARIBO (Applied Robotics for Installation and Base Operations) project for alternative transportation options for on-base wounded Soldier transit (Marshall 2014). The ultimate goal is to provide a means for Soldiers to schedule an on-demand autonomous robotic passenger vehicle to arrive and drive door-to-door to and from the Medical Barracks to the on-base medical facilities. The benefit of this work is to understand the levels of trust that will enhance usage and effective human-robot interaction. Results of this set of studies will provide additional insight into the impact of trust antecedents (e.g., transparency, cueing, human characteristics) on trust development, as well as explore how the trust relationship changes as the human role transitions from a driver, to a safety rider, to a supervisor external to the vehicle, as well as to a passenger. Initial findings advance current trust theory by demonstrating significant relationships with working memory capacity and coping style related to driving, distress, workload, and task performance (Schaefer & Scribner 2015). Future work on this project is two part: (1) understanding the effects of the availability of driver control interfaces (i.e., steering and speed control versus automation engage/disengage buttons) on usability, performance, and trust; and (2) exploring the effects of transparent user interfaces design on trust development and calibration for passengers.

Acknowledgments This research is a continuation of the author's dissertation work supported in part by the US Army Research Laboratory (Cooperative Agreement Number W911-10-2-0016) and in part by an appointment to the US Army Research Postdoctoral Fellowship Program administered by the Oak Ridge Associated Universities through a cooperative agreement with the US Army Research Laboratory (Cooperative Agreement Number W911-NF-12-2-0019). The views and conclusions contained in this document are those of the author and should not be interpreted as representing the official policies, either expressed or implied, of the Army Research Laboratory or the US Government. The US Government is authorized to reproduce and distribute reprints for Government purposes notwithstanding any copyright notation herein. Special acknowledgments hereby include the author's dissertation committee: Drs. Peter A. Hancock, John D. Lee, Florian Jentsch, Peter Kincaid, Deborah R. Billings, and Lauren Reinerman. Additional acknowledgments are made to internal technical reviewers from the US Army Research Laboratory: Dr. Susan G. Hill, Dr. Don Headley, Mr. John Lockett, Dr. Kim Drnec, and Dr. Katherine Gamble.

References

Army Research Laboratory (2012) Robotics Collaborative Technology Alliance Annual Program Plan. U.S. Army Research Laboratory, Aberdeen Proving Ground

Bartneck C, Kulić D, Croft E, Zoghbi S (2009) Measurement Instruments for the anthropomorphism, animacy, likeability, perceived intelligence, and perceived safety of robots. Int J Soc Robots 1:71–81. doi:10.1007/s12369-008-001-3

Beer JM, Fisk AD, Rogers WA (2014) Toward a framework for levels of robot autonomy in human-robot interaction. J Hum Robot Interact 3(2):74–99. doi:10.5898/JHRI.3.2.Beer

Bloomqvist K (1997) The many faces of trust. Scand J Manage 13(3):271–286

Campbell DT, Fiske DW (1959) Convergent and discriminant validation by the multitrait-multimethod matrix. Psychol Bull 56:81–105

Chen JYC, Terrence PI (2009) Effects of imperfect automation and individual differences on concurrent performance of military and robotics tasks in a simulated multi-tasking environment. Ergonomics 52(8):907–920. doi:10.1080/00140130802680773

Desai M, Stubbs K, Steinfeld A, Yanco H (2009) Creating trustworthy robots: lessons and inspirations from automated systems. In: Proceedings of the AISB convention: new frontiers in human-robot interaction, Edinburgh. Retrieved from https://www.ri.cmu.edu/pub_files/2009/4/Desai_paper.pdf

DeVellis RF (2003) Scale development theory and applications, vol 26, 2nd edn, Applied social research methods series. Sage, Thousand Oaks

Donnellan MB, Oswald FL, Baird BM, Lucas RE (2006) The Mini-IPIP scales: tiny-yet-effective measures of the Big Five factors of personality. Psychol Assess 18(2):192–203. doi:10.1037/1040-3590.18.2.192

Fink A (2009) How to conduct surveys: a step-by-step guide, 4th edn. Sage, Thousand Oaks

Gonzalez JP, Dodson W, Dean R, Kreafle G, Lacaze A, Sapronov L, Childers M (2009) Using RIVET for parametric analysis of robotic systems. In: Proceedings of 2009 ground vehicle systems engineering and technology symposium (GVSETS), Dearborn

Groom V, Nass C (2007) Can robots be teammates? Benchmarks in human-robot teams. Interact Stud 8(3):483–500. doi:10.1075/is.8.3.10gro

Hancock PA, Billings DR, Schaefer KE, Chen JYC, Parasuraman R, de Visser E (2011) A meta-analysis of factors affecting trust in human-robot interaction. Hum Factors 53(5):517–527. doi:10.1177/0018720811417254

Jian J-Y, Bisantz AM, Drury CG, Llinas J (1998) Foundations for an empirically determined scale of trust in automated systems (report no. AFRL-HE-WP-TR-2000-0102). Air Force Research Laboratory, Wright-Patterson AFB

Lawshe CH (1975) A quantitative approach to content validity. Pers Psychol 24(4):563–575. doi:10.1111/j.1744-6570.1975.tb01393.x

Lee KM, Park N, Song H (2005) Can a robot be perceived as a developing creature? Effects of a robot's long-term cognitive developments on its social presence and people's social responses toward it. Hum Commun Res 31(4):538–563. doi:10.1111/j.1468-2958.2005.tb00882.x

Lewicki RJ, McAllister DJ, Bies RJ (1998) Trust and distrust: new relationships and realities. Acad Manage Rev 23(3):438–458. doi:10.5465/AMR.1998.926620

Lussier B, Gallien M, Guiochet J (2007) Fault tolerant planning for critical robots. In: Proceedings of the 37th annual IEEE/IFIP international conference on dependable systems and networks, pp 144–153. doi:10.1109/DSN.2007.50

Marshall P (2014) Army tests driverless vehicles in 'living lab.' GCN technology, tools, and tactics for public sector IT. Retrieved from http://gcn.com/Articles/2014/07/16/ARIBO-Army-TARDEC.aspx?Page=1

Matthews G, Joyner L, Gilliland K, Campbell SE, Falconer S, Huggins J (1999) Validation of a comprehensive stress state questionnaire: towards a state "Big Three". In: Mervielde I, Dreary IJ, DeFruyt F, Ostendorf F (eds) Personality psychology in Europe, vol 7. Tilburg University Press, Tilburg, pp 335–350

McAllister DJ (1995) Affect- and cognition-based trust as foundations for interpersonal cooperation in organizations. Acad Manage J 38(1):24–59

McKnight DH, Kacmar CJ, Choudhury V (2004) Dispositional trust and distrust distinctions in predicting high- and low-risk internet expert advice site perceptions. e-Service J 3(2):35–58. Retrieved from http://www.jstor.org/stable/10.2979/ESJ.2004.3.2.35

Merritt SM, Ilgen DR (2008) Not all trust is created equal: dispositional and history-based trust in human-automation interactions. Hum Factors 50(2):194–210. doi:10.1518/001872008X288574

Monahan JL (1998) I don't know it but I like you—the influence of non-conscious affect on person perception. Hum Commun Res 24(4):480–500. doi:10.1111/j.1468-2958.1998.tb00428.x

Nomura T, Kanda T, Suzuki T, Kato K (2004) Psychology in human-robot communication: an attempt through investigation of negative attitudes and anxiety toward robots. In: Proceedings of the 2004 IEEE international workshop on robot and human interactive communication, Kurashiki, Okayama, pp 35–40. doi:10.1109/ROMAN.2004.1374726

Powers A, Kiesler S (2006) The advisor robot: tracing people's mental model from a robot's physical attributes. In: 1st ACM SIGCHI/SIGART conference on Human-robot interaction, Salt Lake City, Utah, USA
Rotter JB (1967) A new scale for the measurement of interpersonal trust. J Pers 35(4):651–665. doi:10.1111/j.1467-6494.1967.tb01454
Rouse WB, Morris NM (1986) On looking into the black box: prospects and limits in the search for mental models. Psychol Bull 100(3):349–363. doi:10.1037/00332909.100.3.349
Sanders TL, Wixon T, Schafer KE, Chen JYC, Hancock PA (2014) The influence of modality and transparency on trust in human-robot interaction. In: Proceedings of the fourth annual IEEE CogSIMA conference, San Antonio
Schaefer KE (2013) The perception and measurement of human-robot trust. Dissertation, University of Central Florida, Orlando
Schaefer KE (2015) Perspectives of trust: research at the US Army Research Laboratory. In: R Mittu, G Taylor, D Sofge, WF Lawless (Chairs) Foundations of autonomy and its (cyber) threats: from individuals to interdependence. Symposium conducted at the 2015 Association for the Advancement of Artificial Intelligence (AAAI), Stanford University, Stanford
Schaefer KE, Sanders TL, Yordon RE, Billings DR, Hancock PA (2012) Classification of robot form: factors predicting perceived trustworthiness. Proc Hum Fact Ergon Soc 56:1548–1552. doi:10.1177/1071181312561308
Schaefer KE, Billings DR, Szalma, JL, Adams, JK, Sanders, TL, Chen JYC, Hancock PA (2014) A meta-anlaysis of factors influencing the development of trust in automation: implications for human-robot interaction (report no ARL-TR-6984). U.S. Army Research Laboratory, Aberdeen Proving Ground
Schafer KE, Sanders T, Kessler TA, Wild T, Dunfee M, Hancock PA (2015) Fidelity & validity in robotic simulation. In: Proceedings of the fifth annual IEEE CogSIMA conference, Orlando
Schaefer KE, Scribner D (2015) Individual differences, trust, and vehicle autonomy: A pilot study. In Proceedings of the Human Factors and Ergonomics Society 59(1):786–790. doi: 10.1177/1541931215591242
Scholtz J (2003) Theory and evaluation of human robot interactions. In: Proceedings from the 36th annual Hawaii international conference on system sciences. doi:10.1109/HICSS.2003.1174284
Steinfeld A, Fong T, Kaber D, Lewis M, Scholtz J, Schultz A, Goodrich M (2006) Common metrics for human-robot interaction. In: Proceedings of the first ACM/IEEE international conference on human robot interaction, Salt Lake City, pp 33–40. doi:10.1145/1121241.1121249
Warner RM, Sugarman DB (1996) Attributes of personality based on physical appearance, speech, and handwriting. J Pers Soc Psychol 50:792–799
Wildman JL (2011) Cultural differences in forgiveness: fatalism, trust violations, and trust repair efforts in interpersonal collaboration. Dissertation, University of Central Florida, Orlando
Wildman JL, Fiore SM, Burke CS, Salas E, Garven S (2011) Trust in swift starting action teams: critical considerations. In Stanton NA (ed) Trust in military teams. Ashgate, London, pp 71–88, 335–350
Yagoda RE, Gillan DJ (2012) You want me to trust a robot? The development of a human-robot interaction trust scale. Int J Soc Robotics 4(3):235–248

Chapter 11
Methods for Developing Trust Models for Intelligent Systems

Holly A. Yanco, Munjal Desai, Jill L. Drury, and Aaron Steinfeld

11.1 Introduction

In just one area of the intelligent systems domain, the number of robot systems has greatly increased over the past two decades. According to a survey, 2.2 million domestic service robots were sold in 2010, and that number was expected to rise to 14.4 million by 2014 (IFR 2011). Not only is the number of robots in use increasing, but the number of application domains that utilize robots is also increasing. For example, self-driving cars have been successfully tested on US roads and have driven over 300,000 miles autonomously (e.g., Thrun 2010; Dellaert and Thorpe 1998). Telepresence robots in the medical industry constitute another example of a new application domain for robots (e.g., Michaud et al. 2007; Tsui et al. 2011).

As the use of such systems increases, there is a push to introduce or add additional autonomous capabilities for these robot systems. For example, the Foster-Miller (now QinetiQ) TALON robots used in the military are now capable of navigating to a specified destination using GPS. The unmanned aerial vehicles (UAVs) deployed by the military are also becoming more autonomous (Lin 2008);

H.A. Yanco (✉)
Computer Science Department, University of Massachusetts Lowell, One University Avenue, Lowell, MA 01854, USA

The MITRE Corporation, 202 Burlington Road, Bedford, MA 01730, USA
e-mail: holly@cs.uml.edu

M. Desai
Google Inc., 1600 Amphitheater Parkway, Mountain View, CA 94043, USA

J.L. Drury
The MITRE Corporation, 202 Burlington Road, Bedford, MA 01730, USA

A. Steinfeld
Carnegie Mellon University, 5000 Forbes Avenue, Pittsburgh, PA 15213, USA

© Springer Science+Business Media (outside the USA) 2016

R. Mittu et al. (eds.), *Robust Intelligence and Trust in Autonomous Systems*,
DOI 10.1007/978-1-4899-7668-0_11

the Global Hawk UAV, for example, completes military missions with little human supervision (Ostwald and Hershey 2007).

Robots are not the only examples of automated systems. IBM's intelligent agent Watson is now being used as an aid for medical diagnosis (Strickland 2013). Additionally, many of the trading decisions in the stock and commodities markets are being made by automated systems. Automation has been in use for decades as autopilot systems in airplanes and as assistants for running factories and power plants.

Utilizing autonomous capabilities can provide benefits such as reduced time to complete a task, reduced workload for people using the system, and a reduction in the cost of operation. However, existing research in the domains of plant, industrial, and aviation automation highlights the need to exercise caution while designing autonomous systems, including robots. Research in human-automation interaction (HAI) shows that an operator's trust of the autonomous system is crucial to its use, disuse, or abuse (Parasuraman and Riley 1997).

There can be different motivations to add autonomous capabilities; however, the overall goal is to achieve improved efficiency by reducing time, reducing financial costs, lowering risk, etc. For example, one of the goals of an autonomous car is to reduce the potential of an accident (Guizzo 2011). A similar set of reasons was a motivating factor to add autonomous capabilities to plants, planes, industrial manufacturing, etc. However, the end results of adding autonomous capabilities were not always as expected. There have been several incidents in HAI that have resulted from an inappropriate use of automation (Sarter et al. 1997). Apart from such incidents, research in HAI also shows that adding autonomous capabilities does not always provide an increase in efficiency. The problem stems from the fact that, when systems or subsystems become autonomous, the operators that were formerly responsible for manually controlling those systems are relegated to the position of supervisors. Hence, such systems are often called supervisory control systems.

In supervisory control systems, the operators perform the duty of monitoring and typically only take over control when the autonomous system fails or encounters a situation that it is not designed to handle. A supervisory role leads to two key problems: loss of skill over time (Boehm-Davis et al. 1983) and the loss of vigilance over time in a monitoring capacity (Endsley and Kiris 1995; Parasuraman 1986). Due to these two reasons, when operators are forced to take over manual control, they might not be able to successfully control the system.

As such systems are developed, it is important to understand how people's attitudes about the technology will influence its adoption and correct usage. A key factor shaping people's attitudes towards autonomous systems is their trust of the system; hence, we are striving to learn the factors that influence trust, whether for all autonomous systems or for particular domains. Without an appropriate level of trust or distrust, depending upon the circumstances, people may refuse to use the technology or may misuse it (Parasuraman and Riley 1997). When people have too little trust, they are less likely to take full advantage of the capabilities of the system. If people trust systems too much, such as when challenging environmental conditions cause the systems to operate at the edge of their capabilities, users are unlikely to monitor them to the degree necessary and therefore may miss occasions when they need to take corrective actions.

Thus it is important to understand how develop appropriate levels of trust prior to designing these increasingly capable autonomous systems. Without understanding the factors that influence trust, it is difficult to provide guidance to developers of autonomous systems or to the organizations commissioning their development. In contrast, a knowledge of the way particular factors influence trust can allow a system to be designed to provide additional information when needed to increase or maintain the trust of the system's user in order to ensure the correct usage of the system.

We have seen that trust of intelligent systems is based on a large number of factors (Desai et al. 2012). In our prior work (Desai et al. 2012, 2013), we have found that the mobile robotics domain introduces some different trust-related factors than have been found in the industrial automation domain. There is some overlap, however: a common subset of trust factors appears in both domains. Given our prior results, we believe that there is a core set of factors across all types of intelligent system domains that has yet to be codified. Further, it may be necessary to identify factors specific to each application domain.

Our ultimate goal is to understand the factors that affect trust in automation across a variety of application domains. Once we have identified the factors, our objective is to develop a core model of trust. In this chapter, we present two methods for identifying factors influencing trust and for building a trust model.

In the first method, we used online surveys of potential system users to identify the factors that most influence people's trust in two domains: automated cars and medical diagnosis systems. Our goal was to determine the factors influencing trust for these domains and compare them to determine the degrees of overlap and dissimilarity. Based upon these findings, we present a method for developing a core trust model.

In the second method, we used a series of human subjects experiments on a real robot to explore the influence of a number of variables upon people's trust of a robot system. Based upon the findings from these experiments, we built a model of trust.

This chapter describes our research methodology and findings for both methods of modeling trust, concluding with a discussion of the pros and cons of each method.

11.2 Prior Work in the Development of Trust Models

Sheridan and Verplank (1978) were among the first researchers to mention trust as an important factor for control allocation. According to their research, one of the duties of the operator was to maintain an appropriate trust of the automated system. However, the first researcher to investigate the importance of trust on control allocation was Muir (1989). According to Muir, control allocation was directly proportional to trust: i.e., the more trust the operator had in a system, the more likely he/she was to rely on it and vice versa. If the operator's trust of the automated system is not well calibrated, then it can lead to abuse (over-reliance) or disuse (under-reliance) on automation. Since this model of trust was first proposed, significant research has been done that indicates the presence of other factors that influence control allocation either directly or indirectly via the operator's trust of the automated system. Some of the factors that are known to influence trust or have been hypothesized to influence trust are explained in brief below.

- Reliability: Automation reliability is one of the most widely researched and one of the most influential trust factors. It has been empirically shown to influence an operator's trust of an automated system (Dzindolet et al. 2003; Riley 1996; deVries et al. 2003). Typically, lower reliability results in decreased operator trust and vice versa. However, some work with varying reliability indicates that the timing of the change in reliability can be critical (Prinzel III 2002).
- Risk and reward: Risk and reward are known to be motivating factors for achieving better performance. Since lack of risk or reward reduces the motivation for the operator to expend any effort and over-reliance on automation reduces operator workload (Dzindolet et al. 2003), the end result for situations with low or no motivation is abuse of automation.
- Self-confidence: Lee and Moray (1991) found that control allocation would not always follow the change in trust. Upon further investigation, they found that control allocation is dependent on the difference between the operator's trust of the system and their own self-confidence to control the system under manual control.
- Positivity bias: The concept of positivity bias in HAI research was first proposed by Dzindolet et al. (2003). They borrowed from the social psychology literature, which points to a tendency of people to initially trust other people in the absence of information. Dzindolet et al. showed the existence of positivity bias in HAI through their experiments. The theory of positivity bias in the context of control allocation implies that novice operators would initially tend to trust automation.
- Inertia: Researchers observed that when trust or self-confidence change, it is not immediately followed by a corresponding change in control allocation (Moray and Inagaki 1999). This delay in changing can be referred to as inertia. Such inertia in autonomous systems can be potentially dangerous, even when the operator's trust is well calibrated. Hence, this is an important factor that warrants investigation to help design systems with as little inertia as possible.
- Experience: In an experiment conducted with commercial pilots and undergraduate students, Riley (1996) found that the control allocation strategy of both populations was almost similar with one exception: pilots relied on automation more than the students did. He hypothesized that the pilots' experience with autopilot systems might have resulted in a higher degree of automation usage. Similar results were found in our work (Desai 2012) when participants familiar with robots relied more on automation than those participants not familiar with robots.
- Lag: Riley (1996) hypothesized that lag would be a potential factor that could influence control allocation. If there is a significant amount of lag between the

operator providing an input to the system and the system providing feedback to that effect, the cognitive work required to control the system increases. This increased cognitive load can potentially cause the operator to rely on the automated system more.

11.2.1 Trust Models

In the process of investigating factors that might influence operator's trust and control allocation strategy, researchers have modeled operator trust on automated systems (e.g., Muir 1987; Lee and Moray 1992; Riley 1996; Cohen et al. 1998; Farrell and Lewandowsky 2000; Moray et al. 2000). Over a period of two decades, different types of trust models have been created. Moray and Inagaki (1999) classified trust models into five categories for which they explain the pros and cons of each type in brief: regression models, time series models, qualitative models, argument based probabilistic models, and neural net models.

Regression models help identify independent variables that influence the dependent variable (in most cases trust). These models not only identify the independent variables but also provide information about the relationship (directly proportional or inversely proportional) between each of the independent variables and the dependent variable and the relative impact of that independent variable with respect to that of other variables. The model presented in Sect. 11.4.3 is an example of a regression model. These models, however, cannot model the dynamic variances in the development of trust and hence must be used only when appropriate (e.g., simply identifying factors that impact operator trust). Regression models can be used to identify factors that impact trust but do not significantly vary during interaction with an automated system, and, based on this information, appropriate steps can be taken to optimize overall performance. This information can potentially be provided to the automated system to allow it to better adapt to each operator. Regression models have been utilized by other researchers (Muir 1989; Lee 1992; Lee and Moray 1992).

Time series models can be used to model the dynamic relationship between trust and the independent variables. However, doing so requires prior knowledge of the factors that impact operator trust. Lee and Moray (1992) used a regression model to initially identify factors and then used a time series model (AutoRegressive Moving AVerage model: ARMAV) to investigate the development of operator trust. Through that model, Lee and Moray found that the control allocation depends on prior use of the automated system and individual biases, along with trust and self-confidence. Using a time series model requires a large enough data set that can be discretized into individual events. For example, in an experiment conducted by Lee and Moray, each participant operated the system for a total of 4 hours, which included twenty-eight individual trials (each 6 minutes long). Qualitative data was collected at the end of each run that might have had a faulty system throughout the run. Unlike most other types of models, time series models can be used online to predict future trust and control allocation and perhaps initiate corrective action if needed. However, to our knowledge, no such models exist.

In qualitative models, the researchers establish relationships between different factors based on quantitative data, qualitative data, and their own observations. As Moray and Inagaki (1999) point out, such models can provide valuable insight into how trust, control allocation, and other factors interact. A model of trust partly based on the human-human model of trust developed by Muir (1989) and the model of human-automation interaction by Riley (1994) takes advantage of two well-established qualitative models. Given the heuristic nature of these models, they cannot be used to make precise predictions about trust and control allocation; however, they can and often have been used to create a set of guidelines or recommendations for automation designers and operators (e.g., Muir 1987; Chen 2009).

Farrell and Lewandowsky (2000) trained a neural net to model the operator's control allocation strategy and be able to predict future actions by the operator. The model, based on connectionist principles, was called CONAUT (Connectionist Model of Complacency and Adaptive Recovery). Their model received digitized information as sets of ten bits for each task. Using that model, the authors predicted that cycling between automatic and manual control could eliminate operator complacency. While such models can accurately model trust and control allocation strategies, they require large data sets. Due to the nature of neural networks, it is not feasible to extract any meaningful explanation about how the model works.

11.2.2 Trust in Human-Robot Interaction (HRI)

Human-Robot Interaction (HRI) is a diverse field that spans from medical robots to military robots to social robots to automated cars. While it would be ideal to create a model of trust that generalizes to all of HRI, it is important to narrow the scope of investigation because we hypothesize that the application domain is a significant factor in the trust model. Various taxonomies have been defined for HRI (e.g., Dudek et al. 1993; Yanco and Drury 2004). One such taxonomy for robots defines the system type by their task (Yanco and Drury 2004). Another possible classification for robots is their operating environment: ground, aerial, and marine robots. The scope of the research described in this paper is limited to remotely controlled unmanned ground robots that are designed for non-social tasks. Unmanned ground robots represent a significant number of robots being developed and hence the contributions of this chapter should impact a significant number of application domains within HRI.

Several application domains within the realm of unmanned ground robots are classified as mobile robots, such as factory robots (e.g., Kiva Systems 2011; CasePick Systems 2011), consumer robots (e.g., iRobot 2011; Neato Robotics 2011), and autonomous cars (e.g., Thrun 2010; Dellaert and Thorpe 1998). However, one of the more difficult domains is urban search and rescue (USAR). USAR robots typically operate in highly unstructured environments (Burke et al. 2004), involve a significant amount of risk (to the robot, operating environment, and the victims), and are remotely operated. These factors that make operating USAR robots difficult also make USAR the ideal candidate for examining different factors that influence trust in HRI.

Along with the models of operator reliance on automation (Riley 1996), the models of trust, the list of known factors, and the impact of these factors on operator trust have been well researched in HAI (e.g., Muir 1989; Moray and Inagaki 1999; Dzindolet et al. 2001). However, the automated systems used for research in HAI and in real world applications differ from the typical autonomous robot systems in HRI and therefore necessitate investigating trust models in HRI. Some of the key differences between typical HAI systems and HRI, along with unique characteristics of HRI relevant to operator trust, are explained in brief below.

- Operating environment: The operating environment of most systems in HAI is very structured and well defined (e.g., automated plant operation or automated anomaly detection). On the other hand, the operating environment for USAR can be highly unstructured (Burke et al. 2004) and unfamiliar to the operator. The lack of structure and a priori knowledge of the environment can limit the autonomous capabilities and can also impact the reliability of the autonomous robots.
- Operator location: When operators are co-located with the autonomous system, it is easy for the operator to assess the situation (e.g., auto-pilots). However, with teleoperated robots, the operator can be up to a few hundred feet or more away from the robot. This physical separation between the robot and the operator makes it difficult to assess the operating environment and can impact the development of trust. While sensors and actuators are not unique to robots, remotely controlling actuators is more difficult with noisy sensors. In most of the experimental methodologies used in HAI, noisy sensors are not used and hence their impact on automation or the operator are not investigated.
- Risk: The level of risk involved in HAI domains varies widely, ranging from negligible (e.g., automated decision aids; Madhani et al. 2002; Dzindolet et al. 2001) to extremely high (e.g., autopilots, nuclear plants). However, the research that does exist mostly involves low risk scenarios (Muir 1989; Riley 1996; Sanchez 2006). In contrast, domains like USAR carry a significant amount of risk that the operator needs to understand and manage accordingly.
- Lag: Unlike HAI, where the input to the system and the feedback from the system is immediate, the delay in sending information to the robot and receiving information from the robot can vary based on the distance to the robot and the communication channel. This delay, ranging from a few hundred milliseconds to several minutes (e.g., in the case of the Mars rovers) can make teleoperating a robot incredibly difficult, forcing the operator to rely more on the autonomous behaviors of the robot.
- Levels of autonomy: Automated systems typically studied in HAI operate at one of two levels of autonomy on the far ends of the spectrum (i.e., completely

manual control or fully automated). In HRI, robots can often be operated at varying levels of autonomy (e.g., Bruemmer et al. 2002; Desai and Yanco 2005).

- Reliability: Due to the nature of noisy and often failure prone sensors used in robotics, the reliability of automated behaviors that rely on those sensors is often lower than typically high reliability levels used for HAI research (Bliss and Acton 2003; Dixon and Wickens 2006).
- Cognitive overload: Teleoperating a remote robot can be a cognitively demanding task. Such demands can impact other tasks that need to be carried out simultaneously. Cognitive load can also result in operators ceasing to switch autonomy modes (Baker and Yanco 2004).

Along with these differences, the experimental methodology used for most of HAI research has either been based on abstract systems, micro-worlds, or low fidelity simulations (Moray and Inagaki 1999). These setups cannot be used to investigate the subtle effects of different characteristics listed above. Hence, a real-world experimental scenario will be used to examine trust in HRI in one of the methods presented in this chapter. Section 11.4.1 explains the details of the experimental methodology along with the different factors that will be examined and a motivation for examining them.

11.3 The Use of Surveys as a Method for Developing Trust Models

While experiments that allow people to use real systems can produce valuable insights into the factors that influence trust, the nature of the experimental procedures do not allow for very large sets of people to be included. To allow for a larger set of people to be queried, we decided to explore the use of surveys for developing trust models. We selected two domains to begin: automotive and medical. Specifically, we focused on driverless cars (e.g., Google Cars) and automated medical diagnoses (e.g., IBM's Watson). There were two dimensions for each survey: the safety criticality of the situation in which the system was being used and name-brand recognition. We designed the surveys and administered them electronically, using Survey Monkey and Amazon's Mechanical Turk. We then performed statistical analyses of the survey results to discover common factors across the domains, domain-specific factors, and implications of safety criticality and brand recognition on trust factors. We found commonalities as well as dissimilarities in factors between the two domains, suggesting the possibility of creating a core model of trust that could be modified for individual domains.

11.3.1 Methodology

We chose the automotive and medical domains for several reasons. The successful completion of over 300,000 miles by Google's driverless car, as well as the rulings in three states and the District of Columbia legalizing the use of driverless cars (Clark 2013), holds much promise for these cars becoming commonplace in the near future. Watson, a question-answering agent capable of referencing and considering millions of stored medical journal articles, is also promising. Little research has been conducted about the public's opinion on IBM's Watson, so the relationship between humans and medical diagnosis agents is uncharted territory.

We felt that the general public could be expected in the future to interact with both automated cars and Watson (in conjunction with their physicians). Thus, we developed computer-based survey instruments that could be administered over the Internet to a wide audience. The surveys resided in Survey Monkey and were accessed via Amazon's Mechanical Turk so that respondents could be paid for their time. The surveys were administered in two rounds, with the first round being exploratory. After making improvements to the surveys, including the addition of factors identified by the initial participants in the "other" category, we released the second round, the results of which are reported in this paper.

Each round of surveys consisted of eight different variations: four for each of the two domains. All of the surveys began with the same demographic questions, including gender, age, computer usage, video game playing, and tendencies to take risks. Then each survey variant had a unique scenario designed to capture differences in public opinions depending on the seriousness of the situation ("safety critical" versus "non-safety critical") and the brand of the automated machine (well-known brand from a large organization versus a brand from an unknown startup). Thus there are four variations for each domain: safety critical and well-known brand (branded); safety critical and unknown brand (non-branded); non-safety critical and branded; and non-safety critical and unbranded.

In the automotive safety critical scenario, the environment was described as high-speed, with lots of traffic. In the non-safety critical scenario, the environment was described as low-speed with little traffic. While one might argue that all driving is safety critical, clearly it is more difficult to ensure safe travel at higher speeds and with more traffic. It is also more difficult to imagine oneself taking over control from such an autonomous system at high speeds in difficult driving conditions.

In the medical safety critical scenario, the task described was to determine diagnoses and treatments of three possible types of cancer. In the non-safety critical scenario, the respondent was given ample information to be certain that the affliction was not life threatening. The three possible afflictions in the non-safety critical scenario include mononucleosis, influenza, or the common cold. Cancer denotes a greater level of importance and urgency whereas the latter situation seems less dire.

In addition to the severity of the situation, we wanted to see whether the brand of the automated machine affected people's trust level as well. For the automotive domain, we explicitly described the automated system as being a Google Car for the two branded surveys. For the medical domain, we specified that Watson was a product of IBM in two survey variants. In the remaining survey variants, we did not label the automated machine as either the Google Car or IBM's Watson; instead, we said that a small, startup company developed the systems. In this way, we hoped to identify the extent to which the reputation of the company influences trust in an intelligent system in these domains.

Each survey in the automotive domain presented a list of 29 factors that could influence a person's trust of an automated system; surveys in the medical domain presented 30 factors. This list of factors was determined initially from a literature search, including the factors from Desai (2012) discussed below in the results section. We started with a shorter list in the initial design of our surveys; we released each of these initial surveys to small sample sizes (25 per survey; 100 in each domain, for a total of 200). Based upon these preliminary results, we added some additional factors, which were identified by respondents in a free-text "other" field. This process resulted in the full list factors for each domain used in the second version of the surveys, some of which were specific to the particular automation domain and others that were common to the two. The results presented in this paper are from the second version of the surveys, with 100 respondents for each of the eight survey variants.

The surveys also included three test questions used to ensure that respondents were actually reading the survey and answering to the best of their ability: "this sentence has seven words in it," "most dogs have four legs," and "the influence of the color of one's shirt" on their trust of an autonomous system. If a respondent answered one or more of these test questions incorrectly, their data was removed from the dataset.

We created each survey on Survey Monkey and utilized Amazon Mechanical Turk to disseminate them to the public. We narrowed our pool to residents of the United States with a minimum age of 18. We paid each respondent $0.90 to complete the survey. This human subjects research was approved by MITRE's IRB.

11.3.2 Results and Discussion

We released 100 HITs (Human Intelligence Tasks) on Mechanical Turk for each of the versions of our surveys. Each survey had 83 questions, similar except for the wording that pertained to branding/not and safety critical/not. After discarding responses that had one or more of the test questions described in Sect. 11.3.1 answered incorrectly, we had 382 responses in the medical diagnosis domain (231 male, 151 female; mean age 31.1 (9.0)) and 355 in the car domain (191 male, 164 female; mean age 35.6 (12.6)).

For the medical domain, we had 91 valid responses for the branded and safety critical version, 101 for branded and not safety critical, 97 for non-branded and safety critical, and 93 for non-branded and not safety critical. For the automotive domain, we had 90 valid responses for the branded and safety critical version, 92 for branded and not safety critical, 82 for non-branded and safety critical, and 91 for non-branded and not safety critical. The gender and age demographics were not significantly different between the survey versions for each domain.

For each of the trust factors, respondents were asked to rank how the factor would influence their trust in the system on a 7 point Likert scale, with 1 meaning "strongly disagree" and 7 meaning "strongly agree." The results for the trust factors were aggregated for the automotive domain and for the medical domain. In Tables 11.1 and 11.2, we present the list of factors sorted on the mean score from the Likert scale; while a Likert scale is not a continuous scale and averaging the responses is not strictly correct, it does allow us to see which factors have greater influence on trust across the respondents. Due to this limitation of a Likert scale, we discuss our results in terms of the top, middle and bottom thirds, rather than a strict ordering based upon the mean.

In both domains, the ability of a system to stay up-to-date, statistics about its past performance, and the extent of the research on the system's reliability are important factors for influencing trust in the system, appearing in the top third in both domains. In the middle third, both domains included the person's own past experience with the system, the reputation of the system, the effectiveness of the system's training and prior learning, and observing a system's failure. These common factors could form the basis of a model of trust for automated systems; of course, we need to expand our work to many other domains in order to discover the true core.

In the bottom third, both domains include the system's possibility of being hacked, the system's user-friendliness, its ability to communicate effectively, the popularity of the system, and the aesthetics of the system. These factors are being judged as unimportant to trust by respondents in both domains. However, there may be some domains where issues related more to user interface and the usability of the system could come into play. For example, in a social robot domain such as companion robots for the elderly, the way the system looks could have a greater influence on the user's trust of the system: a pet-like robot covered in fur might be more trusted than a more machine-like system showing metal and wires, for example.

We found that there are domain specific factors present in the top third of the list. For the medical domain, respondents ranked the accuracy of the diagnosis, verification of the diagnosis, and the doctor's ability to use the machine in the top third. In the automotive domain, reliability also ranked in the top third through several of the factors. In our survey design, we elected to have a number of questions about reliability to determine if there were different aspects of reliability. While we did see some differences, the list of factors could be reduced by using reliability in place of this group of factors; we will do this when we move to the next phase where we ask respondents to rank trust factors in order of importance.

Of note is where the responsibility of system verification and understanding lies between the two domains. In the top third of the factors in the medical domain, we see that people are looking to the doctor to mediate the results of the automated system. However, respondents are relying more on themselves in

Table 11.1 Rankings of the factors that can influence trust of an automated system in the automotive domain

	Automotive domain				
	Rank	Ref	Influence factor	Mean	Std. dev
Top third	1	A	Statistics of the car's past performance	5.98	1.32
	2	B	Extent of research on the car's reliability	5.87	1.39
	3		My own research on the car	5.82	1.33
	4		Existence of error/problem indicators	5.79	1.49
	5		Possibility that the hardware or software may fail	5.69	1.70
	6		Credibility of engineers who designed the car	5.69	1.53
	7	C	The car's ability to stay up-to-date	5.64	1.53
	8		Technical capabilities of the car	5.55	1.55
	9		Your understanding of the way the car works	5.54	1.48
	10		Your past experience with the car	5.53	1.66
Middle third	11	D	Reputation of the car	5.49	1.57
	12		Level of accuracy of the car's routes	5.49	1.48
	13		Amount of current roadway information available to the car (e.g., weather, traffic, construction, etc.)	5.43	1.65
	14	E	Effectiveness of the car's training and prior learning	5.41	1.63
	15		Amount of information that the car can access	5.41	1.64
	16	F	Observing a system failure (e.g., making a wrong turn, running a stop light)	5.37	2.00
	17		Accuracy of the route chosen	5.36	1.60
	18		User's familiarity with the car	5.29	1.58
	19		The reputation of the car manufacturer	5.27	1.66
Bottom third	20		Agreement of routes between car and my knowledge	5.26	1.55
	21		The car's methods of information collection	5.24	1.52
	22	G	Possibility of the car being hacked	5.09	1.94
	23	H	The user-friendliness of the car	5.04	1.69
	24		Amount of verification by your friend of the car's proposed route and driving ability	4.73	1.71
	25		Your friend's training to use the car effectively	4.68	1.99
	26	I	The car's ability to communicate effectively (e.g., accurate grammar, breadth of vocabulary)	4.60	1.88
	27	J	Popularity of the car	3.38	1.74
	28	K	Aesthetics of the car	3.01	1.72

Factors ranked in the same thirds for both the automotive (this table) and medical (Table 11.2) domains are cross-referenced with letters in the "Ref" column. These common factors appearing in the same third of the rankings give evidence that a core model of trust factors could be developed. The other factors, which are common to both domains but ranked in different thirds or which are domain specific, would be the domain specific factors used to customize the core trust model for a particular domain

Table 11.2 Rankings of the factors that can influence trust of an automated system in the medical domain

	Medical domain				
	Rank	Ref	Influence factor	Mean	Std. dev
Top third	1		Accuracy of the diagnosis	6.33	1.04
	2		Level of accuracy of the machine's diagnosis	6.07	1.16
	3	A	Statistics of machine's past performance	6.04	1.20
	4	C	The machine's ability to stay up-to-date	5.97	1.17
	5		Amount of your information available to the machine (e.g., x-rays, physicals, cat scans, etc.)	5.85	1.26
	6		Amount of verification by your doctor of the machine's suggestions	5.84	1.19
	7		Agreement of diagnoses between doctor and machine	5.83	1.32
	8		Doctor's training to use the machine effectively	5.80	1.24
	9		Amount of information that the machine can access	5.79	1.25
	10	B	Extent of research on the machine's reliability	5.79	1.3
Middle third	11	E	Effectiveness of the machine's training and prior learning	5.63	1.35
	12		Technical capabilities of the machine	5.62	1.31
	13		Existence of error/problem indicators	5.52	1.49
	14	D	Reputation of the machine	5.50	1.41
	15		The machine's methods of information collection	5.46	1.29
	16		Possibility that the hardware or software may fail	5.34	1.64
	17		Credibility of engineers who designed the machine	5.31	1.52
	18		Your past experience with the machine	5.25	1.44
	19	F	Observing a system failure (e.g., making an incorrect diagnosis)	5.15	1.88
Bottom third	20		User's familiarity with the machine	5.07	1.52
	21	G	Possibility of the machine being hacked	5.06	1.88
	22		My own research on the machine	5.04	1.52
	23		Your understanding of the way the machine works	5.03	1.59
	24		The reputation of the machine's manufacturer	4.87	1.65
	25		Amount of time the doctor consults other doctors	4.74	1.72
	26	I	The machine's ability to communicate effectively (accurate grammar, breadth of vocabulary)	4.61	1.68
	27	H	The user-friendliness of the machine	4.07	1.66
	28	J	Popularity of the machine	3.77	1.72
	29	K	Aesthetics of the machine	2.64	1.74

Factors ranked in the same thirds for both the automotive (Table 11.1) and medical (this table) domains are cross-referenced with letters in the in the "Ref" column the automotive domain. This responsibility can be demonstrated by the fact that "your understanding of the way the [system] works" ranks in the top third for the automotive domain, but in the bottom third for the medical domain. Models of trust for automated systems will need to take into account whether the system is used directly by an end-user or whether it is utilized by a mediator for the end-user. Other such domains might include automated stock trading systems.

In some of our earlier work (Desai 2012), we also utilized Amazon's Mechanical Turk to determine factors that would influence human-robot interaction for novice robot users. To obtain these results, Desai created a series of videos showing robots moving in a hallway environment, which were watched by the survey respondents. Test questions included the color of the robot shown in the video to ensure that the video had been watched. There were 386 valid responses received.

Desai (2012) reports the top six factors that influence trust of a robot system are reliability, predictability, trust in engineers that designed the robot, technical capabilities of the robot, system failure (e.g., failing sensors, lights, etc.), and risk involved in the operation. The factors in the middle third were error by automation, reward involved in the operation, interface used to control the robot, lag (delay between sending commands and the robot responding to them), and stress. The factors in the bottom third were training, situation awareness (knowing what is happening around the robot), past experience with the robot, size of the robot, and speed of the robot.

While our surveys presented 30 possible factors to respondents and Desai's had a total of 17, we see some similarities between the factors in the top third, most notably reliability (although, as discussed above, our surveys presented several questions about aspects of reliability). We also found that trust in the engineers who designed the system was important to our respondents, largely through the different surveys presented for branded vs. non-branded automated systems.

For both application domains, we found a significant difference in people's trust of the system based upon whether the system was made by a well-known company (Google for the automotive domain; IBM's Watson for the medical domain) vs. a "small, startup company." Our surveys had two questions about branding, to which participants answered on a 7 point Likert scale, with 1 meaning "strongly disagree" and 7 meaning "strongly agree." In the first, participants were asked to rate the statement "I trust the machine's capabilities because it was created by ['IBM', 'Google', or 'a small, startup company']." The second statement asked if the participant's "trust in a fully-autonomous system similar to this machine would decrease if it was created by ['a lesser-known company' for the IBM and Google versions or 'a more established company' such as Google or IBM]." Results are shown in Tables 11.3 and 11.4.

Clearly, given these findings, it will require additional work for designers of automated systems to convince users to trust the systems made by small companies. However, one could note that Google was a small, startup company not long ago. Other factors such as past performance of the system can also be used to assist with the trust of a non-branded automated system.

Table 11.3 Branded vs. non-branded technology: medical domain

	Questions		Mean		
	Brand (Watson)	Non-brand	Brand	Non-brand	T value
Safety critical	I trust the machine's capabilities because it was created by IBM	I trust the machine's capabilities because it was created by a small, upstart company	3.04	2.54	0.006
	My trust in a fully autonomous system similar to this machine would decrease if it was created by a lesser-known company	My trust in a fully autonomous system similar to the machine would decrease if it was created by a more established company such as IBM	3.42	2.81	0.003
Non safety critical	I trust the machine's capabilities because it was created by IBM	I trust the machine's capabilities because it was created by a small, upstart company	3.24	2.67	0.002
	My trust in a fully autonomous system similar to this machine would decrease if it was created by a lesser-known company	My trust in a fully autonomous system similar to the machine would decrease if it was created by a more established company such as IBM	3.46	2.87	0.003

Reputation matters: Significant differences were seen for responses for branded automated systems in the medical domain

Table 11.4 Branded vs. non-branded technology: automotive domain

	Questions		Mean		
	Brand (Google)	Non-brand	Brand	Non-brand	T Value
Safety critical	I trust the car's capabilities because it was created by Google	I trust the car's capabilities because it was created by a small, upstart company	3.19	2.41	0.000
	My trust in a fully autonomous system similar to cars would decrease if it was created by a lesser-known company	My trust in a fully autonomous system similar to cars would decrease if it was created by a more established company such as Google	3.67	2.89	0.001
Non safety critical	I trust the car's capabilities because it was created by Google	I trust the car's capabilities because it was created by a small, upstart company	3.33	2.55	0.000
	My trust in a fully autonomous system similar to cars would decrease if it was created by a lesser-known company	My trust in a fully autonomous system similar to cars would decrease if it was created by a more established company such as Google	3.88	2.82	0.000

Reputation matters, part II: Significant differences were also seen for responses for branded automated systems in the automotive domain

11.3.3 Modeling Trust

A core model of trust can be formulated by looking at the common factors in the top third of the survey results. In the case of this work, the core factors would be statistics about the system's past performance, the extent of research on the system's reliability, and the system's ability to stay up to date. While this last factor will have different meanings in each domain, either referring to the medical information necessary to perform a diagnosis or to road information needed to drive safely and accurately, it is still a core factor that influences people's trust in an automated system.

Factors that rank highly in one domain, but low in the other, could be used as factors to customize the core model of trust for the particular application domain. For example, in the medical domain, we see top ranked factors addressing the accuracy of the system and the doctor's interpretation of the system. In contrast, in the automotive domain, we see more reliance on self-knowledge and the car's ability to convey information.

It is important to note that, given the fact that the users of Amazon's Mechanical Turk skew towards having more education than the average population (Ross et al. 2010), the responses reported in this paper might not be applicable to the general population but instead might only be applicable to the population with an undergraduate degree or greater. We need to conduct an analysis of the data with respect to education level to determine if there are differences between responses for different levels of education. However, despite this potential limitation of our survey population, we believe surveys like ours can identify factors that will influence trust.

While our surveys allowed people to specify the relative importance of the factors, they did not provide people with the means to indicate whether that factor would result in an increase or decrease in trust. Our next step will be to conduct surveys asking people to choose the top factors which influence their trust, ranking them from most to least important. We will also explore the influence that these factors have upon each other; for example, a system's ability to explain its action influences the system's understandability.

We are also expanding this research to other automated system domains. Our methodology will need to change for some of these domains, as we have been relying on people from the population of Mechanical Turk workers. While such people are well qualified to answer questions about cars and doctor's visits, they will be less qualified to answer questions about the use of automated systems in very specialized domains such as the military or power plants. However, we believe that the use of surveys, whether completed by "average" people or people working in specialized domains, will allow us to identify the top factors influencing trust in automated systems in each domain. As we explore more domains, we will be able to identify those factors that are common to many domains; these factors will form the common core of a trust model.

11.4 Robot Studies as a Method for Developing Trust Models

The primary goal of the research described in this section was to create a better understanding of different factors that impact operator trust and control allocation when interacting with an autonomous remote robot. We also wanted to investigate how certain attributes central to remote robot teleoperation (e.g., situation awareness, workload, task difficulty) impact operator behavior. By observing the variations in the different factors and how they affect operator trust and control allocation strategy, a model of operator interaction specifically for teleoperation of an autonomous remote robot was constructed and has been used to create a set of guidelines that can improve the overall system performance.

11.4.1 Methodology

The robot used is an iRobot ATRV-JR platform, shown in Fig. 11.1. The ARTV-JR has differential drive and a wide array of sensors. These sensors include a front facing SICK LMS-200 laser range finder that can scan 180°, a rear facing Hokuyo URG-04LX laser range finder with a field of view of 240°, a Directed Perception PTU-D46-17 pan-tilt unit with a Sony XC-999 camera mounted on it, and a rear facing Canon VC-C4 camera mounted on the back of the robot. The robot also has a 3.0 GHz Intel Core2Duo processor with 4 GB of memory and runs Ubuntu 8.04.

Fig. 11.1 The ATRV-JR used in the robot experiments to explore trust factors in human-robot interaction

It has an 802.11 n radio capable of operating on both the 2.4 and 5.0 GHz range. The client code to control the robot is written in C++ using Player (Gerkey et al. 2003) and compiled using GCC.

Almost all of the prior research in HAI has focused on using two autonomy modes on the far ends of the spectrum. In accordance with this existing research, we decided to provide the participants with two autonomy modes. One of those autonomy modes was at the high end of the autonomy spectrum. Rather than selecting the second autonomy mode to be manual teleoperation mode we decided to opt for a similar autonomy mode where the robot would assist the participants. The key reason was to always keep the participant informed about the robot's behavior, something that would not be possible with a pure manual teleoperation mode. The participants could operate the robot in one of two autonomy modes: robot-assisted mode or fully autonomous mode. The participants were free to select either mode and could switch between them as many times as they wanted. They were also told that there were no benefits or penalties for selecting either mode. When each run was started, no autonomy mode was selected by default, thereby requiring the participants to make an explicit selection. The maximum speed at which the robot moved was the same in both modes and was restricted to approximately 0.12 meters per second. These configurations ensured that the performance of both autonomy modes was similar.

In the fully autonomous mode, the robot ignored the participant's input and followed the hard coded path. The obstacle avoidance algorithm ensured that the robot never hit any object in the course. In the robot-assisted mode, the participant had a significant portion of the control and could easily override the robot's movements, which were based on the path it was supposed to follow. The robot's vectors were calculated the same way in both autonomy modes and were displayed on the user interface (UI) on the laser display to show the participant the robot's planned direction.

Figure 11.2 shows the test course designed for these experiments. The course was approximately 18.3 meters long and had 5 obstacles (boxes) placed about 2.7 meters from each other. The width of the course was 2.4 meters. The clearance on either side of the boxes was 0.9 meters, versus a robot width of approximately 0.7 meters. Therefore the small clearance on either side of the boxes made it difficult to drive.

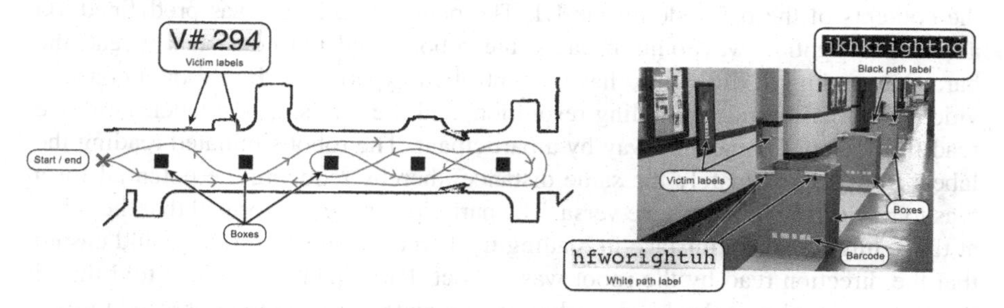

Fig. 11.2 The test course used in the experiments

The course had moderate foot traffic, as it was placed in a hallway in a public building; however, we found that people walking through the area were able to do so in a way that did not interrupt the experiment. Additionally, during each experiment, one of our researchers was with the robot at all times, so could ask people to avoid the robot, if necessary.

The robot started and ended each run at the same location. For each run, the participants had to follow a preset path. Since we planned five runs, we designed five different paths (also referred to as maps) based on the following criteria:

- The length of each map must be the same (~61 meters).
- The number of U-turns in a map must be the same (3 U-turns).
- The number of transitions from the left side of the course to the right and vice versa must be the same (three transitions).

Since the maps were similar in difficulty and length, they were not counter-balanced. Instead, the maps were selected based on a randomly generated sequence. A sample map is shown in green in Fig. 11.2.

Each box on the course had text labels to provide navigational information to the participants. Text labels were placed on top of the boxes to indicate the path ahead. Since the boxes were wide, similar labels were placed on both edges of the face as shown in Fig. 11.2, to make it easy for the participants to read the labels as the robot moved past the boxes. The labels indicated one of three directions 'left', 'right', or 'uturn'. These directions were padded with additional characters to prevent the participants from recognizing the label without reading them.

Two sets of labels were necessary to prevent the participants from driving in an infinite loop. Figure 11.2 shows the two types of labels that were used. The labels with a white background (referred to as white labels) were followed for the first half of the entire length and then the labels with a black background (referred to as black labels) for the second half. The transition from following the white labels to black labels was indicated to the participants via the user interface (UI).

The boxes also had barcodes made from retro-reflective tapes that the robot could read using its laser rangefinder (Fig. 11.2). While the robot did not actually use these barcodes in the experiments (the localized pose of the robot was used instead to encode the paths), the participants were told that the robot reads the barcodes to determine the path ahead, just like they read the labels. The robot displayed the contents of the bar code on the UI. The path for each run was predefined via a set of navigation waypoints because the robot could not consistently read the barcodes, making it difficult to have a controlled experiment. Based on a constant video compression rate, sampling resolution, and the font size, the labels could be read from about 0.9 meters away by a participant. The robot simulated reading the labels from approximately the same distance, thereby reducing the potential for a bias to rely on the robot or vice versa. The participants were informed that the robot at times might make a mistake in reading the barcodes and that they should ensure that the direction read by the robot was correct. Participants were also told that if the robot did make a mistake in reading the barcode, it would then proceed to pass

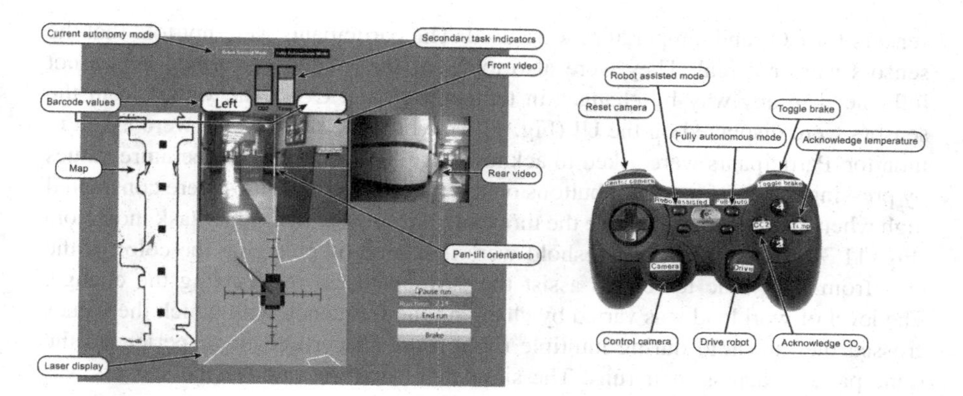

Fig. 11.3 The user interface for the robot

the next box on the incorrect side, resulting in the participant being charged with an error on their score (described below).

The course also had four simulated victims. These victims were represented using text labels like the one shown in Fig. 11.2. The victim tags were placed only on the walls of the course between 0.8 and 1.8 meters from the floor. The victim locations were paired with the paths and were never placed in the same location for any of the participant's five runs. While there was a number associated with each victim, the participants were told to ignore the number while reporting the victims. Whenever participants found a new victim, they were told to inform the experimenter that they have found a victim. They were explicitly instructed to only report victims not reported previously. The experimenter noted information about victims reported by the participants and also kept track of unique victims identified.

Figure 11.3 shows the UI utilized for controlling the robot. The video from the front camera was displayed in the center and the video from the back camera was displayed on the top right (mirrored to simulate a rear view mirror in a car). The map of the course with the pose of the robot was displayed on the left. The distance information from both lasers was displayed on the bottom around a graphic of the robot just under the video. There were vectors that originate from the center of the robot and extend out. These vectors indicated the current magnitude and orientation of the participant's input via the gamepad and the robot's target velocity. The participant's vector was displayed in light gray and the robot's vector was displayed in blue.

The participants provided input using the gamepad shown in Fig. 11.3. Participants could drive the robot, control the pan tilt unit for the front camera, select the autonomy modes, turn the brakes on or off, re-center the camera, and acknowledge the secondary tasks.

The participants were asked to drive the robot as quickly as they could along a specified path, while searching for victims, not hitting objects in the course, and responding to the secondary tasks. To create additional workload, simulated sensors for CO_2 and temperature were used. The participants were not told that the sensors were not real. They were also told that the robot's performance was not influenced in any way by changes in temperature and CO_2. The values from the sensors were displayed on the UI (Fig. 11.3), which the participants were asked to monitor. Participants were asked to acknowledge high CO_2 and temperature values by pressing the corresponding buttons on the gamepad. The values were considered high when their values are above the threshold lines on the secondary task indicators (Fig. 11.3); values over the threshold were indicated by changing the color of the bars from light blue to red, to assist the participants in recognizing the change. The level of workload was varied by changing the frequency with which the values crossed the threshold during multiple robot runs; all participants experienced the same patterns across their runs. The simulated sampling rate for the sensors was kept steady.

In the feedback (Sect. 11.4.2.2), reduced task difficulty (Sect. 11.4.2.3) and long-term (Sect. 11.4.2.4) experiments, the simulated sensor readings were removed from the interface. In their place, participants were asked at regular intervals (every 25 s) whether their trust in the robot had increased, stayed the same, or decreased. The answers given were plotted over time during the run. We then defined the area under the trust curve (AUTC) as a metric that could be used to measure on-line trust, as opposed to end of run measures such as were used by Muir and Jian.

Using higher levels of automation can reduce workload and hence is desirable, especially under heavy workload from other tasks. To prevent participants from using high levels of autonomy all the time, regardless of the autonomous system's performance, it is typical to introduce some amount of risk. Hence, in line with similar studies (e.g., Riley 1996; Lee and Moray 1992; Dzindolet et al. 2002), the compensation was based in part on the overall performance. The participants could select a gift card to a local restaurant or Amazon.com. The maximum amount that the participants could earn was $30. Base compensation was $10. Another $10 was based on the average performance of five runs. The last $10 was based on the average time needed to compete the five runs, provided that the performance on those runs was high enough.

The performance for each run was based on multiple factors, with different weights for each of these factors predetermined. The participants are told there was a significant penalty for passing a box on the incorrect side, regardless of the autonomy mode. If the participants passed a box on the wrong side, they were heavily penalized (20 points per box). In addition to the loss of score, participants were told that time would be added based on the number of wrong turns they took, but the specific penalties were not revealed. For the first box passed on the wrong side, no additional time was added, to allow participants to realize that the reliability of the system had dropped. For the second incorrect pass, 60 s were added, with an additional 120 s for the third and an additional 240 for the fourth, continuing with a cumulative increase. Finding the victims was also an important task, so 10 points were deducted for each victim missed.

The scoring formula was not revealed to participants, although they were told about the factors that influence their score. The score for each run was bounded between 0 and 100. If the score was 50 or more, the participants were eligible for a time bonus; if they completed all of the runs in an average of under 11:45 min, they received an additional $10. If they had a score of 50 or more and averaged between 11:45 and 15 min, they received a $5 bonus. Participants were told about this interdependence between score and time, which was designed to prevent participants from quickly running through the course, ignoring the tasks, while also providing a significant motivation to perform the task quickly.

At the end of each run, the score was calculated and the participants were informed about the amount of compensation that could be received based only on that run. At the end of five runs, the average compensation was calculated and given to the participant.

There were three sets of questionnaires. The pre-experiment questionnaire was administered after the participants signed the consent form; it focused on demographic information (i.e., age, familiarity with technology similar to robot user interfaces, tendency towards risky behavior, etc.). The post-run questionnaire was administered immediately after each run; participants were asked to rate their performance, the robot's performance, and the likelihood of not receiving their milestone payment. Participants were also asked to fill out previously validated trust surveys (Muir 1989; Jian et al. 2000) and a NASA Task-Load Index (TLX) questionnaire (Hart and Staveland 1988) after each run. After the last post-run questionnaire, the post-experiment questionnaire was administered, which included questions about wanting to use the robot again and its performance. These human subjects studies were approved by the University of Massachusetts Lowell's IRB.

After participants signed the informed consent form, they were given an overview of the robot system and the task to be performed. Then participants were asked to drive the robot through the trial course in fully autonomous mode. The experimenter guided the participants during this process by explaining the controls and helping with tasks if necessary. The trial course was half the length of the test course. Once participants finished the first trial run, they were asked to drive the robot again through the same course in the robot-assisted mode. Since there were multiple tasks that participants needed to perform, we decided to first show them the fully autonomous mode, as that would be a less overwhelming experience. Once the participants finished the second trial run, they were asked to fill out the post-run questionnaire. While the data from this questionnaire was not used, it allowed participants to familiarize themselves with it and also helped to reinforce some of the aspects of the run that they needed to remember.

After the two trial runs, the participants were asked to drive the robot for five more runs. In each run, a different map was used. During these runs the reliability of robot autonomy was either held high throughout the run or was changed, according to four pre-planned reliability configuration, shown in Fig. 11.4. The changes in reliability were triggered when the robot passed specific points in the course. These locations were equal in length and there were no overlaps. For all four patterns, the robot always started with high reliability. The length of each low reliability span was about one third the length of the entire course. Using different dynamic patterns for reliability allowed us to investigate how participants responded to a drop in reliability at different stages and how the changes influenced control allocation.

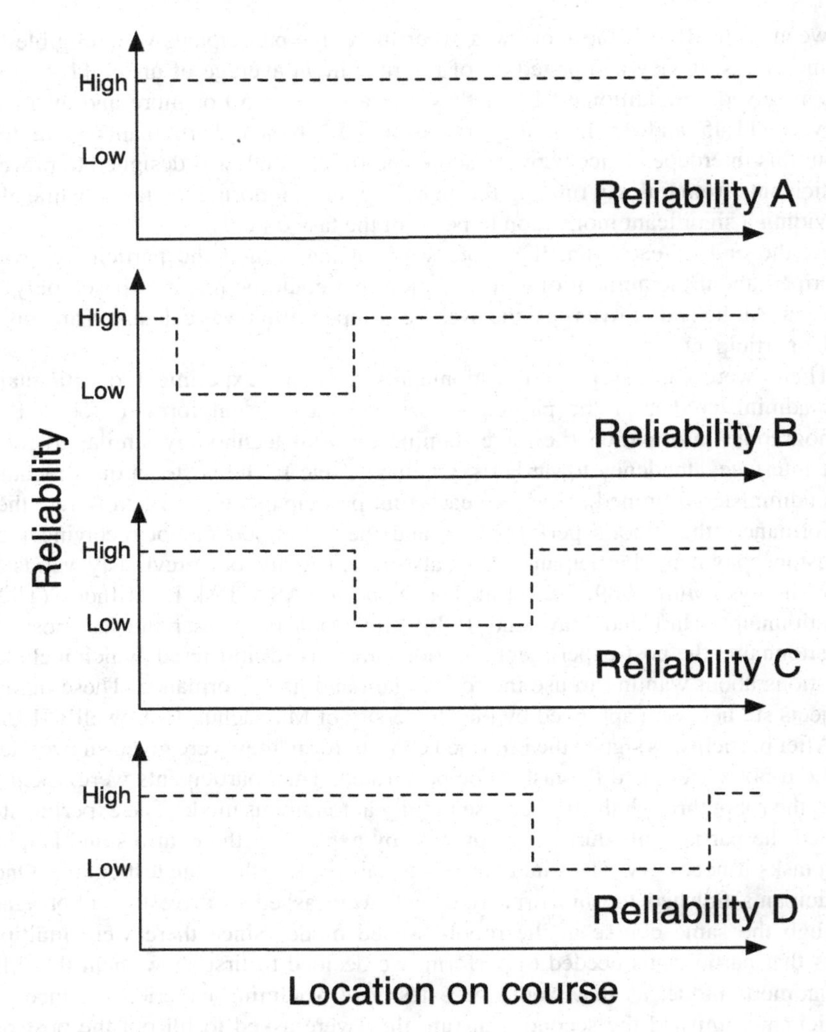

Fig. 11.4 Reliability configurations for the robot's runs

Every participant started with a baseline run under full reliability (Reliability A in Fig. 11.4). Then, the four reliability profiles were counter-balanced for the remaining four runs.

The methodology explained in this section was utilized for all of the experiments. Since multiple factors (e.g., reliability, situation awareness, long-term use) needed to be investigated, it was not feasible to design a within-subjects experiment. Hence, a between-subjects experiment was designed. The overall concept was to conduct multiple experiments, each with two independent variables (e.g., reliability and situation awareness). The dependent variables were the operator's trust and the control allocation strategy. To discern the influence of reliability and other factors being investigated, a baseline experiment with dynamic reliability (DR) as the only independent variable was conducted first. Data from that experiment was used as a baseline for comparison with data from other experiments.

11.4.2 Results and Discussion

Using the methodology described in the prior section, we conducted experiments to determine how a number of factors would influence an operator's trust and control allocation strategy: lowering situation awareness (SA), providing feedback about the robot's confidence in its current operations, reducing task difficulty, and long-term interaction on operator trust and control allocation. This section presents qualitative models based on the impact of those factors as determined by the experiments (for the full results of the experiments, see Desai 2012) as well as a set of guidelines, proposed to help better design autonomous robot systems for remote teleoperation and to improve system performance during operation. These models are presented in the context of the Human interaction with Autonomous Remote Robots for Teleoperation (HARRT) model described in the next section.

11.4.2.1 Reducing Situation Awareness (SA)

Figure 11.5 shows the impact of comparing our baseline dynamic reliability (DR) experiment with our low SA experiment (LSA) where the user interface was modified to impact the participant's SA.

As the participants' SA was reduced, it increased their workload. We suspect the increase in workload was due to the additional effort (cognitive and otherwise) required to maintain the minimum required level of SA. Additionally, lowering SA makes the task of remote teleoperation more difficult, which could also increase workload. The combination of increased workload and poor SA increased the time needed to finish the task.

We suspect that lowering SA forced participants into relying more on the fully autonomous (FA) mode. Higher reliance on the FA mode improved the control allocation strategy, since the ideal control allocation strategy required the participants to rely more on FA than the robot-assisted (RA) mode. While the increase in trust was unexpected, it can be explained by the higher reliance on FA for a task that was difficult to perform manually.

Lowering SA also reduced the participants' rating of the robot's performance, even though there was not a significant difference in performance. We suspect this was due to two reasons: poor SA made it difficult to correctly judge the robot's performance and the participants could have blamed the robot for providing inadequate information needed for teleoperation.

Guidelines based on the SA model, shown in Fig. 11.5, are as follows.

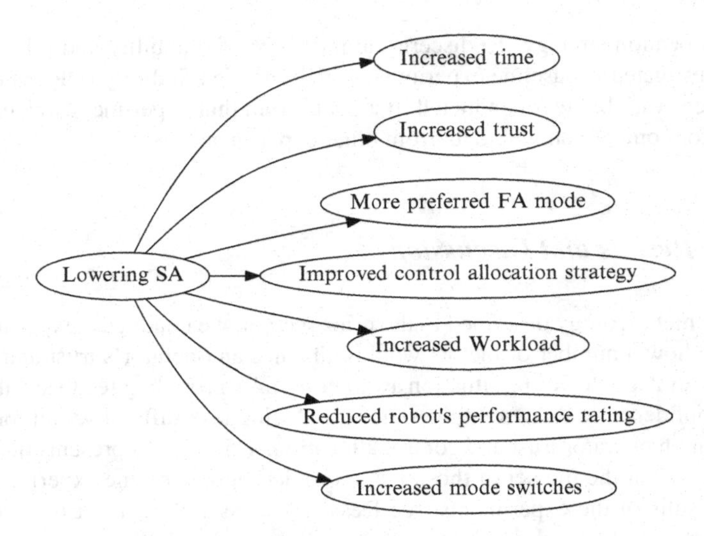

Fig. 11.5 The impact of reducing situation awareness (SA) on different factors. All of the effects shown are based on significant differences between the Low Situation Awareness (LSA) and Dynamic Reliability (DR) experiments

Guideline 1: *Reduced SA leads to higher reliance on autonomous behaviors.* Intentionally reducing SA to force operators to rely on autonomous behaviors is not recommended as a design strategy due to the other undesirable side effects. However, such influence does remain a possibility, but should only be exercised when absolutely necessary, since doing so can potentially impact safety and performance.

Guideline 2: *Suspend or defer non-critical tasks when SA is reduced.* Even with higher reliance on automation, the workload is expected to increase, so tasks that are not critical should be suspended or deferred to offset the increased workload and to prevent an overall detrimental impact on performance.

Guideline 3: *Switch functions unaffected by reduced SA to automation.* Functions not impacted by reduced SA can be switched over to automation in an attempt to reduce workload.

Guideline 4: *Educate operators about SA.* Operators associate robot performance with SA and therefore operators must be informed (during training or during the interaction) that low SA does not necessarily impact the robot's performance.

11.4.2.2 Providing Feedback

Figure 11.6 shows the results of comparing results of the baseline Real-Time Trust (RT) experiment with that of the Feedback (F) experiment where the participants were provided with feedback concerning the robot's confidence in its own sensors.

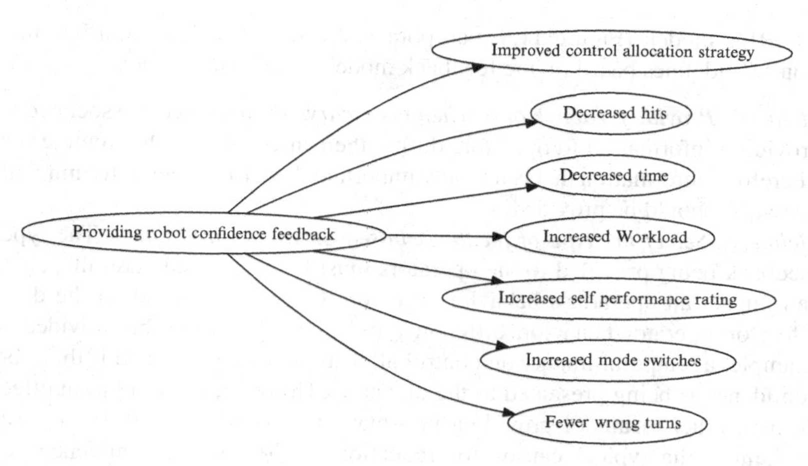

Fig. 11.6 The impact of providing feedback on different factors. All of the effects shown are based on significant differences between the Feedback (F) and Real-Time Trust (RT) experiments

Providing information about the robot's confidence in its own sensors and decision making to the participants increased their workload, as they were given additional information that needed to be processed. Also, participants reacted to the change in robot's confidence by aggressively changing autonomy modes and therefore increased the number of autonomy mode switches. We suspect these autonomy mode changes were another reason that resulted in an increase in workload.

However, increased autonomy mode switches and better robot supervision due to the variations in the robot's confidence resulted in a better control allocation strategy, which in turn led to better performance. Despite the better performance, the participant's trust of the robot did not increase; we suspect this lack of increase in trust was due to the type of feedback provided to the participants.

It is often conjectured that providing feedback should improve an operator's trust in the system by helping operators better align their mental model with that of the system's architecture and operation. However, in this case, the information provided to the participants could not have helped achieve more synchronized mental models. We suspect this discrepancy occurred because no information was provided that could sufficiently explain why the robot made a mistake in reading the labels. Providing such information requires feedback that provides details about the robot's internal processes. For example, informing the user that the robot cannot read labels accurately at certain angles would explain the decrease in the robot's confidence and help the operators better understand the robot's internal operation. The feedback also provided negative information to the participants. It informed the participants that the robot's confidence was medium, low, or at best functioning as intended.

Providing feedback seems to directly impact workload and the operator's control allocation strategy and the impact of feedback on other attributes aligned with the HARRT model. (Figure 11.9 incorporates all of the models described in this section.) Guidelines based on the feedback model are described below.

Guideline 5: *Provide feedback only when necessary.* There is a cost associated with providing information to operators during their interaction with a remote robot. Therefore, information that is not only important, but also essential for immediate operation should be provided.

Guideline 6: *Select the type of feedback based on the desired effect.* The type of feedback being provided to the operators must be considered carefully, since it can impact an operator's behavior. The corollary is, that based on the desired effect on operator behavior, different types of feedback can be provided. For example, a temporal impact on control allocation can be expected if the robot's confidence is being presented to the operators. However, if a long-term effect is desired, other means of providing information must be selected. For example, explaining the typical causes for reduction in the robot's confidence could provide the operators with better understanding of the robot and result in a permanent effect. Guideline 5 must be considered while doing so.

11.4.2.3 Reducing Task Difficulty

Figure 11.7 shows the results of comparing data from the baseline Real-Time Trust (RT) experiment with that of the Reduced Difficulty (RD) experiment where the complexity of the teleoperation task was reduced.

With the teleoperation task easier to perform, we expected the participants to not rely on the fully autonomous mode as much, and, consequently, a poor control allocation strategy was expected. However, the control allocation strategy improved along with an increase in autonomy mode switches. We suspect the reduced difficulty of the teleoperation task reduced the participants' workload and allowed

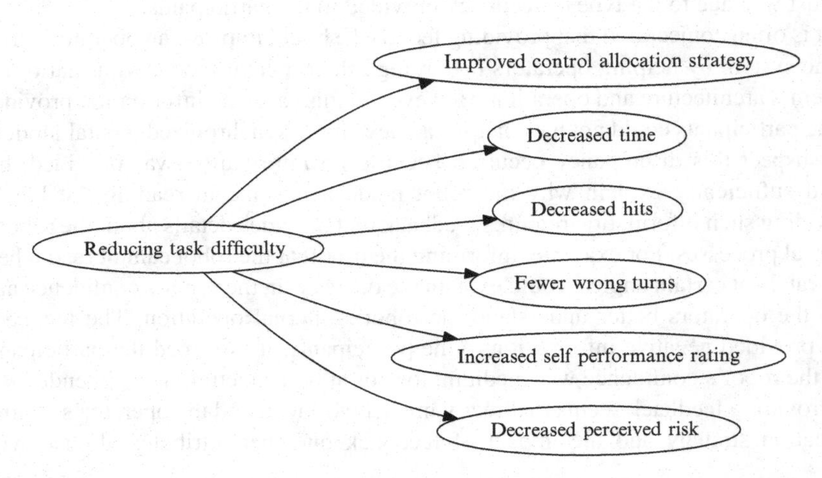

Fig. 11.7 The impact of reducing task difficulty on different factors. All of the effects shown are based on significant differences between the Reduced Difficulty (RD) and RT experiments

them to better observe the robot's performance in the fully autonomous mode. This better robot supervision allowed them to switch autonomy modes appropriately and improve the control allocation strategy. We suspect that improvement in supervision and the resulting increase in autonomy mode switches increased the workload enough to offset the initial reduction in workload due to the easier task.

The easier teleoperation task and the better robot supervision improved performance and safety by reducing the number of hits, reducing the time needed to finish, and reducing the number of wrong turns. Reducing the difficulty of the task seems to primarily impact an operator's control allocation strategy. The impact on other attributes aligns with the HARRT model. Guidelines based on the reduced difficulty model are described below:

Guideline 7: *Tasks with reduced difficulty result in better robot supervision and no reduction in workload.* If the difficulty of the task reduces during an interaction or for interactions that involve a relatively easy remote robot teleoperation task, operators should be expected to allocate the additional available cognitive resources towards better supervision of the robot's behavior or secondary tasks.

Guideline 8: *Do not expect operators to assume manual control for easier tasks.* Operators will not necessarily opt for lower autonomy modes, at least in scenarios involving multiple tasks or a relatively high workload. While a reduction in the difficulty of the task will improve performance and safety, the operator's trust of the system will not be affected.

11.4.2.4 Long-Term Interaction

The long-term interaction experiment (LT) was conducted to investigate if an operator's trust and control allocation strategy change over a longer period of time. We looked for trends to incorporate into the model and a set of guidelines. Another goal of the LT experiment was to investigate if there is a difference between operators who are familiar with robots and those who are not.

Interestingly, no significant differences were found between sessions two through six for any attribute. This lack of a difference between sessions and the significant similarities found between sessions indicates that an operator's behavior during initial interaction can predict his or her behavior over the short term.

With respect to the impact of familiarity with robots, several significant differences were found. Figure 11.8 shows the impact familiarity with robots has on operator behavior. It shows that while there was not a difference in performance, participants who were familiar with robots trusted them less and had an increased workload, perhaps due to feeling the need to execute better robot supervision, in accordance with the HARRT model. The better robot supervision in turn positively affected their control allocation strategy and also is consistent with the HARRT model. Figure 11.9 shows the familiarity model incorporated into the HARRT model and guidelines based on the reduced long-term and familiarity model are described below:

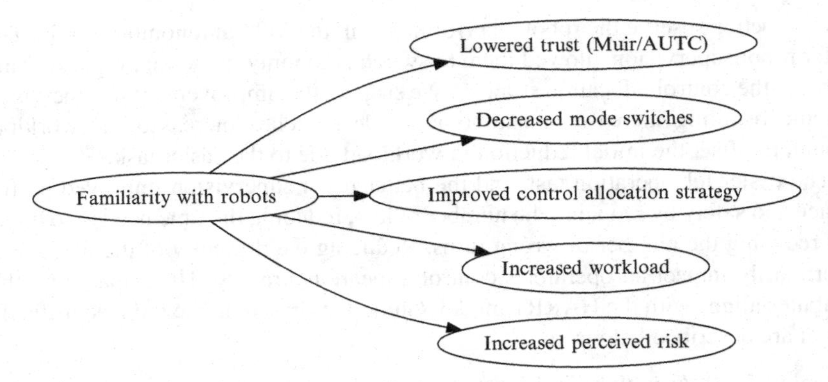

Fig. 11.8 The impact of familiarity with robots on different factors. All of the effects shown are based on significant differences between the two participant groups in the Long-term (LT) experiment

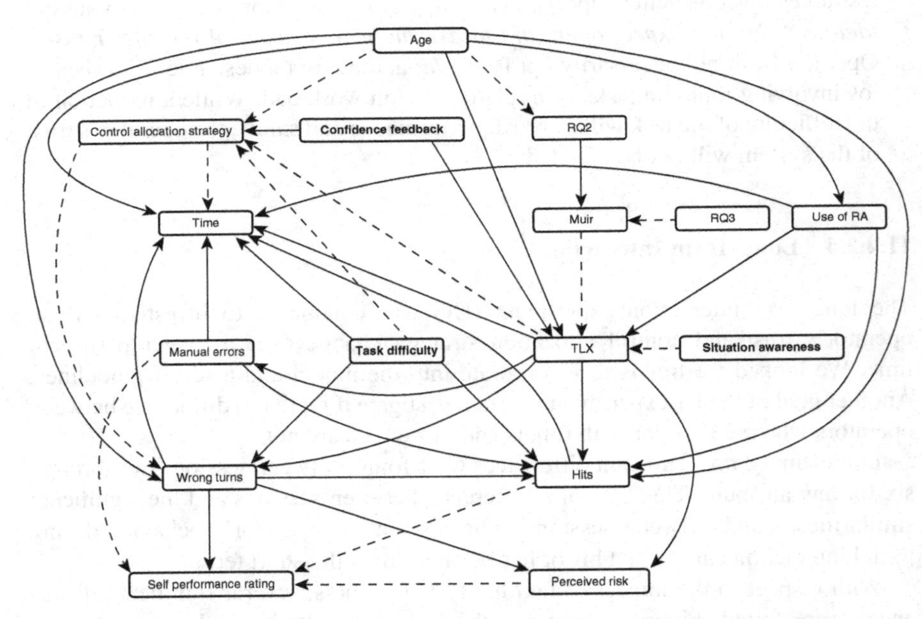

Fig. 11.9 The original human and autonomous remote robot teleoperation (HARRT) model augmented with all of models described in Sect. 11.4.2. The dashed and solid arrow lines indicate an inverse relationship or a proportional relationship, respectively. RQ2 is the second question about risk from Grasmick et al. (1993): "Sometimes I will take a risk just for the fun of it." RQ4 is the fourth question from Grasmick et al. (1993): "Excitement and adventure are more important to me than security." Participants were asked to answer these questions using a 6-point Likert scale

Guideline 9: *Initial operator behavior does not change over the short term.* It is possible to quickly assess and predict an operator's behavior over a longer period of time, based on their initial interactions with the robot.

Guideline 10: *Familiarity with robots does not impact performance.* Familiarity with robots should not be interpreted as or confused with expertise in remote robot teleoperation. While familiarity with robots impacts trust, it does not impact performance.

11.4.2.5 Impact of Timing of Periods of Low Reliability

Periods of low reliability early in the interaction not only have a more immediate detrimental impact on trust, but that effect lasts throughout the interaction as it also impedes the recovery of trust. Since the experimental setup was designed to require participants to rely more on the fully autonomous mode, the impact of decreased trust on other parameters was not as noticeable. However, for most balanced operations, the impact on trust would also be accompanied by a similar impact on control allocation, performance, and workload. Guidelines based on the impact of periods of low reliability early in the interaction are described below:

Guideline 11: *Operator's initial interactions must always be stable.* The implications of the timing data are that initial segments of every interaction must be stable and reliable. If needed, this experience should be facilitated by conducting a short, controlled interaction.

Guideline 12: *In the event of a reliability drop early in the interaction, corrective measures must be taken.* These steps (e.g., providing information explaining the cause for the reduction in reliability) must essentially minimize or prevent erratic operator behavior due to confusion or other factors. There are costs associated with these preventive steps, along with other implications associated with different measures, so caution must be exercised while selecting corrective measures.

11.4.2.6 Impact of Age

As people grow older, their attitude towards risk changes: they are willing to take fewer risks (e.g., Mather et al. 2009). We also found a significant correlation with age to answers to several questions about risk asked of participants. This unwillingness to take on more risk is shown in robot use through the fact that they prefer some autonomy modes and do not switch out of their comfort zone as often. Attitudes towards risk change with age, but so does the view or the definition of risk. It was often mentioned by the older participants that the compensation did not matter to them as much. However, it must also be said that they were still motivated to perform well. The inertia in control allocation exhibited by the older participants could potentially also have increased their workload and ultimately performance. Guidelines based on the impact of age are described below:

Guideline 13: *Know your target audience.* It is important to take into account the different population groups that will be interacting with a robot. Understanding the motivations of the operators can help explain their view on potential risks and better predict their behavior.

Guideline 14: *Accommodate operators of all ages.* Due to a higher probability of poor control allocation and poor performance for older operators, more time should be spent training them. To counteract the inertia observed, additional steps can also be taken. However, caution must be exercised to ensure that these steps do not increase their workload. For the other end of the age spectrum, given their tendency to take more risk, the risks involved in the scenario must be explained carefully. Since the younger population has the ability to better manage workload and better robot management, it should be easier to influence their control allocation strategy if needed.

11.4.3 Modeling Trust

Using the experimental methodology, multiple experiments were conducted to examine the impact of different factors on operator trust and control allocation. These factors were selected based on different criteria. Some factors were selected based on the results of the initial surveys (i.e., reliability and risk). In fact, to better model real world scenarios, we ensured that dynamic reliability and risk were inherent in all of the experiments. Other factors like situation awareness (SA) and reduced task difficulty (RD) were selected based on their significance to the remote robot teleoperation task and also on our observations of other experiments involving remote robot teleoperation. Factors like feedback and long-term interaction were selected based on conjectures and commonly held beliefs. For example, it is often assumed that providing feedback to the operator should increase their trust of the robot and improve performance.

The results from these experiments showed interesting, sometimes unexpected, but overall insightful data. Using that data we were able to find different attributes that are relevant to human interaction with remote autonomous robot and the mediating relationships between them.

These results were used to create the Human interaction with Autonomous Remote Robots for Teleoperation (HARRT) model a regression based model, shown in Fig. 11.9. Based on the HARRT model and the specific experiments, guidelines were proposed that should help improve overall performance by better managing the different tradeoffs (e.g., workloads, situation awareness, feedback) to influence operators' control allocation strategy. These results also highlight some of the differences between HAI and HRI. For example, a primary difference between HAI and HRI was the lack of direct correlation between trust and control allocation, a result always observed in HAI research.

11.5 Conclusions and Future Work

Our ultimate goal is to build models of the factors that influence people's trust in automated systems, across many domains, building a common core model of trust for automated systems and identifying factors specific to particular domains. Such models will serve to inform the designers of automated systems, allowing the development of systems that address the key factors for developing and maintaining a person's trust of an automated system. This chapter presents some of our initial work towards this goal, identifying the factors that most influence people's trust of automated cars, medical diagnosis systems, and remotely operated robot systems.

We have presented two methods for developing different types of models of trust in automation. The HARRT model is a regression model that used data from extensive experimentation. However, given the limitations of regression modeling, the HARRT model is not able to be used dynamically to predict current levels of user trust during the use of the robot. It is best used as a predictive model of how trust will evolve. However, it could be modified into a trust model that could be run dynamically, which would provide the robot with the means to determine the moment-by-moment modifications in its behavior that are necessary to elicit appropriate trust levels from the user. In contrast, the survey-based method results in a list of factors that are important to generating trust. Because the survey-based model is not designed for real-time execution, it is best used prior to system design to generate requirements that are trust-related.

The choice of the modeling approach for any given situation can thus be based on the time and resources available, and whether real-time adjustments in the robot's behavior are desired so as to elicit the optimal level of trust at any given point. The survey-based modeling method is more appropriately used when a large number of potential respondents are available; for example, the user pool consists of a large segment of the general population. If the system is to be used by specialized populations (e.g., doctors, first responders), it is best to conduct studies within those user groups. The characteristics of each modeling approach are summarized in Table 11.5.

We are planning to create an example of an executable HARRT model as part of our future work. This work will involve developing alternative behaviors when pre-identified conditions occur. For example, if the robot determines that a sensor

Table 11.5 Characteristics of HARRT and survey-based modeling approaches

Characteristic	HARRT	Survey-based model
Time and cost to create	High	Low
Method to create	Experimentation with system	Surveys
Format	Directed graph	List of factors
Real-time execution possible?	Yes, with modifications	No
Potential use	Modify robot's behavior in real time to elicit appropriate trust	Requirements generation

is becoming unreliable, this condition may trigger the robot to provide an error message and modify the user interface layout to make an alternative sensor's readouts more salient. This will be a successful strategy if the user trusts the unreliable sensor less and the alternative sensor more, as compared to earlier trust levels.

Additional future work will use a survey-based approach to examine the effects of mediated versus unmediated autonomy on the relative importance of the trust factors. By mediated, we mean that there is a human expert user who interprets the robot's results or who actually operates the robot on behalf of the end user (that is, the user who is benefiting from the robot's work). The use of IBM's Watson by an oncologist represents a mediated use from the standpoint of the cancer patient. We hypothesize that the trust factors will be rated differently from users in mediated versus unmediated situations.

Acknowledgements This research has been supported in part by the National Science Foundation (IIS-0905228 and IIS-0905148) at the University of Massachusetts Lowell and Carnegie Mellon University, respectively, and by The MITRE Corporation Innovation Program (Project 51MSR661-CA; Approved for Public Release; Distribution Unlimited; 15-1753).

Munjal Desai conducted the research described in this chapter while a doctoral student at the University of Massachusetts Lowell. Michelle Carlson of The MITRE Corporation assisted with the design and analysis of the survey-based research. Hyangshim Kwak and Kenneth Voet from the United States Military Academy assisted with the survey-based work during their internships at The MITRE Corporation. Many people in the Robotics Laboratory at the University of Massachusetts Lowell have assisted with the robot testing over several years, including Jordan Allspaw, Daniel Brooks, Sean McSheehy, Mikhail Medvedev, and Katherine Tsui. At Carnegie Mellon University, robot testing was conducted with assistance from Christian Bruggeman, Sofia Gadea-Omelchenko, Poornima Kaniarasu, and Marynel Vázquez.

All product names, trademarks, and registered trademarks are the property of their respective holders.

References

Baker M, Yanco H (2004) Autonomy mode suggestions for improving human-robot interaction. IEEE Int Conf Syst Man Cybernet 3:2948–2953

Bliss JP, Acton SA (2003) Alarm mistrust in automobiles: how collision alarm reliability affects driving. Appl Ergon 34(6):499–509

Boehm-Davis DA, Curry RE, Wiener EL, Harrison L (1983) Human factors of flight-deck automation: report on a NASA-industry workshop. Ergonomics 26(10):953–961

Bruemmer DJ, Dudenhoeffer DD, Marble JL (2002) Dynamic autonomy for urban search and rescue, AAAI Mobile Robot Workshop

Burke JL, Murphy RR, Rogers E, Lumelsky VL, Scholtz J (2004) Final report for the DARPA/NSF interdisciplinary study on human-robot interaction. IEEE Trans Syst Man Cybernet Part C 34(2):103–112

CasePick Systems (2011) http://www.casepick.com/company. Accessed 30 Dec 2011

Chen JY (2009) Concurrent performance of military tasks and robotics tasks: effects of automation unreliability and individual differences. In: Fourth annual ACM/IEEE international conference on human-robot interaction, pp 181–188

Clark M (2013) States take the wheel on driverless cars, USA Today http://www.usatoday.com/story/news/nation/2013/07/29/states-driverless-cars/2595613/

Cohen MS, Parasuraman R, Freeman JT (1998) Trust in decision aids: a model and its training implications. Command and control research and technology symposium

Dellaert F, Thorpe C (1998) Robust car tracking using Kalman filtering and Bayesian templates. In: Intelligent transportation systems conference, pp 72–83

Desai M (2012), Modeling trust to improve human-robot interaction, Ph.D. thesis, University of Massachusetts, Lowell

Desai M, Medvedev M, Vazquez M, McSheehy S, Gadea-Omelchenko S, Bruggeman C, Steinfeld A, Yanco H (2012) Effects of changing reliability on trust of robot systems. In: Seventh annual ACM/IEEE international conference on human-robot interaction

Desai M, Kaniarasu P, Medvedev M, Steinfeld A, Yanco H (2013) Impact of robot failures and feedback on real-time trust. In: Eighth annual ACM/IEEE international conference on human-robot interaction

Desai M, Yanco H (2005) Blending human and robot inputs for sliding scale autonomy. In: IEEE international workshop on robots and human interactive communication, pp 537–542

deVries P, Midden C, Bouwhuis D (2003) The effects of errors on system trust, self-confidence, and the allocation of control in route planning. Int J Hum Comput Stud 58(6):719–735

Dixon SR, Wickens C (2006) Automation reliability in unmanned aerial vehicle control: a reliance-compliance model of automation dependence in high workload. Hum Factors 48(3):474–486

Dudek G, Jenkin M, Milios E, Wilkes D (1993) A taxonomy for swarm robots. In: IEEE/RSJ international conference on intelligent robots and systems, pp 441–447

Dzindolet M, Pierce L, Beck H, Dawe L, Anderson B (2001) Predicting misuse and disuse of combat identification systems. Military Psychol 13(3):147–164

Dzindolet M, Pierce L, Beck H, Dawe L (2002) The perceived utility of human and automated aids in a visual detection task. Hum Factors 44(1):79–94

Dzindolet MT, Peterson SA, Pomranky RA, Pierce LG, Beck HP (2003) The role of trust in automation reliance. Int J Hum Comput Stud 58(6):697–718

Endsley M, Kiris E (1995) The out-of-the-loop performance problem and level of control in automation. Hum Factors 37(2):381–394

Farrell S, Lewandowsky S (2000) A connectionist model of complacency and adaptive recovery under automation. J Exp Psychol 26(2):395–410

Gerkey B, Vaughan R, Howard A (2003) The player/stage project: tools for multi-robot and distributed sensor systems. In: Eleventh international conference on advanced robotics, pp 317–323

Grasmick H, Tittle C, Bursick R Jr, Arneklev B (1993) Testing the core empirical implications of Gottfredson and Hirschi's general theory of crime. J Res Crime Delinq 30(1):5–29

Guizzo E (2011) How Google's self-driving car works. In: IEEE spectrum http://spectrum.ieee.org/automaton/robotics/artificial-intelligence/how-google-self-driving-car-works

Hart SG, Staveland LE (1988) Development of NASA-TLX (Task Load Index): results of empirical and theoretical research. Human Mental Workload 1(3):139–183

International Federation of Robotics (IFR) (2011) Statistics about service robots. http://www.ifr.org/service-robots/statistics/. Accessed 30 Dec 2011

iRobot (2011) iRobot Roomba. http://www.irobot.com/roomba. Accessed 30 Dec 2011

Jian J, Bisantz A, Drury C (2000) Foundations for an empirically determined scale of trust in automated systems. Int J Cognit Ergon 4(1):53–71

Kiva Systems (2011) Kiva Systems. http://www.kivasystems.com/. Accessed 30 Dec 2011

Lee J (1992) Trust, self-confidence and operator's adaptation to automation, PhD thesis, University of Illinois at Urbana-Champaign

Lee J, Moray N (1991) Trust, self-confidence and supervisory control in a process control simulation. In: IEEE international conference on systems, man, and cybernetics, Charlottesville, pp 291–295

Lee JD, Moray N (1992) Trust, control strategies and allocation of function in human-machine systems. Ergonomics 31(10):1243–1270

Lin P (2008) Autonomous military robotics: risk, ethics, and design. Technical report, Defense Technical Information Center

Madhani K, Khasawneh M, Kaewkuekool S, Gramopadhye A, Melloy B (2002) Measurement of human trust in a hybrid inspection for varying error patterns. Hum Factors Ergon Soc Annual Meeting 46:418–422

Mather M, Gorlick MA, Lighthall NR (2009) To brake or accelerate when the light turns yellow? Stress reduces older adults' risk taking in a driving game. Psychol Sci 20(2):174–176

Michaud F, Boissy P, Corriveau H, Grant A, Lauria M, Labonte D, Cloutier R, Roux M, Royer M, Iannuzzi D (2007) Telepresence robot for home care assistance. AAAI spring symposium on multidisciplinary collaboration for socially assistive robotics

Moray N, Inagaki T (1999) Laboratory studies of trust between humans and machines in automated systems. Trans Inst Meas Control 21(4–5):203–211

Moray N, Inagaki T, Itoh M (2000) Adaptive automation, trust, and self-confidence in fault management of time-critical tasks. J Exp Psychol 6(1):44–58

Muir BM (1987) Trust between humans and machines, and the design of decision aids. Int J Man-Mach Stud 27(5–6):527–539

Muir BM (1989) Operators' trust in and use of automatic controllers in a supervisory process control task. PhD thesis, University of Toronto

Neato Robotics (2011) Neato XV-11. http://www.neatorobotics.com/. Accessed 30 Dec 2011

Ostwald P, Hershey W (2007) Helping Global Hawk fly with the rest of us. Integrated communications, navigation, and surveillance conference, April/May

Parasuraman R (1986) Vigilance, monitoring, and search. In: Boff K, Thomas J, Kaufman L (eds) Handbook of perception and human performance: cognitive processes and performance. Wiley, New York

Parasuraman R, Riley V (1997) Humans and automation: use, misuse, disuse, abuse. Hum Factors 39(2):230–253

Prinzel III LJ (2002) The relationship of self-efficacy and complacency in pilot-automation interaction. Technical report, Langley Research Center

Riley V (1994), Human use of automation. PhD thesis, University of Minnesota

Riley V (1996) Operator reliance on automation: theory and data. In: Parasuraman R, Mouloua M (eds) Automation and human performance: theory and applications. Lawrence Erlbaum, Mahwah

Ross J, Irani I, Silberman M, Zaldivar A, Tomlinson B (2010) Who are the crowd workers?: shifting demographics in Amazon Mechanical Turk. In: ACM CHI conference on human factors in computing systems extended abstract, pp 2863–2872

Sanchez J (2006) Factors that affect trust and reliance on an automated aid. PhD thesis, Georgia Institute of Technology

Sarter N, Woods D, Billings C (1997) Automation surprises. Handbook Hum Factors Ergon 2:1926–1943

Sheridan TB, Verplank WL (1978) Human and computer control of undersea teleoperators. Technical report, Department of Mechanical Engineering, Massachusetts Institute of Technology

Strickland GE (2013) Watson goes to med school. IEEE Spectrum 50(1):42–45

Thrun S (2010) What we're driving at. http://googleblog.blogspot.com/2010/10/what-were-driving-at.html

Tsui K, Norton A, Brooks D, Yanco H, Kontak D (2011) Designing telepresence robot systems for use by people with special needs. International symposium on quality of life technologies

Yanco HA, Drury J (2004) Classifying human-robot interaction: an updated taxonomy. IEEE Int Conf Syst Man Cybernet 3:2841–2846

Chapter 12
The Intersection of Robust Intelligence and Trust: Hybrid Teams, Firms and Systems

W.F. Lawless and Donald Sofge

12.1 Introduction

Our goal with autonomy is to control hybrid teams (arbitrary combinations of humans, machines and robots). Traditional approaches to social models treat interdependence as a problem to be removed to improve the replication of experiments (Kenny et al. 1998), or one to be resolved before teams can be controlled (e.g., Jamshidi 2009). In contrast, we consider interdependence to be a resource that teams can use to *solve* ill-defined problems (IDP). But predictability and replicability are lost as a consequence. Unlike swarms, machine learning and game theory approaches to interdependence, we conclude that hybrid teams, like human teams, cannot be controlled directly to solve IDPs; instead, they can be indirectly controlled with self-governance (Lawless et al. 2013). How to improve control is the goal of our chapter.

The difference in approaches is foundational. Traditional approaches assume a complete, "God's-eye view" of reality, implying that whatever information can be sensed can be collected to model social reality (Rand and Nowak 2013, p. 415). In contrast, the physics of interdependence (which we model with Signal Detection Theory) precludes completeness, limiting the information that can be sensed or collected either by machines or humans, physically constraining meaning and situational awareness. Our approach generates better models of social reality with

W. Lawless (✉)
Paine College, 1235 15th Street, Augusta, GA, USA
e-mail: wlawless@paine.edu

D. Sofge
Naval Research Laboratory, 4555 Overlook Avenue SW, Washington, DC, USA

© Springer Science+Business Media (outside the USA) 2016

R. Mittu et al. (eds.), *Robust Intelligence and Trust in Autonomous Systems*,
DOI 10.1007/978-1-4899-7668-0_12

concrete conclusions; e.g., incomplete social information causes uncertainty; social autonomy cannot occur without social interaction; and autonomy is a resource when benefits exceed interaction costs (Coase 1960).

In addition, we plan to continue to perform research on the autonomy of hybrid teams. Our research is centered around modeling tradeoffs with Fourier pairs from signal detection theory (SDT); social interdependence or bistability (i.e., multiple states); multitasking; and Nash equilibria. Individuals multitask poorly (Wickens 1992); teams and firms exist to multitask (Ambrose 2001; e.g., a baseball team multitasks as its members play different positions). Unlike traditional game-theoretic models which promote cooperation but not social governance (e.g., Rand and Nowak 2013), Nash equilibria are one of the primary tools of self-governance where a society multitasks by exploiting the competition naturally existing between the orthogonal (bistable) beliefs of groups in processing the signals or information they emit or receive to solve the IDPs that improve social welfare (Lawless et al. 2013).

Game theory models of social reality do not attempt to be "a good representation of that world" (Rand and Nowak 2013, p. 416). They assume that only one view of reality is possible. In contrast, bistability assumes that two orthogonal interpretations spontaneously arise simultaneously in every social situation (e.g., Republicans and Democrats often come to differing or orthogonal interpretations of reality). Thus, our bistable models better capture existing social reality (e.g., Lawless et al. 2013). This result has important implications, as when attempting to reduce tragic decisions; e.g., during the time that Department of Energy (DOE) contaminated the environment with widespread radioactive wastes, now costing up to $200 Billion to remediate, DOE was self-regulated (Lawless et al. 2008), whereas today, its decisions are competitively challenged by numerous State and Federal agencies, and the public, yet the quality of its decisions has improved dramatically (Lawless et al. 2014).

12.1.1 Background

Biologists approach the studies of nesting agents with an open mind about a nest's welfare, an approach made difficult in the study of human teams, groups and systems by the cognitive biases that have impeded the development of a new mathematics of interdependence to replace game theory (e.g., Barabási 2009). For example, from Helbing (2013):

> ... we need to 'think out of the box' and require a paradigm shift towards a new economic thinking characterized by a systemic, interaction-oriented perspective inspired by knowledge about complex, ecological, and social systems.

Interdependence theory is needed for the efficient and effective control of autonomous hybrid teams (Lawless et al. 2013), an arbitrary combination of humans, machines and robots, in preparation for a rapidly approaching future with computational teams. (e.g., "Smart Drones", from Keller (2013)). But since the introduction of interdependence into game theory almost 70 years ago, it has floundered (Schweitzer et al. 2009) likely because its assumptions have never been validated; e.g., in games, cooperation is valued over competition, but in real life, cooperation between two competitors, like Apple and Google, is known as collusion (Lawless et al. 2011); interdependence is treated as a static or repeated static phenomenon, e.g., like in the movies; and, more relevant, with its folk theorem, the choices made by a team are determined by a simple sum of the individual choices of its members.

These biases extend to the social science of teams, where interdependence is a nuisance to be removed to produce the statistically necessary independence among subjects (i.i.d.) to be able to replicate an experiment (e.g., Kenny et al. 1998). The study of interdependence is encumbered further by the cost in collecting an increasingly large number of teams as team size increases in order to reach statistical significance. To reduce costs, most studies of teams focus on small three-member groups, usually concluding that cooperation among members produces superior results compared to competition (e.g., Bell et al. 2012); while we agree, groups with three to six members are hampered in generating competition (Kerr and MacCoun 1985), a problem not only for say juries weighing alternative decisions, but also for the theory of groups. In field studies, for example, Hackman (2011) concluded that conflict (competition) in teams made them more creative; supporting Hackman, in our study of citizen groups advising the Department of Energy (DOE) on the cleanup of its wide-spread radionuclide contamination across its complex, we found that observing conflict is sufficiently entertaining to hold an audience's attention as a group generates the information needed to decide on a course of action, that competition among viewpoints produced more concrete recommendations that advanced DOE's cleanup compared to consensus decisions (Lawless et al. 2008); and in the lab, that the larger the group size, the more conflict and interdependence it generates, along with better decisions (Lawless et al. 2014).

In sum, the poor state of team theory impedes generalizations from teams to higher orders of organization; compared to systems that enforce cooperation, competitive checks and balances significantly improve social well-being; and a new method must be found to replace the costs of running experiments with larger group sizes, which in addition raises questions about the applicability of laboratory experiments to real problems (Kerr and MacCoun 1985). Also, when the institutions (cooperative social structures) that allow the existence of competition to improve social well-being themselves arbitrarily change the interpretations of their own rules, transactional uncertainty increases (e.g., the dueling editorials about the U.S. Administration's treatment of JP Morgan by Freeman (2013) and Eavis (2013) illustrate the value of competition in generating information; in addition, their content implies that the Administration's change in its rules made the business transactions by JP Morgan more uncertain).

12.2 Theory

Interdependence is the bidirectional effect of a group on the individual, ranging from a minimum for independence (e.g., the individuation process of converting a team into a collection of individuals), to a maximum with the disappearance of the individual into a group (e.g., the deindividuation process of converting individuals into a mob or swarm). Interdependence causes teams to form into organizations (independent person A does *x*; person A sells item *x* to anyone, say independent person B who combines *x* and *y*; person B sells combined *xy* item to anyone, say independent person C who adds *z*; etc.; these actors are performing independent tasks until two or more actors interdependently depend on each other, forming multi-tasking communication channels between a team's members with the flow of information, objects and material; they can form an organization when the money–energy—generated is greater than their losses—entropy; in Coase (1937)); but interdependence causes uncertainty (on the assembly line, normally person A is paid to do *x* and person B is paid to expect *x* to combine it with *y*; with success, a team becomes trusted; but when person A does something unexpected, person B is left confused about what to do, illustrating both interdependence and uncertainty). Measurements of uncertainty cause incompleteness; e.g., religion, politics or sororities all require specific actions and beliefs for membership—to get to the upper echelon of a group requires that an agent master its rituals and beliefs, implying that for trust to increase among fellow members, the leader must become a "true" believer; but when the agent fully adopts these beliefs, an incomplete view of reality is formed; i.e., it becomes less able to predict its competitor's actions reducing trust in the competitor (as may happen during a hostile merger; e.g., the hostile bid for Cadbury by Kraft; in Cimilluca et al. 2009).

Subsequently a meeting between counterparts from opposing groups illuminates incompleteness. Namely, political party A meets opposing political party B for a discussion which highlights their mutual uncertainty. Take for example the strong views of the need for austerity held in Germany that creates a state of incompleteness in social reality over the future of Greece and the European Union; e.g., Fichtner and Smoltczyk (2013) wrote:

> "But then I get a call from (former US Treasury Secretary) Timothy Geithner," says Schäuble, "and he says, 'You do know that we wouldn't have made the decision to allow Lehman Brothers to go bankrupt if we had been asked 24 hours later, don't you?'' Schäuble shrugs his shoulders and falls silent. He cradles his head in his hands and narrows his eyes, using body language to ask: "Well, what do you do in that situation? What's the right thing to do? What isn't? What's going to blow up in your face tomorrow?"

Or consider what happens when computer firm member A of Apple disagrees with a colleague at Apple (Vogelstein 2013):

> The pressure to meet Jobs's deadlines was so intense that normal discussions quickly devolved into shouting matches. Exhausted engineers quit their jobs—then came back to work a few days later once they had slept a little. Forstall's chief of staff, Kim Vorrath, once slammed her office door so hard it got stuck and locked her in, and co-workers took more

> than an hour to get her out. "We were all standing there watching it," Grignon says. "Part of it was funny. But it was also one of those moments where you step back and realize how [expletive] it all is."

Or suppose a member A of computer firm Google meets member B of computer firm Apple, pleading for Apple to support its product, inadvertently illustrating the disagreement existing between the two firms; Dilger (2013) wrote:

> It also explains why Google's chairman Eric Schmidt continued to suggest the potential for Apple to give up on its own maps and simply adopt Google's as late as April, far after there was any hope in such a scenario actually occurring.

The result of conflict is information (from information theory; Conant 1976); afterwards, with mutually acceptable structures, illustrating socially appropriate cooperation among competing groups, information can be converted into actions that improve social welfare (increasing free energy), characterized as knowledge (generating low entropy). Consequently, we assert that an institutionalized conflict center, which we call a Nash equilibrium (NE), is a social asset that helps those societies evolve that can manage an NE, compared to those that cannot or would not (compare night satellite photos of the USA with Cuba; or South Korea with North Korea; or Germany with Russia; see Fig. 12.1). We also assert that the knowledge from social transactions cannot be generated other than by social interaction (from Coase (1960)).

Fig. 12.1 A night photo of North Korea and surrounding nations. From http://earthobservatory.nasa.gov/IOTD/view.php?id=83182

12.3 Outline of the Mathematics

12.3.1 *Field Model*

Putting uncertainty aside until later, the effects of a community matrix *A* can be measured in the field. Assume that competition for resources occurs within and between groups; that, unlike the inability of individuals to multitask (Wickens 1992), multitasking is the purpose of a group (Ambrose 2001). The optimal group multitasks seamlessly, generating a baseline entropy for stable organizations that we initially, but incorrectly, set to zero, noting that, similarly, stable knowledge implies zero entropy (Conant 1976). We justify this assumption at this time by observing that, compared to functional groups, an individual is less able to survive. That is, a collective of individuals is in a higher state of average uncertainty or agitation than the same individuals independently performing the identical actions but as part of a group using coordination to multitask.

Competition between groups increases cooperation within groups (Bowles 2012). Given *A* as an operator that serves as a community matrix of, for example, possible cooperators from a tribe's ingroup working together to multi-task, or competitors in an outgroup, let a_{ij} represent the effects of *agent-i* on *agent-j*, the opposite for a_{ji} (May 1973; for a review of ingroup-outgroup effects, see Tajfel 1970; for a review of tribal effects, see Chagnon 2012). The strength of cooperation to multi-task can be measured by the state of interdependence in community matrix *A*, where interdependence is the effect that a group has on the choices and behaviors of its members; we designate interdependence as ρ:

$$\rho = \left(MS_{G/T} - MS_{S/G/T}\right) / \left(MS_{G/T} + (n-1)\, MS_{S/G/T}\right), \qquad (12.1)$$

$MS_{G/T}$ is the sum of the mean squares from the group on a measurement of an arbitrary factor, *T*, such as a culture, an issue, or a problem that is a group's focus as it assigns roles that produce multitasking; $MS_{S/G/T}$ is the aggregated contribution from the individuals on a measurement of factor *T*; and *n* represents the number of members in a group being measured (from Kenny et al. (1998, p. 235)). At its extremes, ρ ranges from −1 as multitasking goes to zero when the group is replaced by a collection of independent individuals, or to +1 as multitasking replaces the individual with slavish subservience to a group's efforts, like groupthink or authoritarianism.

With (12.1), we build *A* and convert it into an orthogonal matrix. Let *A* be a symmetric matrix with potential eigenvalues $\lambda_1, \ldots \lambda_n$. If *Q* is an orthogonal matrix with real values, and if $Q^{-1} = Q^t$ (i.e., if the inverse of *Q* equals its transpose), then the row vectors (or column vectors) are orthogonal, and Q^tAQ diagonalizes symmetric matrix *A* into its eigenvalues.

Let *A* be an operator on a social object, ψ, within its internal zone of influence; ψ could be an agent or a team, etc.; and let ψ be a column vector that represents the state of the social object as operator A transforms state vector ψ into a matrix. When ψ is represented on two sides of an equation as:

$$A\,|\psi> = x|\,\psi> \qquad (12.2)$$

then x is a scalar that is the eigenvalue, λ, or characteristic of A, and ψ becomes an eigenvector or eigenfunction. The usual way to solve for the eigenvalue, λ, is with an iterative process: $A\psi - \lambda I\psi = (A - \lambda I)\,\psi = 0$, where I is the identity matrix (i.e., $\begin{bmatrix} 1 & 0 \\ 0 & 1 \end{bmatrix}$).

The outer product of two state vectors is an operator; and the outer product of two eigenvectors is a special operator or projector, P, that projects an unknown state vector, ψ, into an eigenfunction and eigenvalue. Eigenfunctions form a basis that is orthonormal; i.e., given eigenfunctions ψ and ϕ and $< \psi|\phi >$ as the inner product of the two eigenfunctions, then $< \psi|\phi > = \psi_1\phi_1 + \psi_2\phi_2 + \cdots = \delta_{ij}$, where δ_{ij} as the Kronecker delta equals to 1 when $i = j$, otherwise 0. Thus, in our model, all state vectors are normalized, their inner product summing to 1 because their eigenvectors are equal (i.e., $cos\ 0° = 1$); it also means that the dot product of the two elements of a bistable or orthogonal vector is 0, and that the probabilities of measuring interdependent (or bistable) factors always sums to 1. This causes classical measurement uncertainty; i.e., when the probability of one bistable factor goes to one, the other goes to zero (e.g., the argument by Freeman (2013) and Eavis (2013) over the increased transactional uncertainty for JP Morgan caused by the Administration).

If ψ was a simple column vector representing the state of its independent elements, putting aside manipulations to find the eigenvalues, there would be little ambiguity in constructing conceptual models or in understanding them based on what amounts to a convergent, rational process. Assuming that intuition is a stable interpretation of reality, conceptual difficulties arise and intuition fails when interdependence (groupiness) is introduced. Beginning with simple bistability, ψ becomes a superposition of two orthogonal but non-factorable states, such as an observation and an action; a republican and a democrat; or a single tribal ingroup and outgroup (e.g., Lawless et al. 2011). Putting time evolution aside, we gain insight into a static situation by letting $|0>$ be the name of a column vector that represents one of the orthogonal factors that forms a basis, and $|1>$ the other factor (e.g., we arbitrarily set observation to $|0> = \begin{bmatrix} 1 \\ 0 \end{bmatrix}$, and action to $|1> = \begin{bmatrix} 0 \\ 1 \end{bmatrix}$); similarly, we could let a single person oscillate between being a conservative, represented by $|0>$, and a liberal, represented by $|1>$ (and vice versa); or ingroup A versus outgroup B (Tajfel 1970). Two orthogonal vectors $|0>$ and $|1>$ form a basis in 2-D (i.e., $cos\ 90° = 0$).

12.3.2 *Interdependence*

To model a group in a state of interdependence, we introduce the tensor product of independent elements, for example, $|0> \otimes |0>$, represented as |00>; and $|1> \otimes |1>$, represented as |11>. The basis for a 2-agent system becomes $\{|00>, |01>, |10>, |11>\}$. Factorability means independent objects, implying that any separable vector space V by tensor decomposition into basis elements is not interdependent. i.e., given state vector $|\psi>$ in a system, where $V = V_1 \otimes V_2 \otimes \cdots \otimes V_n$, the state $|\psi>$ is separable *iff* $|\psi> = V_1 \otimes V_2 \otimes \cdots \otimes V_n$. Otherwise, $|\psi>$ is in an interdependent state. An example of a non-factorable state is:

$$|\psi> = \frac{1}{\sqrt{2}}(|00> + |11>), \tag{12.3}$$

To prove, let $(a_1|0> + b_1|1>)(a_2|0> + b_2|1>) = \frac{1}{\sqrt{2}}(|00> + |11>)$. However, for this equation, no combination of a's and b's exists such that a_1b_2 and a_2b_1 are both zero. Moreover, trying to break (12.3) into separable elements not only loses information from this state of interdependence, but also the inability to factor (12.3) means that the measurement of interdependence produces two incomplete states that cannot be recombined to reproduce the original state of interdependence.

12.3.3 *Incompleteness and Uncertainty*

An individual's beliefs might be altered by new information but confirmation bias makes it unlikely that contradictory new information will be judged objectively by a committed believer (Darley and Gross 2000), or even appreciated by a neutral believer (for reasons discussed below). Avoidance of entering into states of cognitive dissonance keeps most important attitudes and beliefs of humans stable indicating that internal conflict is necessary to change strongly held beliefs (Lawless et al. 2013). Together, these two biases make it unlikely that a follower of one political view (e.g., conservative or liberal) would entertain an opposing viewpoint, especially when entertaining such a view threatened power, status or access to resources under control. An example would be the US government shutdown that happened in October 2013, decried by both political parties for entirely different reasons. First from the *New York Times* (Weisman and Peters 2013):

> "You don't get to extract a ransom for doing your job," Mr. Obama said in the White House briefing room as the clock ticked to midnight. ... [oppositely] "I talked to the president tonight," the speaker said on the House floor. He summed up Mr. Obama's remarks as: "I'm not going to negotiate. I'm not going to negotiate."

And from the *Wall Street Journal* (Hook and Peterson 2013):

> Senate Majority Leader Harry Reid (D., Nev.) rejected the move, saying he wouldn't enter negotiations until the House agreed to reopen the government by extending its funding for several weeks. "We like to resolve issues, but we will not go to conference with a gun to our head," Mr. Reid said on the Senate floor. ... [oppositely] Republicans denounced Senate Democrats for refusing to negotiate. "Our hope this evening is we will be able to put reasonable people in a room," said House Rules Committee Chairman Pete Sessions (R., Texas).

To simplify what constitutes a complexity of its own, assume there are teams of ideologues on either side of an issue, and that all others are swing voters ensconced in the neutral camp. For those in a political swing camp, we postulate that both views are held simultaneously in an indeterminate state of interdependence. For a single social agent in a superposition of orthogonal factors (opposed beliefs; or beliefs and actions), we propose:

$$|\psi >= a|0 > +b|1 >, \tag{12.4}$$

with the basis for a single agent written as {|0>, |1>}, where $|a * a'| = a * a = a^2$ (here a' is the complex conjugate that we use to represent the oscillations caused by an illusion) gives the probability of a social object being found in state |0>, with b^2 giving the probability of being in state |1>. But, for an individual, this state vector is factorable, suggesting that the oscillating (conflicting) perspectives for independent neutral individuals may be simply aggregated separately to reconstruct elements of the oscillation.

While (12.4) is easily factored; breaking apart a bistable state of superposition leads to a loss of information, producing incompleteness about the interdependent state. The effect of measuring a in (12.3) produces incomplete information about the measurement of b, thus constituting the measurement problem. An excellent example is illustrated by the recent difficulty with predicting elections during 2014–2015 (Morrissey 2015):

> In attempting to explain his failure to predict the Conservative landslide in the UK election, calling it a crises [i.e., the existence of interdependence in data], Silver also missed forecasting the "Scottish independence referendum", with the "no" side projected to win by just 2 to 3 percentage points. In fact, "no" won by almost 11 percentage points ... [and] that Republicans were favored to win the Senate in the 2014 U.S. midterms, they nevertheless significantly underestimated the GOP's ... margins over Democrats were about 4 points better than the polls in the average Senate race ... [and] Pre-election polls badly underestimated Likud's performance in the Israeli legislative elections earlier this year, projecting the party to about 22 seats in the Knesset when it in fact won 30. ...

12.4 Evidence of Incompleteness for Groups

The function of a group is to multitask. Multitasking with agent-1 and agent-2 forces them to focus on their individual tasks to manage the work-flows and communications between them to constitute the elements of a multitask, reducing the information available to them about their own performances.

12.4.1 The Evidence from Studies of Organizations

First, Bloom et al. (2007) found that the estimation by managers of their firm's performance was unrelated to their firm's actual performance. Second, a significant association between training and performance was found by Lawless et al. (2010) that no association existed between the book knowledge of air combat skills and combat outcomes. Third, uncertainty in the observations of better-run organizations was found to become noise (Lawless et al. 2013). Fourth, despite that most mergers fail (Andrade and Stafford 1999), they are often pursued to offset a vulnerability, to gain a new technology, to remove a competitor, but also to transform a business model for an organization that is failing.

In sum, as Galton discovered when a crowd of independent individuals was able to accurately estimate the weight of an ox, groups that process all of the available information are more likely than any one individual to be correct. But when a group acts as one under maximum groupiness, it loses its ability to process all of the external information; the tradeoff is that the group becomes better at cooperating to multitask to derive the solution of a well-defined problem.

12.4.2 Modeling Competing Groups with Limit Cycles

We postulate that at the level of individuals and groups, there is a constant competition to focus on the orthogonal functions for observation and action, orthogonal views like conservatism and liberalism, or orthogonal membership in tribe *A* or tribe *B*. The competition between these orthogonal functions results in limit cycles (May 1973; see Fig. 12.2).

Limit cycles depend on the free flow of neutrals to different (ideological, commercial, scientific, etc.) belief positions (the central tenet of capitalism). Limit cycles can be suppressed under authoritarian rule. In a dictatorship, social stability is maintained by censoring information (May 1973); i.e., by forcibly setting *a* or *b* to zero. But while social control is gained by an autocrat, in that incomplete information governs, the opportunity for mistakes increases dramatically (e.g., from Lawless et al. (2013, 2014)): DOE's mismanagement of nuclear wastes prior to 1985; China's air and water contamination today; the USS Vincennes shoot-down of an Iranian airbus in 1988, killing all aboard; and the USS Greeneville's collision with a Japanese fishing boat, killing nine of its tourists and crewmembers.

Compared to a collection of independent individuals, we had initially assumed that the entropy (*S*) is set to zero for a perfect team, the driving motivation to form a tribe. We now justify this assumption in the limit as follows (and contradict it

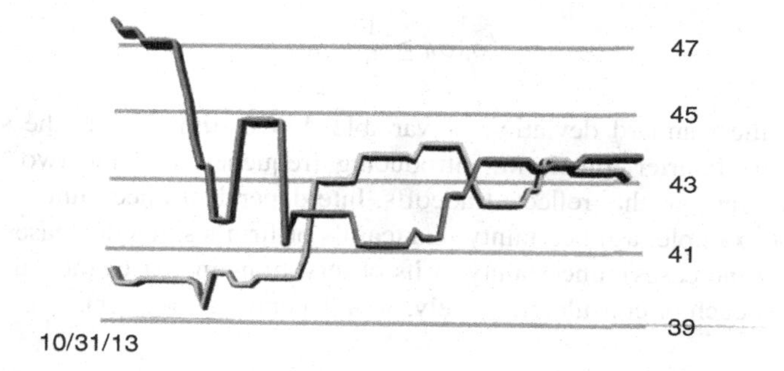

Fig. 12.2 Instead of a limit cycle (i.e., portraying N_1 versus N_2; in May (1973)), we display the data with N over time, t. In the forthcoming campaign for Representatives and Senators for the Democratic (blue) and Republican (green) Party, the data represents the "Generic Congressional Vote" obtained by averaging poll numbers collected and published by Real Clear Politics (collected from October 31, 2013 to January 6, 2014; this, past and current data can be found at http://www.realclearpolitics.com/epolls/other/generic_congressional_vote-2170.html). Three limit cycles are shown of decreasing magnitude: From 11/25–12/17; from 12/17–12/25; and from 12/25–12/30

later). Transaction costs are lower for individuals inside of a firm performing the same functions as for those same individuals multitasking in a firm (Coase 1960; Bowles 2012). This cost differential motivates the six-sigma processes designed to reduce waste in a firm, but if it becomes an overriding management goal, it impedes the tradeoffs a firm must make to find the new sources of free energy needed to adapt or to innovate (Christensen 2011), unexpectedly generating more entropy in a changing environment (May 1973), possibly setting a firm up to fail as its competition increases.

Equation 12.2 does not allow us to capture tradeoffs. To do this for two operators, A and B, we write:

$$[A, B] = AB - BA. \tag{12.5}$$

When two operators representing two different tribes have the same eigenvalue, then the operators commute: $[A, B] = AB - BA = 0$. With agreement between two erstwhile competitors, the combined social system is stable, no oscillations occur, nor do limit cycles exist (the goal of an autocracy). But when disagreement arises between two competitors, their two operators do not commute, giving:

$$[A, B] = iC, \tag{12.6}$$

where C is a measure of the gap or distance in reality between A and B. However, as multitasking improves, the tradeoffs between each group's focus on their tasks interfere with their meta-perspectives on how best to change or optimize tasks to improve performance (Smith and Tushman 2005), motivating tradeoffs that may or may not be efficacious:

$$\sigma_A \sigma_B \geq \frac{1}{2} \tag{12.7}$$

where σ_A is the standard deviation of variable A over time, σ_B is the standard deviation of its Fourier transform, introducing frequency, and the two together forming a Fourier pair that reflects tradeoffs. Interdependent uncertainty generates tradeoffs. For example, as uncertainty in a team's or firm's skills decreases (i.e., as a team's skills increases), uncertainty in its observations increases (accounting for the value of a coach or consultant, namely, an independent observer).

12.5 Gaps

The number 1/2 in (12.7) we liken to a "gap" in reality; gaps can exist over time or distance. Equations 12.2 and 12.5 reflect the existence of this "gap" in social reality, C, that permits social dynamics to operate. Social dynamics derive from the challenges to claims, social illusions (Adelson 2000) and irrationality (Kahneman 2011), feeding limit cycles (May 1973). For time gaps, the evidence indicates that the conscious awareness of signals takes about 500 ms, but under decision-making, it can extend to several seconds (up to 7 s) before a human's consciousness becomes aware of its "desire" to switch to a new choice (Bode et al. 2011) that can then be articulated by the human brain's running narrator (Gazzaniga 2011), the latter often construed as "free will". For distance gaps, at a minimum, say between two beliefs, cosine 0° is one; at a maximum, cosine 90° is zero.

Gaps are needed to create a state of superposition over a claim, to process the challenges that establish the oscillations between claims, giving observers time and space to process sequentially the information derived from opposing perspectives. But the human motivation is to believe that knowledge processing is too cumbersome, leading to the various illusions such as that a merger reduces overcrowding in a collapsing market, when the market itself may be ending; that six-sigma processes safely improve profits (but see Christensen 2011); or that market returns improve by chasing market leaders. The motivation in these and other illusions is to ignore, reduce or replace the "gaps" in reality with a rational approach (one without gaps) instead of an emphasis (focus) on problem solutions.

Despite the accumulating evidence against the traditional model, it remains rational (e.g., Bayesian). Silver (2012) concluded that the brain forms and continually updates a set of Bayesian "priors" learned over a lifetime used retrospectively to interpret new data that corresponds to its environment. But Silver's technique of aggregating polling data copies Galton's insight. The more important question is why Democrats and Republicans look at the same data but interpret it differently at the same time, thereby generating bistable illusions, conflict and oscillations. Numerous examples exist; e.g., R.A. Fisher, the esteemed statistician and evolutionary biologist, argued against the new evidence that smoking cigarettes would cause cancer; but Fisher was a smoker (Stolley 1991), likely the cause of his bias against accepting the new evidence.

12.6 Conclusions

We have argued that interdependence combines with cognitive dissonance to make those of us who adopt strong beliefs act to suppress both our internal cognitive narrator, known as confirmation bias (Darley and Gross 2000) but also the alternative views of our ingroup, forming the ingroup-outgroup bias (Tajfel 1970). When these beliefs are unchallenged, they give the illusion of stable reality; but when challenged, they drive the oscillations of social behavior between competing individuals, teams, tribes, or firms across a system. Thus, the presence of alternative views in the decision process is not only the end of certainty that motivates tradeoffs ((12.4) and (12.7), respectively), but these Nash equilibria are also the source of information that competition generates for observers to process that preclude, reduce or mitigate tragedies (e.g., unlike Communist China during its great famine in the 1950s, no modern democracy has ever suffered from famine; in Sen (2000); and unlike Nazi Germany during the 1930s–1940s, no modern democracy has ever started a war against another democracy; in Wendt (1999)). To defend an individual, Chagnon (2012) concluded that people find safety in numbers of their own. However, although not very popular to any single tribe of Republicans or Democrats, competing religions or different races, nonetheless, it is the competition for the strongest idea that has become the pillar of free speech that forms the foundation of modern societies (Holmes 1919).

Without competition, incomplete information impedes social evolution. But with conflict and its management, indirect control of hybrid teams may be feasible. Social uncertainty spontaneously generates interdependence, just as interdependence generates social uncertainty. Both require a space-time "gap" in reality that promotes competition as neutrals (e.g., citizens, courts and administration) sort through the interpretations when they are free to make the best choice, switching back when a choice does not pan out, forming limit cycles that provide indirect social control. Social information must remain incomplete, forcibly true under the dictatorships that attempt to maintain direct control (May 1973/2001), but inescapably true in democracies with working checks and balances. However, unlike dictatorships which thrive on the failure of social change, the never ending search for completeness in democracies leads to social evolution. All things considered, social (political) predictions made in democracies are more likely to be wrong than those made in dictatorships.

Finally, we began by setting the baseline entropy for well-functioning teams to zero. We need to revise it to underscore the cognitive difficulty implied by (12.3) for two or more agents multitasking together in a state of superposition. Equation 12.3 suggests on the one hand how a team or an organization can perform at a high level, but also why on the other hand they are incomplete witnesses of their performance. How can agents generate the data in Fig. 12.2 for (12.3) or (12.5)? We suspect that a conflict center creates additive or destructive interference among superposed neutrals; that winning a debate or selling more computer products on 1 day somewhat suppresses a conflict center's complementary element, producing stable results; that a tie causes no movement in the results; and that a more competitive counterattack from a previously failing candidate or firm creates the return arm in the results that builds a limit cycle to exploit the gaps in reality.

In sum, we have reached new conclusions about robust intelligence and trust. With the bistable agent model, we conclude that the robust intelligence necessary to succeed appears unlikely to occur for a collection of agents acting independently, reducing autonomy and thermodynamic effectiveness (e.g., lower productivity). By multitasking (MT) together, a team of agents is more effective than the same agents independently performing the same tasks. However, for a team of bistable agents, bias reduces its robust intelligence and autonomy. Instead, robustness requires a MT team observing *Reality* to contradict an opposing team (two teams of MTs best capture situational awareness), implicating the value of competition in determining *Reality*, but these two teams are also insufficient for robustness. Robust intelligence requires three teams: two opposing MT teams to construct *Reality* plus a neutral team of freely moving bistable independent agents attracted or repelled to one or another team to optimize the thermodynamic forces that determine team effectiveness and efficiency. Thus, given two competitive teams, adding a third team to constitute a spectrum of neutral bistable agents able to invest freely (properties, ideas, works), act freely (joining and rejecting either team), and observe freely makes the greatest contribution to robust intelligence, to mitigating mistakes, and to maximizing effective and efficient autonomy. However, a reliable, valid metric is still needed. Moving from subjective measures (e.g., surveys, questionnaires) to a countable metric of individuals joining or leaving teams establishes a Hilbert space, our end result with three teams.

References

Adelson EH (2000) Lightness perceptions and lightness illusions. In: Gazzaniga M (ed) The new cognitive sciences, 2nd edn. MIT, Cambridge

Ambrose SH (2001) Paleolithic technology and human evolution. Science 291:1748–1753

Andrade G, Stafford E (1999) Investigating the economic role of mergers (Working Paper 00-006). Cambridge

Barabási AL (2009) Scale-free networks: a decade and beyond. Science 325:412–413

Bell BS, Kozlowski SWJ, Blawath S (2012) Team learning: a theoretical integration and review. In: Kozlowski SWJ (ed) The oxford handbook of organizational psychology, vol 1. Oxford Library of Psychology, New York

Bloom N, Dorgan S, Dowdy J, Van Reenen J (2007) Management practice and productivity. Quart J Econ 122(4):1351–1408

Bode S, He AH, Soon CS, Trampel R, Turner R, Haynes JR (2011) Tracking the unconscious generation of free decisions using ultra-high field fMRI. PLoS One 6(6):e21612

Bowles S (2012) Warriors, levelers, and the role of conflict in human evolution. Science 336:876–879
Chagnon NA (2012) The Yanomamo. Wordsworth, New York
Christensen C (2011) How pursuit of profits kills innovation and the U.S. forbes, economy. http://www.forbes.com/sites/stevedenning/2011/11/18/clayton-christensen-how-pursuit-of-profits-kills-innovation-and-the-us-economy/
Cimilluca D, Rohwedder C, McCracken J (2009) Cadbury Sneers at Kraft's Hostile Bid. U.S. food giant takes unsweetened, $16 billion offer directly to confectioners' holders, Wall Street J, from http://www.wsj.com/articles/SB10001424052748704402404574524952121366382
Coase R (1937) The nature of the firm. Economica 4:386
Coase R (1960) The problem of social costs. J Law Econ 3:1–44
Conant RC (1976) Laws of information which govern systems. IEEE Trans Syst Man Cybernet 6:240–255
Darley JM, Gross PH (2000) A hypothesis-confirming bias in labelling effects. In: Stangor C (ed) Stereotypes and prejudice: essential readings. Psychology Press, Philadelphia, p 212
Dilger DE (2013) Data bites dogma: apple's iOS ate up android, Blackberry U.S. market share losses this summer, AppleInsider. http://appleinsider.com/articles/13/10/05/data-bites-dogma-apples-ios-ate-up-android-blackberry-us-market-share-losses-this-summer
Eavis P (2013) Despite its cries of unfair treatment, JP Morgan is no victim, New York Times. http://dealbook.nytimes.com/2013/09/30/despite-cries-of-unfair-treatment-jpmorgan-is-no-victim/
Fichtner U, Smoltczyk A (2013) Architect of austerity: schäuble's search for a way forward, spiegel online international. http://www.spiegel.de/international/germany/how-german-finance-minister-schaeuble-navigates-the-euro-crisis-a-924526.html
Freeman J (2013) Opinion: looting JP Morgan. Wall Street J http://www.wsj.com/video/opinion-looting-jp-morgan/C4804E6F-4FFC-4A7F-AD95-5BEC6E7F8931.html
Gazzaniga MS (2011) Who's in charge? Free will and the science of the brain. Ecco, New York
Hackman JR (2011) Six common misperceptions about teamwork. Harvard Business Review http://blogs.hbr.org/cs/
Helbing D (2013) How and why our conventional economic thinking causes global crises (discussion paper). EconoPhysics Forum, ETH Zurich
Holmes OW (1919) Dissent: Abrams v. United States
Hook J, Peterson K (2013) Government shuts down as Congress misses deadline. Senate rejects house bill to delay part of health law; Obama gives notice to workers, Wall Street J http://www.wsj.com/articles/SB10001424052702304373104579107051184641222
Jamshidi M (2009) Control of system of systems. In: Nanayakkara T, Sahin F, Jamshidi M (eds) Intelligent control systems, vol 2. Taylor & Francis, London (Chapter 8)
Kahneman D (2011) Thinking, fast and slow. MacMillan (Farrar, Straus & Giroux), New York
Keller B Op-Ed (2013) Smart drones. New York Times. http://www.nytimes.com/2013/03/17/opinion/sunday/keller-smart-drones.html?_r=0
Kenny DA, Kashy DA, Bolger N (1998) Data analyses in social psychology. In: Gilbert DT, Fiske ST, Lindzey G (eds) Handbook of social psychology, vol 1, 4th edn. McGraw-Hill, Boston, pp 233–265
Kerr NL, MacCoun RJ (1985) The effects of jury size and polling method on the process and product of jury deliberation. J Pers Soc Psychol 48:349–363
Lawless WF, Whitton J, Poppeliers C (2008) Case studies from the UK and US of stakeholder decision-making on radioactive waste management. ASCE Pract Period Hazard Toxic Radioact Waste Manage 12(2):70–78
Lawless WF, Rifkin S, Sofge DA, Hobbs SH, Angjellari-Dajci F, Chaudron L, Wood J (2010) Conservation of information: reverse engineering dark social systems. Struct Dyn 4(2):1–30
Lawless WF, Angjellari-Dajci F, Sofge DA, Grayson J, Sousa JL, Rychly L (2011) A new approach to organizations: stability and transformation in dark social networks. J Enterprise Transformation 1:290–322

Lawless WF, Llinas J, Mittu R, Sofge D, Sibley C, Coyne J, Russell S (2013). Robust intelligence (RI) under uncertainty: mathematical and conceptual foundations of autonomous hybrid (human-machine-robot) teams, organizations and systems. Struct Dyn 6(2):1–35
Lawless WF, Akiyoshi M, Angjellari-Dajcic F, Whitton J (2014) Public consent for the geologic disposal of highly radioactive wastes and spent nuclear fuel. Int J Environ Stud 71(1):41–62
May RM (1973/2001) Stability and complexity in model ecosystems. Princeton University Press, Princeton
Morrissey E (2015) Nate Silver, David Axelrod agree: yes, there is a crisis in polling, Hot Air http://hotair.com/archives/2015/05/08/nate-silver-yes-there-is-a-crisis-in-polling/
Rand DG, Nowak MA (2013) Human cooperation. Cognit Sci 17(8):413–425
Sen A (2000) Development as freedom, Knopf
Schweitzer F, Fagiolo G, Sornette D, Vega-Redondo F, Vespignani A, White DR (2009) Economic networks: the new challenges. Science 325:422–425
Silver N (2012) The signal and the noise. Why so many predictions fail—but some don't. Penguin, New York
Smith WK, Tushman ML (2005) Managing strategic contradictions: a top management model for managing innovation streams. Organ Sci 16(5):522–536
Stolley PD (1991) When genius errs: RA Fisher and the lung cancer controversy. Am J Epidemiol 133:416
Tajfel H (1970) Experiments in intergroup discrimination. Scientific American 223(2):96–102
Vogelstein F (2013) And then steve said, 'Let There Be an iPhone', New York Times Magazine http://www.nytimes.com/2013/10/06/magazine/and-then-steve-said-let-there-be-an-iphone.html
Weisman J, Peters JW (2013) Government shuts down in budget impasse. New York Times http://www.nytimes.com/2013/10/01/us/politics/congress-shutdown-debate.html?_r=0
Wendt A (1999) Social theory of international politics. Cambridge University Press, Cambridge
Wickens CD (1992) Engineering psychology and human performance, 2nd edn. Merrill, Colombus

Printed in the United States
By Bookmasters